International Conference on Recent Trends in Physics

(ICRTP 2012)

Journal of Physics: Conference Series Volume 365

Indore, India
4-5 February 2012

ISBN: 978-1-62276-782-3
ISSN: 1742-6588

Printed from e-media with permission by:

Curran Associates, Inc.
57 Morehouse Lane
Red Hook, NY 12571

proceedings.com

Some format issues inherent in the e-media version may also appear in this print version.

Printed by Curran Associates, Inc. (2013)

Institute of Physics
Dirac House, Temple Back
Bristol BS1 6BE UK

Phone: 44 1 17 929 7481
Fax: 44 1 17 920 0979

techtracking@iop.org

Additional copies of this publication are available from:

Curran Associates, Inc.
57 Morehouse Lane
Red Hook, NY 12571 USA
Phone: 845-758-0400
Fax: 845-758-2634
Email: curran@proceedings.com
Web: www.proceedings.com

TABLE OF CONTENTS

INVITED TALKS

CONDENSED MATTER PHYSICS

LASER PHYSICS

Journal of Physics: Conference Series **365** (2012) 011001 doi:10.1088/1742-6596/365/1/011001

Preface

The International Conference on Recent Trends in Physics (ICRTP2012) took place in Indore, India, on 4–5 February 2012. The conference was hosted by the School of Physics, Devi Ahilya University, Indore – 452001.

The objective of the conference was to provide a platform for interaction among scientists, teachers, researchers and students, and to share their ideas, thoughts and scientific findings in various areas of physics.

ICRTP2012 attracted a total of 130 abstracts submitted by scientists from 7 different countries. The conference included an inaugural talk and 13 Invited talks. Aside from regular oral contributions, 118 posters were presented.

A particular highlight of the conference was a special session for oral presentations by young PhD students. The aim of this session was to provide a platform for the budding scientists to present their recent findings in the presence of their community. The two best oral presentations, judged by a special three-member committee, were awarded prizes. Similarly the two best posters, judged by a committee of five experts, were also awarded prizes.

It is our pleasure to thank the members of the International Advisory Committee and Local Organizing Committee for their invaluable help, especially for their proposals for invited talks. A total of 82 papers were submitted to be considered for publication and 59 papers were accepted for inclusion in the proceedings. All the papers were reviewed, and we wish to thank to all the referees for their support.

We are grateful to the University Grants Commission (UGC), New Delhi, India for their financial support of the organization of ICRTP2012. Support from Marketing Centre (India), Indore and Transolutions, Indore is also appreciated.

Finally, we express our sincere thanks to our university administration for their continuous support. Special thanks go to all the faculty members, administrative staff and students of the School of Physics for their tireless efforts in organizing ICRTP2012.

Indore, 30 April 2012

Shashank N Kane*
Ashutosh Mishra
Anup Kumar Dutta
Pratima Sen
Guest Editors

School of Physics, Devi Ahilya University, Indore – 452001, India
*E-mail address: kane_sn@yahoo.com

Peer review statement

All papers published in this volume of *Journal of Physics: Conference Series* have been peer reviewed through processes administered by the proceedings Editors. Reviews were conducted by expert referees to the professional and scientific standards expected of a proceedings journal published by IOP Publishing.

XXIst Century Ferrites

F. Mazaleyrat , K. Zehani, A. Pasko, V. Loyau and M. LoBue
SATIE, UMR CNRS 8029, ENS de Cachan, 61, Avenue du President Wilson,
94235 Cachan, France

E-mail: mazaleyrat@satie.ens-cachan.fr

Abstract. Ferrites have always been a subject of great interest from point of view of magnetic application, since the fist compass to present date. In contrast, the scientific interest for iron based magnetic oxides decreased after Ørsted discovery as they where replaced by coil as magnetizing sources. Néel discovery of ferrimagnetism boosted again interest and leads to strong developments during two decades before being of less interest. Recently, the evolution of power electronics toward higher frequency, the downsizing of ceramics microstucture to nanometer scale, the increasing price of rare-earth elements and the development of magnetocaloric materials put light again on ferrites. A review on three ferrite families is given herein: harder nanostructured $Ba_{2+}Fe_{12}O_{19}$ magnet processed by spark plasma sintering, magnetocaloric effect associated to the spin transition reorientation of W-ferrite and low temperature spark plasma sintered Ni-Zn-Cu ferrites for high frequency power applications.

1. Introduction

1.1 A brief history of ferrites

Ferrite is the oldest magnetic material known, together with iron, but the only terrestrial native magnetic material, since natural metallic iron is from meteoritic origin. Furthermore, magnetic virtue of iron is observable only in presence of a appreciable magnetic field, i.e. a loadstone, or natural magnet, the only source of magnetic field known until Ørsted's discovery in 1819. The load stone is mainly composed of magnetite, Fe_3O_4 nowadays classified as a semi-hard ferromagnetic, which is formed by cooling of magma coming out from earth liquid coat, in atmosphere. Contrarily to common thinking, magnetite is not efficiently magnetized by the earth magnetic field, but lightning impacts that produce very strong currents do. Indeed, load stones are mostly found on the top of volcanoes and magnetized in circles. The attracting power of loadstone on iron was reported in 685 BC by Thales of Miletus. During the early middle-age, Chinese made relatively inefficient compasses from spoon shaped magnetite stones, but the modern compass made out from a magnetized hard-steel needle pivoting on a brass pin was known about Xth century in China and Arabic word although is precise origin (Arabia or China) is still controversial. However, the real efficient navigation compass was invented in Genova, from an Arabic one, improved by putting it in a brass box, which damps oscillations by producing eddy currents. Improvement in understanding of magnetism was brung by a French military engineer, Pierre de Maricourt (known as Petrus Peregrinus, the vehicular language of science was Latin until the XVIIIth century) in 1269: he defined north and south poles and showed that breaking a magnet creates new poles on the broken surfaces (one would say now div**B**=0!). Then it was not before 1600, when the doctor of Queen Elisabeth the first, William Gilbert, observed the direction of a compass around magnetite bar and sphere, plotting for the first time what we call now a magnetic vector field. Until the 1820's, the loadstone was used mainly for magnetizing compass needles and scientific shows for aristocrats and bourgeois in the Europeans universities and curiosity cabinets. The most famous exception is the young uneducated bookbinder Michael Faraday, who used a magnetite bar to make the first prototype electrical motor in the aim of experimenting the reciprocal effect of Arago's rotation. Faraday in his magnetic classification (ferro, para, diamagnetism) obviously ranked magnetite in the ferromagnetic bodies class from macroscopic observation. Then, magnetite disappeared from scientific preoccupations since coils feed by Volta stacks and soon after electrical machines, where better magnetic field sources and hard steels better permanent magnets. When Pierre Weiss (a French political refugee in Switzerland when Alsace was occupied by Prussians) established the mean field theory based on Langevin classical statistical paramagnetism, he discovered that magnetic metals poorly obey his theory as the Ms vs T slope in the low temperature range was too steep in his theory. However, he and other physicists had a great confidence on the mean field theory since it was describing very well the Ms vs T curve of magnetite. Ironically, the fact magnetite is not ferromagnetic was demonstrated by Weiss' most famous student, Louis Néel in 1948. Interestingly, while most of magnetic alloys were known and used in industry (Fe-Si, Fe-Co, Fe-Ni, Cr-Co hard steels, AlNiCo), ferrites where still limited to iron ferrite, namely magnetite having the spinelle structure and formula $M^{2+}Fe_2O_4$ (M=Fe). Néel's theory and high frequency electronics development boosted ferrite research and created ferrite industry. Great efforts have been done in order improve ferrite properties by softening (M=Mg), hardening (M=Co), increasing magnetisation ($M=Zn_{0.5}+Mn_{0.5}$) or making them electrical insulators (M=Cu or Ni). With industrial development of radiocommunications and power electronics, two main families of cubic ferrites emerges in the 50's, in one hand, the $(Mn,Zn)Fe_2O_4$ for LF-MF band having a high induction, high permeability and low losses and, in the over hand, (Ni,Zn) Fe_2O_4 for HF-VHF band showing high resistivity [2]. One should also mention Li-ferrites and Y-garnets used in UHF, the later being the only transparent magnetic material known. In parallel, a great advance in hard materials was realized in Japan in 1955 with the discovery of hexagonal ferrites, from which the common M type, $BaFe_{12}O_{19}$ showing the magnetoplumbite structure, and the less known Y, W and Z type, used in the UHF-SHF band having slightly different structures. Since the beginning of the 60's, only gradual improvements have been obtained in ferrite industry: we are now trying to show three directions that can undergo strong evolution in the near future.

1.2 Why nanostructured dense ferrites?

Magnetism is mainly governed by three kind of energy:

- exchange energy, a quantum mechanical term originating from electrostatic interaction and depending on atom to magnetic electrons shell diameter (a/2r), represented by the exchange constant

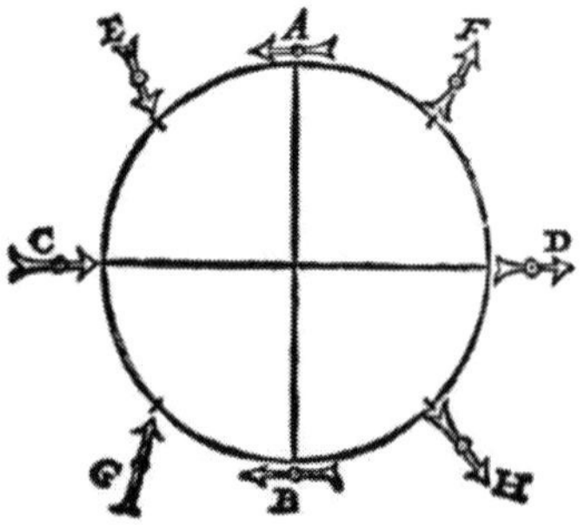

Figure. 1: Gilbert's drawing of field around a spherical magnet.

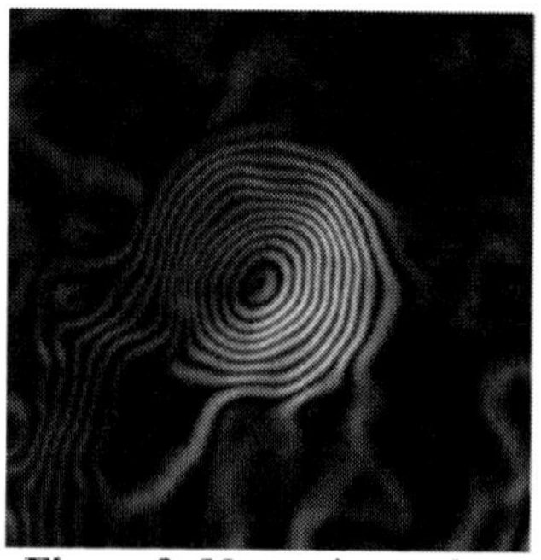

Figure.2: Vortex image in a 70 nm permalloy particle obtai-ned by electron holography.

Figure.3: Domain structure in a Fe-Si grain oriented steel sheet obtai-ned by magneto-optical Kerr effect.

Table 1: Structural parameters and magnetic properties of dense nanostructured hexaferrite magnets obtained by SPS [9]

Elaboration conditions	Grain size (±10 nm)	Density (%)	M_S (Am^2/kg)	M_R/M_S	μ_0H_C (T)
Conventional	>1000	> 95	62	0.51	0.26
Calcinated 900°C	60	-	55	0.51	0.56
SPS 800°C/0min	70	76	66	0.57	0.51
SPS 800°C/13min	84	88	66	0.57	0.49
SPS 800°C/20min	77	86	62	0.58	0.49

$50<A<200$ pJm^{-1}, which forces spins to be parallel (when $a/2r>1$) or anti-parallel (when $a/2r<1$) at atomic scale;
- magnetocrystalline energy, represented by the anisotropy constant $100<K<10^6$ Jm^{-3} linked to spin orbit coupling and lattice symmetry which forces spin to be aligned within an easy axis;
- magnetostatic energy, proportional to the squared saturation magnetisation $0.3<J_S<2.5$ T, a classical term described by Maxwell equations acting at any distance from atomic to macroscopic scale, which tends to create a close structure of magnetic moments.

Minimisation of the total energy for a given particle as a function of its size, shows that in coarse grain materials, the magnetization organizes itself in a so-called domain structure where spin are parallel inside the domains and anti-parallel from one domain to its neighbour (eventually perpendicular in cubic materials), these domain being separated by a region where spins turns helically, the Bloch's wall, which thickness is $L_w=(A/K)^{1/2}$.

In soft materials, where the magneto-cristalline energy is low compared to magnetostatic, a crystal is multi-domain when its size is roughly over $3L_w$ typically about 100 nm. The single-domain (uniformly magnetized) configuration is energetically favourable when exchange dominates the magneto-crystalline energy, which corresponds to a particle more or less smaller than 8Lex, with $Lex=(2\mu_0A/J_S^2)^{1/2}$. In between, the favoured state is a vortex structure where the spins curls around an easy-axis [4,5]. As domain walls motion is always a dissipative phenomena, the single domain state in soft magnets is favourable to high frequency applications.

In hard materials, the magnetocrystalline terms is very strong ($3L_w$ is typically few nm) and the uniformly magnetized configuration is favoured for particles smaller than $36\mu_0(AK)^{1/2}J_S^{-2}$. Because the mechanism of magnetic switching involves mainly phenomena related to the walls (pinning of the wall on defects, nucleation of the walls), the single domain state in hard magnets increases the coercivity, in principle close to the limit given by the so called Stoner-Wohlfarth theory $\mu_0H_K = 2K/M_S$ [6,7].

In this paper, we will show how these XXth century compounds should evolved with modern chemical and elaboration processes, dragged by the needs of innovative technological applications. In the first part, it will be shown how giant coercivity nanostructured Ba ferrites can be obtained. In the second part, magnetocaloric effect associated to the spin reorientation transition will be addressed, and the third one will describe the properties of Ni-Zn-Cu and the realisation of magnetic components.

2. Giant coercivity of nanostructured M-type Ba hexaferrite

Enhencement of the coercivity of M ferrite has been first obtained by replacement of Ba with Sr. However, this improvement is not spectacular and the coercivity remains below 0.3 T in isotropic magnets and 0.35 in anisotropic ones. Several authors have studied the effect of rare-earth elements (La substitution for Ba, and Co, Ni respectively [3]) and more recently in Sr ferrites, resulting in a corcivity enhencement by about 20% with a maximum value of 0.42 T for $Sr_{0.8}La_{0.2}Fe_{11.8}Co_{0.2}O_{19}$ [8]. Our approach has been completely different: starting from the fact coercivity is inversly dependent on particle size, downsizing of grain down to nanometer range should increase noticeably that figure, a well known fact observed in Ba ferrite powders. Indeed, according to Néel and Stoner, the upper limit of the coercivity is $\mu_0H_K = 1.7$ T, for perfectly oriented particles, and 0.48 H_K for an assembly of

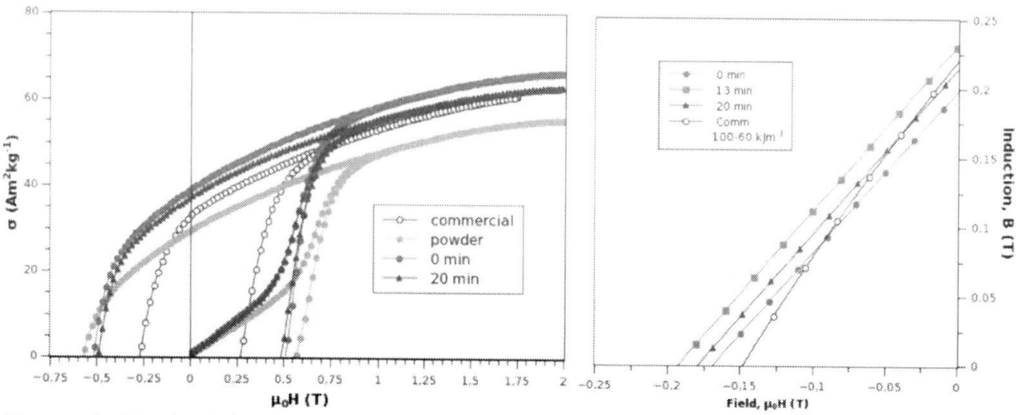

Figure. 4 : Hystéresis loops of nanostructured hexaferrite magnets obtained by SPS (left) and second quandran B vs. H loops (right). Lines are iso-BH curves from 60 Jm^{-3} (bottom) to 100 Jm^{-3}.

non-interacting uniaxial particles with isotropic distribution of easy axes. Several authors have succeed to produce Ba ferrite nanoparticles by soft chemical route and they have found coercivities as high as 0.58 T for a mean grain size about 100 nm [9,10] . As the anisotropy field is 1.7 T, the theoretical value for an isotropic magnet is 0.82 T, so that they reached 70% of the upper limit. However, applications requires solid materials with high density in order to maximize the magnetic induction. This densification is currently obtained by sintering the powder at temperatures about 1000°C and above, yielding grain growth and in turn a dramatic drop in coercive field.

In a previous work [11], sol-gel route was used to produce particles smaller than the single domain limit (about 200 nm). After reaction at 900°C, the coercivity for the powder is over 0.5 T. Then the problem to be solved is the grain growth, which was done by using the spark plasma sintering process (SPS). Basically, the technique consist in pulsing an electrical current in a powder closed in a die between to pistons, these part being made from conducting refractory material, usually graphite, under neutral atmosphere or vacuum. The process is capable to reduce the temperature by 100-200°C compared to standard sintering and to keep more or less the grain size of the powder. For metallic powders SPS is actually a Joule heating with locally high temperature at the grains contact points, which, with the help of pressure makes the process very fast. There was a controversial on the process involved in ceramics since they are usually insulating. Actually, even if the ferrites can have resistivity as high as 10^6 Ωm, this figure rapidly goes down with temperature because of the semi-conducting behavior and drops by several orders of magnitude above 500°C [2]. The result is that after sintering at 800°C (heating and cooling rates about 200°C/min) for different times, a more than 80% dense material can be obtained with a coercivity close to 0.5 T (see Table1 and Fig. 4), a value never reached before for a hexaferrite anisotropic magnet, which makes it very strong against demagnetisation. The B vs H curve in the second quadrant (see Fig. 4) is almost linear owing to the high coercivity, and the energy product $(BH)_{max}$ which is usually considered as the mertit factor for magnets, reaches 90 kJm^{-3} , a valuable enhancement compared to usual value arroud 60.

3. Magnetocaloric effect associated to spin transition reorientation in W-type Ba hexaferrite

The magnetocaloric effect (MCE) referrers to an exchange of heat between a magnetic material and a medium, upon magnetization or demagnetisation. In particular, when the temperature is close to an order/disorder magnetic transition (typically a Curie point), the application of magnetic field can drive the material from disorder (paramagnetic) to magnetic order by orienting spin towards the field, producing heat release to the medium. Inversely, demagnetization drives back the material to magnetic

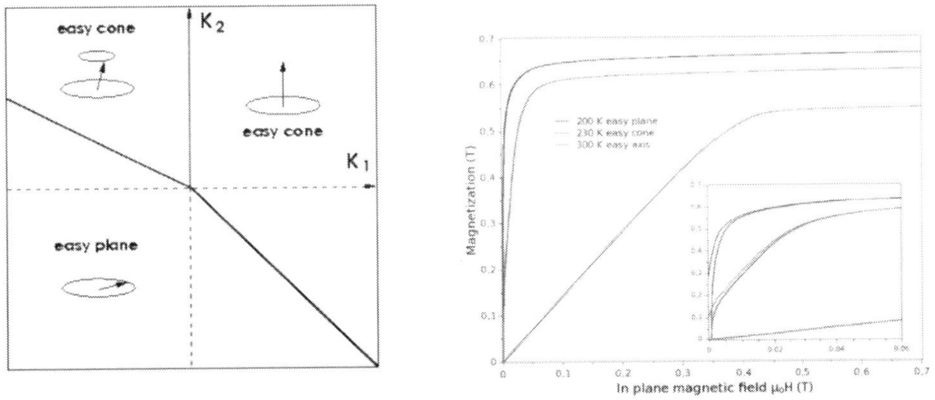

Figure. 5: the three states of the magnetization in W-ferrite as a function of the relative value of the anisotropy constants and the magnetization curves in the different states when temperature changes.

disorder and heat is absorbed by the material. This thermodynamic cycle is quite similar to the vapour to liquid transition upon pressure exploited in classical cooling systems. The MCE at room temperature was first studied by Weiss on pure gadolinium [12], but can be experimented also on Cr substituted Fe-based amorphous ribbons [13] or any king of alloys having a close to 300 K Curie point such as Fe-Ni-Cu alloys [14]. However, the MCE is not only associated to order/disorder transition (ferro to paramagnetic, i.e. Curie point) but to any change in order yielding a drastic change in the net magnetization without hysteresis. The term giant magnetocaloric effect usually refers to a first order magnetic transition produced by a crystallographic lattice symmetry change : this change can be either ferro to para like in LaFeSi [15] or ferro to antiferro like in CoMnSi [16] or CoMnGeP [17], both yielding a large entropy change due to the metamagnetic transition. Also, a much less known magnetocaloric effet is that associated to spin reorientation transition (SRT), as it has been shown by Belov in 1978 on W-type hexaferrite [18].

W-type hexaferrite is a complex structure composed of M spinel sublattice having the composition $BaCo_xZn_{1-x}Fe_{16}O_{27}$. In this particular compound, the Co contain, x, controls the first and second order anisotropy constants (K_1 and K_2 in Fig. 5) and hence the type of anisotropy and its direction. As these constants evolve in different ways with temperature, the magnetic anisotropy also change with temperature.

Magnetisation curves performed on single crystals of W-ferrite with x=0.7 shows a change in shape across the SRT estimated around 230 K for the easy plane to cone, and about 260 K for cone to axis SRT. The magnetocaloric effect is usually measured applying Maxwell relation to the M(T) curve as in [17], however this method is naturally inaccurate because the computation of dM/dT. A more complex but safer method is the direct measurement of entropy change. This has been done using a Peltier cell based calorimeter [19]. LoBue et al. Shown that this material exhibits a maximum entropy change of $\Delta s = 0.2\ Jkg^{-1}K^{-1}$ [20]. This value is 20 times smaller than usual magnetocaloric materials but with an almost 15-fold reduced field. Also, a key feature of SRT MCE (compared to ferro/para transition) is that the transition can be driven by the field even several ten of K above the zero field SRT, so it can be used in a wide range of temperature [21]. Thus the question is to determine whether or not these W-ferite have practical interest compared to LaFeSi and to find a global criterion for comparison. Indeed, if LaFeSi exhibits high Δs, the field necessary to reach this value is above 1.5 T, a value that exceeds the residual magnetisation of stronger magnets. As a consequence, Halbach type magnets, 'magic ring' type, have to be used to produce that field.

Let's consider a Halbach ring with inner and outer diameter a and b, respectively, and length L, which cavity is completely filled with the magnetocaloric material. The field in the cavity is given by

Table 2: Comparison between MC materials and device dimensions.

	Δs ($Jkg^{-1}K^{-1}$)	M_0H (T)	F ($Jm^{-3}K^{-1}$)	RE mass /1000 cm^3	V (cm^{-3})	V_{MC} (cm^{-3})	RE mass (kg)
Gd	4	1.6	3200	7.9	1000	153	4.4
LaFeSi	6	1.6	4700	1,3	670	102	2.3
W-ferrite	0.2	0.1	900	0	2350	3050	1.1
NdFeB	-	1.4	-	2.4	-	-	-

$H=M_R\ln(a/b)$. The total volume of the system is given by $V=\pi b^2L$ and the energy exchanged with the fluid is $E=\rho V_{MC}\ \Delta s(T_2-T_1)$, with $V_{MC}=\pi a^2L$. From these simple equations, one can easily derive the energy per total volume of the system (including magnet) per kelvin:

$$F=\frac{E}{V(T_2-T_1)}=\rho\Delta s e^{-2H/M_S} \tag{1}$$

On the same basis, if one takes the cooling power corresponding to a 1000 cm^3 system (magnet+MC material) based on gadolinium as a reference, one can compute the volume of equivalent system based on other MC material and NdFeB magnets. In Table 2, systems based on different material are presented. These results show clearly that if refrigerators based on W-ferrite would be bigger compared to RE-based ones, they would need at least twice less RE elements. Since single crystals can't be use for application purpose and because magnetization by domain wall motion should be avoided, next development step will be to produce nanosize oriented W-ferrites by means SPS.

4. NiZn spinel ferrites for power integration.

Ni-Zn ferrite are well known and widely used for signal application in high frequency due to there high resistivity, but usually not for power applications. The reason is that losses in these materials are considered too high for this kind of applications. Copper substitution in Ni-Zn ferrite was introduced by Nomura in 1992 with the aim of reducing the sintering temperature [22]. Indeed, the high mobility of copper ions makes possible to reduce the sintering temperature significantly (1000°C or even less) that makes possible co-firering with silver instead of Ag-Pd, a more resistive and expensive alloy. Later on, it was proved that by cobalt doping, the material undergoes a spin transition reorientation near room temperature [24] around which the losses are minimized under inductions of several 10 mT above 1 MHz, allowing to use that material in high frequency power applications instead of Mn-Zn [25].

Properties of $Ni_{0.195}Cu_{0.2}Zn_{0.599}Co_{0.006}Fe_2O_4$ synthesized by the classical solid route (basic oxides mixing, calcination, milling) and sintered by means of SPS are reported in Table 3. All samples, even sintered at 650°C, are solid and show porosity of 7 to 3%. As SPS is done using graphite dies, the atmosphere

Table 3: Structural and magnetic properties of NiZnCu ferrite sintered at different temperatures (5 min heating/cooling plus 10 min plateau time).

Sintering		Grain size	Density	Resistivity (Ωm)		ε_R (@1 MHz)		μ_R	f_R
(°C)	min	(nm ±20)	(%)	sintered	burnt	sintered	burnt		MHz
650	10	175	93	10^3	3.10^5	17	5	18	200
700	10	180	94	30	7.10^4	45	18	40	100
750	10	220	96	60	10^3	60	37	150	33
800	10	280	97	100	400	60	40	210	24

must be neutral. As a consequence, heating oxides in contact with carbon produce carbon deposition on the surface of samples and possibly reduction. The carbon deposition is detected by XRD on as-sintered samples and evidenced by the low conductivity. A subsequent carbon burning step has to be performed on all samples at 500°C in air for 20 min. The result is a dramatic increase in resistivity reaching values similar to standard annealing in air for the lowest temperatures. Above 700°C, the resistivity remains one or two orders of magnitude below, presumably because of reduction in the core of the sample, which can't be re-oxidized because the porosity is closed. The core reduction is indicated by the high permittivity which DC value can be as high as 6000 for samples annealed at 800°C due to internal barrier layer capacitance (IBLC) phenomenon [26].

The permeability is increasing with sintering temperature and at the same time, the resonance frequency (i.e. the frequency at which the imaginary part of the permeability is maximum) is decreasing accordingly. The merit factor $\mu_R.f_R$ is between 3.6 and 5 GHz, the latter value being comparable to that obtained by classical process [25]. Usually, the increase of the permeability upon sintering temperature is explained by the increase in grain size. Indeed, the area of Bloch walls is enhanced accordingly leading to an upraise in permeability. However, in our case, the grains are small enough to prevent formation of Bloch walls in a grain. Part of the explanation ties to the effect of porosity according to Lorenz relation

$$\chi_{eff} = \frac{(1-p)\chi}{1+pN\chi} \qquad (2)$$

As a computation can show, this effect is still not sufficient to explain the change in permeability. It is well known that due to its half-filled 3d shell, Fe^{3+} has a week spin-orbit coupling and thus gives almost no magnetocrystalline anisotropy and magnetostriction. Thus, in this material, anisotropy and magnetostriction are both due to Ni^{2+} only and negative. As we have shown, SPS produce reduction of Fe^{3+} into Fe^{2+} which not only results in increasing the conductivity but also give a positive contribution to both anisotropy and magnetostriction. As a results, there is a net reduction of the anisotropy of the lattice producing the observe enhancement of the permeability. As consequence, the reduction of Fe^{3+} is not only a drawback but also an advantage if it is well controlled in order to optimize the properties globally. Finally SPS can be used to produce in a single step magnetic cores. Usually SPS samples are made in the shape of flat cylinder and then drilled to make a ring for magnetic measurement and realize a component (inductor, transformer). It is however better to produce directly a toroid since drilling is ceramics is a tricky process. Using the mould shape sketch in Fig. 6, we have been able to make ring in one step (see picture) and we have shown that the magnetic

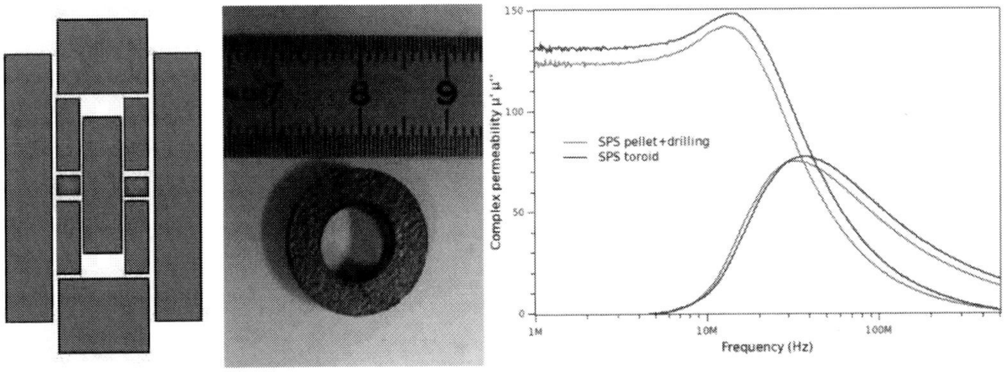

Figure. 6 : Realization of toroids directly by SPS. Left, mould geometry; centre picture of toroid (scale is cm); right, comparison between permeability spectra of a toroid directly made by SPS and another made by drill a SPS pellet made from the same powder in same conditions (750°C, 110 min)

properties are very similar to that measured on drilled samples produced exactly in the same conditions. The last step will be now to make directly a monolithic component by sintering the coil in the magnetic material.

5. Conclusion

Although ferrites are known since long and can be seen as old materials, they have been in constant evolution in order to satisfy the the demand of industry in terms of properties. One of the main advantage of ferrites is their adaptability: indeed, because of the accommodation capacity of different sites (tétra, octa and eventually bipiramidal) substitutions with many different ions (divalent in general but also mixed-valence) is possible and due to the localized electron nature of their magnetism, the effect of these substitutions is relatively predictable and usually show quite linear dependence on the substitution rate. For all these reasons, it is expected that ferrites will have a strong development in the XXIst century not only in the three directions explored herein, Ni-Zu-Cu spinels for high frequency, W and M-type hexaferrites for magnetocaloric refrigeration and hard magnets, but also Z and Y-type hexaferrites for microwave antennas and the multiferroic bismuth ferrite BiFeO that shows ferromagnetic as well as ferroelectric behaviours [27, 28], biomedical applications [29] as well as spin electronics devices such as magnetic tunnel junctions [30] and spin filters [31].

References

[1] Néel P, 1948 *Ann. de Phys.* **3**137
[2] Smit J, Wijn H, 1959 *Ferrites (Eindhoven: Philips Technical Library)*
[3] Wijn H 1970), 547*in Landolt & Bornstein III/4b, (Springer-Verlag, New York,*
[4] Hytch H et al. 2003 *Phys. Rev Lett.* **91** 257207
[5] Mazaleyrat F et al. 2009 *J. Alloys and Compounds* **483** 473
[6] Néel L 1947 *Comptes Rendus Acad. Sci. Paris* **224** 94
[7] Stoner E , Wohlfarth E 1948 *Trans. Roy. Soc. A* **240** 599
[8] Tenaud P et al. 2004 *J. Alloys and Compounds* **370** 331
[9] Pankhurst Q. et al., 1996 *J. Magn. Magn. Mat.* **155** 104
[10] Zhong W, et al., 1997 *J. Magn. Magn. Mat.* **168** 196
[11] Mazaleyrat F, Pasko A, Bartok A , LoBue M 2011 *J. Appl. Phys.* **109** 07A708
[12] Urbain G, Weiss P, Trombe F, 1935 Compt. Rend. **200** 2132
[13] Franco V, Conde C F , Conde A, Kiss L F 2007 *Appl. Phys. Lett.* **90** 052509
[14] Bozorth R M 1951, *Ferromagnetism (van Nostrand Co Inc.)*
[15] Fujita et al. 2003 *Phys. Rev. B* **67** 104416
[16] Niziol S, et al. 1982 *J. Magn. Magn. Mater.* **27** 281
[17] Gercsi Z, Hono K, Sandeman K G 2011 *Phys. Rev. B* **83,** 174403
[18] Belov K. P. et al. 1977 *Sov. Phys. JETP* **45** 307
[19] Naiden E P and Zhilyakov S M 1997 *Russ. Phys.* **J. 40**, 86974
[20] LoBue M. , et al. 2012 *J. Appl. Phys.* **111,** 07A905
[21] Basso V, Sasso C P , and Kupferling M 2010 Rev. Sci. Instrum. 81 113904
[22] Nomura T 1992 *Proc. Int. Conf. On Ferrites, ICF-6, Japan* 1206-1213
[23] Lebourgeois R et al. Proc. Int. Conf. On Ferrites, ICF-9, USA (2005) 791
[24] Lucas A , Lebourgeois R, Mazaleyrat F, Labouré E 2010 *Appl. Phys.Le*tt. **97** 182502
[25] Lucas A,Lebourgeois R, Mazaleyrat F, Labouré E , 2011 *J. Magn. Magn. Mat.* **323** 735
[26] Guillemet-Fritch S. et al.2008 Adv. Mat. **20** 551
[27] Wang J et al. 2003 Science **299** 1719
[28] Kumar A et al. this volume, paper CM58
[29] Fortin J.P et al. 2007 *J. Am. Chem. Soc.* **129** 2628
[30] Zutic I,Fabian J, Das Sarma S 2004 *Rev. Modern Phys.* **76** 323
[31] Luders U et al. 2006 *Appl. Phys. Lett.* **88** 082505

X-ray absorption fine structure (XAFS) spectroscopy using synchrotron radiation

B. D. Shrivastava
School of Studies in Physics, Vikram University, Ujjain

rashmibasant@gmail.com

Abstract. The X-ray absorption fine structure (XAFS) spectra are best recorded when a highly intense beam of X-rays from a synchrotron is used along with a good resolution double crystal or curved crystal spectrometer and detectors like ionization chambers, scintillation counters, solid state detectors etc. Several synchrotrons around the world have X-ray beamlines dedicated specifically to XAFS spectroscopy. Fortunately, the Indian synchrotron (Indus-2) at Raja Ramanna Centre for Advanced Technology (RRCAT) at Indore has started operation. A dispersive type EXAFS beamline called BL-8 has been commissioned at this synchrotron and another beamline having double crystal monochromator (DCM) is going to be commissioned shortly. In Indian context, in order that more research workers use these beamlines, the study of XAFS spectroscopy using synchrotron radiation becomes important. In the present work some of the works done by our group on XAFS spectroscopy using synchrotron radiation have been described.

1. Introduction

X-ray absorption fine structure (XAFS) spectroscopy refers to the details of how X-rays are absorbed by an atom at energies near and above the core-level binding energies of that particular atom. At certain energy, the absorption increases drastically and gives rise to an absorption edge. Although most of the absorption spectrum is quite smooth, oscillatory features called X-ray absorption fine structure (XAFS) is found directly above an edge. This fine structure is intrinsically quantum mechanical phenomenon that is based on the X-ray photoelectric effect, in which an X-ray photon incident on an atom within a sample is absorbed and liberates an electron from an inner atomic orbital. The "photoelectron" wave scatters from the atoms around the X-ray absorbing atom, creating interferences between the outgoing and scattered parts of the photoelectron wave function. These quantum interference effects cause an energy-dependent variation in the X-ray absorption probability, which is proportional to the X-ray absorption coefficient, a measurable quantity. Traditionally, this fine structure is split into two energy regions. The first, termed as X-ray absorption near edge structure (XANES), occurs in the region from the edge to approximately 40 eV above the edge, while the second, termed as the extended X-ray absorption fine structure (EXAFS), extends from 40 eV to 1000 eV above the edge. The higher energy portions of EXAFS spectra are dominated by single-scattering events within only some tenths of nm (5-6 Å) from the absorber, and contain information specific to the local structure surrounding the absorber. On the other hand, XANES spectra, being centered close to the absorption jump, are dominated by multiple-scattering events extending a few nm from the

Published under licence by IOP Publishing Ltd

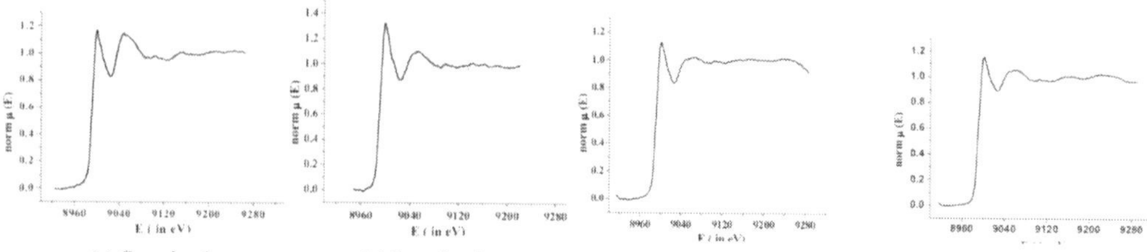

(a) Complex 1 (b) Complex 2 (c) Complex 3 (d) Complex 4

Figure 1. Normalized EXAFS spectra at the K-edge of copper in the four complexes.

absorber. XANES is strongly sensitive to formal oxidation state and coordination chemistry of the absorbing atom, while the EXAFS is used to determine the distances, coordination number, and species of the neighbors of the absorbing atom. As a result, XAS provides a practical and simple way to determine the chemical state and local atomic structure for a selected atomic species. The significant advantage of X-ray absorption spectroscopy over the X-ray crystallography is that the local structural information around the element of interest can be obtained easily and quickly even from disordered samples, such as powders and solutions. To be specific, single crystals of samples are not required in XAS spectroscopy as are necessary in the case of X-ray crystallography. This versatility allows it to be used in a wide variety of disciplines: physics, chemistry, biology, biophysics, medicine, engineering, environmental science, materials science, and geology [1].

2. Transition from X-ray tubes to synchrotron sources

X-ray absorption spectroscopy (XAS) has been extensively used in the past to obtain information about molecular structure viz., the valency, bond type, ionic charges, coordination stoichiometry etc. Many research centers in Indian universities and institutes have been doing extensive research work in this field since 1950 or so. At all of these centers, the laboratory X-ray spectroscopic set-ups were in use which employ photographic method of registration of spectra. These types of set-ups comprise of low power (0.5kW - 3kW) X-ray tubes and Cauchois-type curved mica crystal spectrographs employing X-ray films as detectors. After obtaining a number of spectrograms, the analog and digital spectral records were obtained with the help of microphotometers. These analog and digital spectral records were used to be analyzed manually. The results about X-ray absorption edge energies, edge structures, near edge structures and extended fine structures were being reported. The data was used to be generally analyzed qualitatively and empirically to yield useful information about molecular structure. The data was also analyzed through some established relations to yield information about valency, effective nuclear charge, coordination type, average bond length etc. To our knowledge, such photographic EXAFS data has not been analyzed by Fourier transforming and fitting with theoretical standards. Only recently, we have made such an attempt [2, 3]. Unfortunately, nearly all of these centers ceased to work in 1980s because research workers around the world had started using X-rays from synchrotron instead of X-ray tubes for recording the X-ray absorption spectra.

The experimental X-ray absorption spectroscopic data obtained from X-ray tube sources is plagued by noise and systematic errors. It wasn't until the 1970s that high brilliance synchrotron radiation began to be used to obtain absorption spectra. Experimental techniques have continued to develop over the years, and due to the exponential increase in brilliance over time, high quality XAFS data is now routinely collected at a variety of second and third generation synchrotron sources. Over the past three decades, the technique of XAFS has made great strides toward the goal of providing such information. The existence of intense new synchrotron X-ray sources alone was not enough to achieve this goal, even though such facilities spurred considerable progress. In addition, the full success of the XAFS technique must be attributed in large part to advances in theory, which have led ultimately to a highly quantitative understanding of the phenomena.

X-ray absorption spectroscopic measurements with synchrotron radiation are generally carried out on two different types of beamline set-ups. In the first type of XAFS beamline set-up, a double crystal

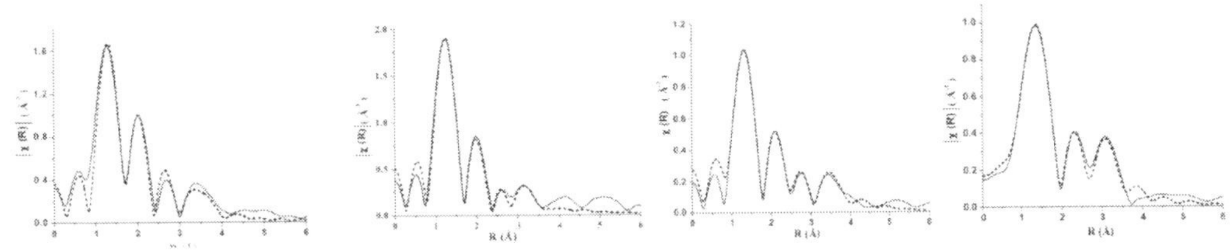

(a) Complex **1** (b) Complex **2** (c) Complex **3** (d) Complex **4**

Figure 2. Fourier transformed EXAFS data for the four complexes. Solid lines are experimental data and dashed lines are modeled fit

monochromator (DCM) is used which selects a particular energy from the incident synchrotron beam. This type of set-up is generally used at all of the synchrotrons throughout the world. In the second type of XAFS beamline set-up, a bent crystal polychromator is used, to select a band of energy from the white synchrotron beam, which is horizontally dispersed and focused on the sample. Fortunately, the 2 GeV Indus-2 synchrotron, at Raja Ramanna Centre for Advanced Technology (RRCAT) at Indore, India has started operation and the second type of EXAFS beamline set-up namely, BL-8 dispersive EXAFS (DEXAFS) beamline, has been commissioned. Another beamline having double crystal monochromator (DCM) is going to be commissioned shortly. In Indian context, in order that more research workers use these beamlines, the study of X-ray absorption spectroscopy using synchrotron radiation becomes important.

3. Analysis of EXAFS data using theoretical model

The data obtained from EXAFS beamline is generally analyzed using available computer software packages like *Athena* and *Artemis* (available from website www.xafs.org). These programs include *AUTOBK* for background removal, *FEFF6L* for generation of the theoretical EXAFS models, and *FEFFIT* for parameter optimization of the model. A brief description of the recommended analysis procedure is given below. [4]

For normalizing the μ(E) vs. E spectra, a linear function is regressed to the pre-edge region, and a linear or quadratic function is regressed to the post-edge region. Normalized μ(E) spectra are produced by subtracting the pre-edge line from the entire data spectrum and then dividing the spectrum by the step height. μ(E) data is then converted to χ(k) data. The χ(k) data is then Fourier transformed in *Artemis*. The distance found in Fourier transform is about 0.2 Å - 0.5 Å shorter than the actual distance due to energy dependence of the phase factors in sine function of the EXAFS equation (given below). Hence, to extract the information about the various parameters that can be determined from EXAFS data, the data has to be compared with theoretical model. This can be done through the software *Artemis*. The program *Artemis* has a special interface called "Atoms" that converts crystal structure information into a cluster of atoms and provides a list of atom positions in x, y, z coordinates for use in FEFF computations. This trial structure is given as input to the FEFF code to generate a theoretical EXAFS model.

The theoretical model is written as a sum of paths of the contribution from all scattering paths of photoelectron that travels from the absorbing atom and then is scattered from one or more neighbouring atoms and finally returns to the absorbing atom.

$$\chi(k) \equiv \sum_i \left[N_i S_0^2 F_{\mathrm{eff}_i}(k)/kR_i^2\right]\cdot \mathrm{Sin}\left[2kR_i + \phi_i(k)\right]\exp(-2\sigma_i^2 k^2)\exp[-2R_i/\lambda(k)]$$

$$\text{with } R_i = R_{0i} + \Delta R_i \quad \text{and} \quad k^2 = [2m_e(E - E_0 + \Delta E_0)/\hbar]$$

In the above equations, the terms $F_{\mathrm{eff}_i}(k)$, $\varphi_i(k)$ and $\lambda(k)$ are the effective scattering amplitude of the photoelectron, the phase shift of the photoelectron and the mean free path of the photoelectron,

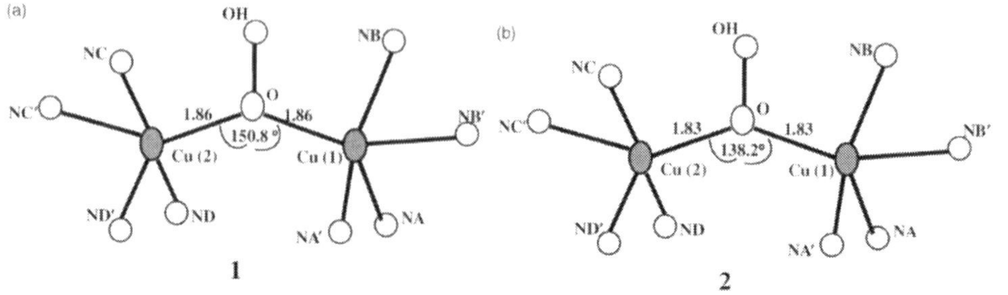

Figure 3. Coordination geometry about the copper (II) ions in complexes **1** and **2** as deduced from the EXAFS data analysis and following Haddad et al [7].

respectively, all of which can be calculated by a computer program such as FEFF. The term R_i is the half path length of the photoelectron (i.e., the distance between the absorber and a coordinating atom for a single-scattering event). The value of R_{0i} is the half path length used in the theoretical calculation which can be modified by ΔR_i. Each path has several adjustable parameters optimized by the computer code FEFFIT to fit the data. These parameters include the passive electron reduction factor (S_0^2), the number of identical paths (N_i), the relative mean-square displacement of the atoms included in path (σ_i^2), an energy shift for each path (ΔE_0), and a change in the path length (ΔR_i).

The theoretical model is adjusted as needed (e.g., with different atom types) until the best possible fit is obtained between theoretical and experimental spectra. The parameters that are often determined from a fit to the EXAFS spectrum affect either the amplitude of the EXAFS oscillations (N, S_0^2, σ^2) or the phase of the oscillations (ΔE_0 and ΔR).

In the following, some of the works done by our group on X-ray absorption spectroscopy using synchrotron radiation have been described.

4. Calibration of dispersive EXAFS beamline at RRCAT, Indore:

Before doing any XAFS measurement on the recently developed energy dispersive EXAFS beamline at the Indus-2 synchrotron source, RRCAT, Indore, calibration of the particular setting of the polychromator has to be done. We have outlined a method which should be used for finding the dispersion and then for calibrating the experimental spectra [5]. This method involves taking the position of the first maximum of the derivative spectra of the two standards, for determining their edge positions and energies. Also, the performance of the beamline for the particular setting should be checked by recording the EXAFS of a standard and by fitting the EXAFS data with a theoretical model. By recording the K-absorption spectra of copper metal foil and a copper complex on this beamline, it has been shown by analysis and extracting information about the various parameters that the performance of the beamline is quite satisfactory [2, 3].

5. Application of EXAFS spectroscopy using synchrotron radiation - EXAFS study of binuclear hydroxo-bridged copper (II) complexes

As an example of the application of EXAFS, a study carried out by us on copper complexes [6] is being presented here. The EXAFS spectra of binuclear monohydroxo-bridged copper(II) complex $[(bpy)_2Cu\text{-}OH\text{-}Cu(bpy)_2](ClO_4)_3$ (**1**), its analogous complex $[(phen)_2Cu\text{-}OH\text{-}Cu(phen)_2](ClO_4)_3$ (**2**) and dihydroxo-bridged copper(II) complex $[Cu_2(\mu\text{-}OH)_2(bipy)_2]SO_4.5H_2O$ (**3**), its analogous complex $[Cu_2(\mu\text{-}OH)_2(phen)_2]SO_4.5H_2O$ (**4**) (where bpy and phen are 2,2'-bipyridine and 1,l0-phenanthroline, respectively) have been recorded at dispersive EXAFS beamline BL-8 on Indus-2 synchrotron, RRCAT, Indore.

The energies of the K-edge of copper are found to be 8989.69 eV, 8990.01 eV, 8988.30 eV and 8988.11 eV for complexes **1**, **2**, **3** and **4**, respectively. Taking the energy of copper metal K-edge as 8980.5 eV, the chemical shifts are found to be 9.19 eV, 9.51 eV, 7.80 eV and 7.61 for complexes **1**, **2**,

Table 1(a) . The EXAFS fitting results for complexes 1 and 2. The XRD results for complex 1 (from Haddad *et al.* [7]) are also given.

Atomic pair	Complex 1					Complex 2		
	EXAFS results				XRD results	EXAFS results		
	N	R(Å)	delr(Å)	σ^2(Å^{-2})	R	N	R(Å)	σ^2(Å^{-2})
Cu1-O1	1	1.86	-0.06	0.0101 ± 0.0026	1.92	1	1.83	0.0134 ± 0.0017
Cu1-NA'	1	2.18	0.17	0.0123 ± 0.0099	2.00	1	2.09	0.0091 ± 0.0033
Cu1-NB	1	2.18	0.17	0.0123 ± 0.0099	2.01	1	2.10	0.0091 ± 0.0033
Cu1-NB'	1	2.21	0.17	0.0123 ± 0.0099	2.04	1	2.13	0.0091 ± 0.0033
Cu1-NA	1	2.36	0.12	0.0123 ± 0.0099	2.24	1	2.10	0.0091 ± 0.0033
Cu1-Cu2	1	3.60	-0.04	0.0014 ± 0.0049	3.64	1	3.42	0.0087 ± 0.0100

Table 1(b) . The EXAFS fitting results for complexes 3 and 4. The XRD results for complex 3 (from Casey *et al.* [8]) are also given.

Atomic pair	Complex 3					Complex 4		
	EXAFS results				XRD results	EXAFS results		
	N	R(Å)	delr(Å)	σ^2(Å^{-2})	R	N	R(Å)	σ^2(Å^{-2})
Cu1-O1	1	1.91	-0.01	0.0082 ± 0.0065	1.92	1	1.86	0.0082 ± 0.0065
Cu1-O2'	1	1.91	-0.01	0.0082 ± 0.0065	1.92	1	1.86	0.0082 ± 0.0065
Cu1-N1	1	1.98	-0.01	0.0082 ± 0.0065	1.99	1	1.94	0.0082 ± 0.0065
Cu1-N2	1	1.99	-0.01	0.0082 ± 0.0065	2.00	1	1.94	0.0082 ± 0.0065
Cu1-O4	1	2.29	-0.09	0.0066 ± 0.0059	2.37	1	2.38	0.0157 ± 0.0063
Cu1-Cu2	1	2.76	-0.08	0.0075 ± 0.0028	2.84	1	2.78	0.0046 ± 0.0022

3 and **4**, respectively. These values of the chemical shifts suggest that copper is in +2 oxidation state in all of these four complexes.The EXAFS data have been analyzed and theoretical models have been generated for the complexes **1** and **3**, and then fitted to their experimental EXAFS data to obtain the bond-lengths, coordination numbers and thermal disorders. As the crystallographic data for the complexes **2** and **4**
are not available in literature, we have determined the structural parameters for these complexes by fitting their experimental EXAFS data with the same theoretical models which were generated for their corresponding analogous complexes, i.e., **1** and **3** respectively.

5.1.Complex 1

For the analysis of the EXAFS data of **1**, the input parameter R_{bkg}, that determines the maximum frequency of the background, was set to 1.0 Å. Fourier transform was performed over k-range: k_{min}=2.30 Å-1, k_{max}=8.38 Å^{-1}. Theoretically modeled data was fitted in the R-space to the experimental data using kw = 2. Fitting was performed for the coordination shells in the R range of 1.0 - 5.0 Å. In the fitting procedure, we have used the first 16 paths obtained from FEFF calculations. The value of goodness-of-fit parameter, i.e., reduced chi-square (χ_ν^2) is 76.31. The results obtained from fitting are given in table 1(a), which gives the local structure parameters for complex 1 obtained from the analysis. The S_0^2 value obtained is 1.57 ± 0.31. ΔE_0 value is also reasonable, i.e., 5.43 eV ± 1.75. Also, the σ^2 values are quiet high as the measurements have been performed at room temperature.

Our results for the bond distances in **1**, with ΔR ranging from -0.04 to 0.17, are given in the table 2(a) and these distances are in agreement with the crystallographic bond distances reported by Haddad et al. [7]. The Cu-O-Cu bridging angle calculated from the obtained distances is 150.8° which is in the range of 135.4° - 165.6° reported for binuclear monohydroxo-bridged complexes. The value of this angle as reported by Haddad et al. is 141.6°. Hence, the theoretical model generated by us for **1** appears to be quite reasonable. Using the present results the inner coordination spheres around the two copper (II) centers have been depicted in figure 3(a). This figure is based on the figure given by Haddad et al. but shows the distances and the bridging angle obtained by us in the present study.

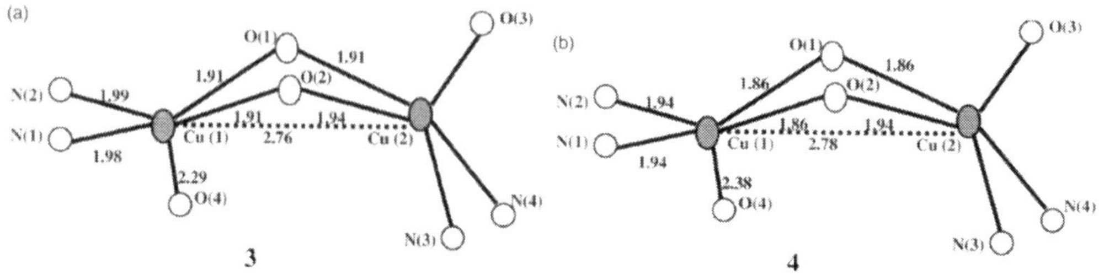

Figure 4. Coordination geometry about the copper (II) ions in complexes **3** and **4** as deduced from the EXAFS data analysis and following Casey et al [8].

5.2 Complex 2

As the theoretical model generated by us for **1** fits well to its experimental EXAFS data and as **1** and **2** are analogous to each other, we have used the same theoretical model generated for **1** to fit the experimental EXAFS data of **2**. The data analysis procedure used for **1**, as explained above, has also been applied to **2** and the input fitting parameters were kept the same during fitting. The S_0^2 value obtained is 1.17 ± 0.24. ΔE_0 value is also reasonable, i.e., -1.25 eV ± 2.78. The structural parameters so obtained for **2** are reported in table 2(a). The Cu-O-Cu bridging angle calculated from the obtained distances is 138.2° which is in the range of 135.4° - 165.6° reported for binuclear monohydroxo-bridged complexes. Using the obtained results, the inner coordination spheres around the two copper (II) centers in **2** have been depicted in figure 3(b).

5.3 Complex 3 and 4

The crystallographic data is available for **3** [8], using which the theoretical model for the complex has been generated. This theoretical model is then fitted to its experimental EXAFS data to obtain the structural parameters for **3** and the results obtained from fitting are given in table 1(b). The results have been found to be comparable with the crystallographic results. The Cu-O-Cu bridging angle calculated from the obtained distances comes out to be 91.8°, while the range reported for binuclear dihydroxo-bridged complexes is 95.6° - 104.1°. The value of this angle as reported by Casey et al. [8] is 97.0°.

As the crystallographic data for **4** is not available in the literature, we have applied the same method as in the case of **1** and **2**, i.e, we have used the theoretical model generated for **3** to fit the experimental EXAFS data of its analogous complex **4**. The structural parameters so obtained for **4** are also reported in table 1(b). The Cu-O-Cu bridging angle calculated from the obtained distances is 96.71° which is in the range of 95.6° - 104.1° reported for binuclear dihydroxo-bridged complexes.

Using the present results the inner coordination spheres around the two copper (II) centres have been depicted in figures 4(a) and 4 (b) for complexes **3** and **4**, respectively. Figure 4(a) is based on the figure given by Casey et al. but shows the distances and the bridging angle obtained by us in the present study.

The metal-metal coordination number in the four complexes has been found to be unity (table 1 (a) and (b)) which points to the presence of a stable binuclear complex. Thus, on the basis the analysis of the EXAFS data, the four complexes have been shown to be binuclear, i.e., they contain two metal atoms per cluster.

Thus, it has been demonstrated that if crystal structure is available for a complex from X-ray crystallographic studies (which require single crystals), structural parameters like bond-length, coordination number and Debye-Waller factor can be determined for any of its analogous complex from EXAFS studies (which do not require single crystals).

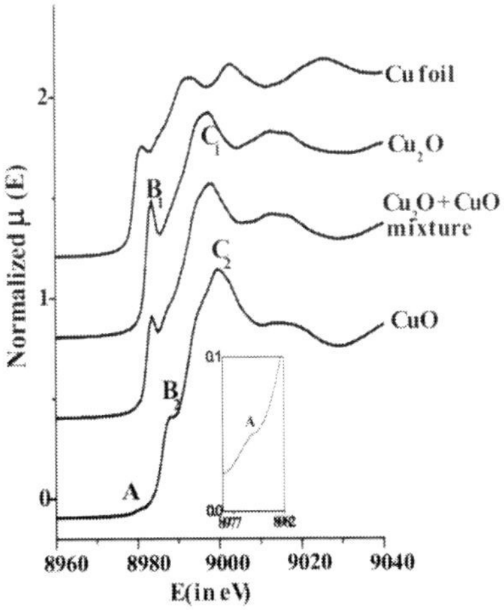

Figure 5. XANES spectra of copper metal, cuprous oxide, cupric oxide and their mixture,

6. Application of XANES spectroscopy using synchrotron radiation - A comparative study of the methods of speciation using X-ray absorption fine structure

As an example of the application of XANES, study carried out by us on speciation of copper compounds in their mixture is being presented here. XANES spectrum reflects the local structure around the absorbing atom and provides information about oxidation state of an excited atom and the coordination symmetry. By taking standards of well-defined chemical species, analysis of XANES can be used to determine metal speciation, i.e., determination of the chemical forms along with the relative quantity of the different species in a given sample. For speciation using XANES, the commonly used methods are: principal component analysis (PCA), target transformation (TT) and linear combination fitting (LCF). Other methods which can be used are: methods based on derivative spectra, method based on the relative position of the absorption edge, residual phase analysis (RPA) and normalized difference absorption edge spectra analysis (NDAES).

We have made comparative studies of the above mentioned different methods of speciation and discussed their relative merits by taking example of mixture of cuprous oxide and cupric oxide in a specific ratio [9, 10]. The X-ray absorption fine structure (XAFS) spectra for these standards and the mixture were recorded using the EXAFS set-up at the synchrotron wiggler beamline 4–1 at the Stanford Synchrotron Radiation Laboratory (SSRL), Stanford, California, USA.

The normalized (energies normalized to the elemental Cu edge, baseline, and background corrected) XANES spectra of copper metal, cuprous oxide, cupric oxide and their mixture are shown in figure 5. The edge position, defined as the maximum point of the first derivative function in the rapidly rising edge step of the absorbance vs. energy plot, have been found to be 8979.0, 8980.6 and 8983.6 eV for Cu foil, Cu_2O and CuO, respectively. The different characteristic features obtained in the present investigation in the XANES region for two copper oxides, which are used as standards, are also shown in figure 5. These features are important as they are ultimately used for speciation purpose.

In the present study, for PCA and target transformation, *SIXPack* data analysis software has been used. (Details can be found at www.ssrl.slac.stanford.edu/swebb/sixpack.htm). LCF has been performed using *Athena*. The position of the edge corresponding to the point having $\mu x = 0.5$ in the normalized spectrum has been determined using *Athena*. The difference spectra and the derivative spectra for the analysis were also constructed using *Athena*. EXAFS data analysis program *Artemis* has been used for optimizing the theoretical model to the measured spectrum.

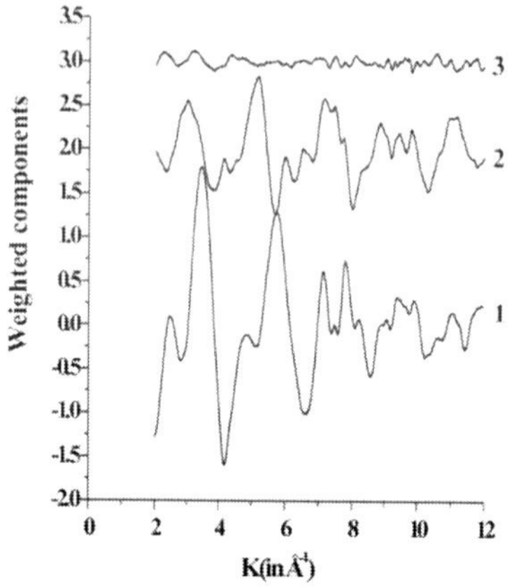

Figure 6. First three components from PCA calculation performed in the EXAFS region and weighted by eigenvalues.

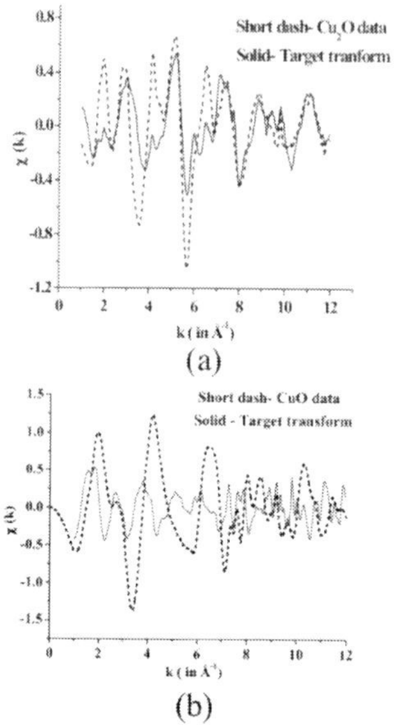

Figure 7. Predicted targets obtained through target transformation of the Cu Kedge EXAFS spectra for (a) Cu_2O and (b) CuO

By applying PCA and target transformation, the number and identity of components in the mixture have been confirmed. Then, five methods have been used by us for determining the proportions of the CuO and Cu_2O in the mixture. The methods of speciation and the results obtained are discussed below. *6.1 Principal component analysis (PCA)*: PCA determines the minimum number of significant components required to satisfactorily regenerate the data matrix, using a reduced space. PCA assumes that the absorbance in a set of spectra can be mathematically modeled as a linear sum of individual Figure 5. XANES spectra of copper metal, cuprous oxide, cupric oxide and their mixture, components, called factors, which correspond to each one of the species present in the sample, plus noise. The two main outputs from PCA are eigenvalues and eigenvectors. PCA has been performed in both XANES and EXAFS regions. However, the results of PCA have been presented in figure 6 only for the EXAFS region in the range 2 to 12 Å^{-1}. The figure shows the different components weighted by their eigenvalues. It can be seen from the figure that the first two components clearly dominate the spectra and the third component appears to be noise.

Table 2. Results of speciation

S.No.	Method of speciation	Percentages of components in the mixture		
		Cu_2O	CuO	Max. % error
1.	LCF	73	27	± 2 %
2.	NDAES	71	29	± 5 %
3.	RPA	70	30	± 5 %
4.	Der. Cu(I)*	74	26	± 2 %
5.	Edge**	79	21	± 5 %
As prepared		75	25	-

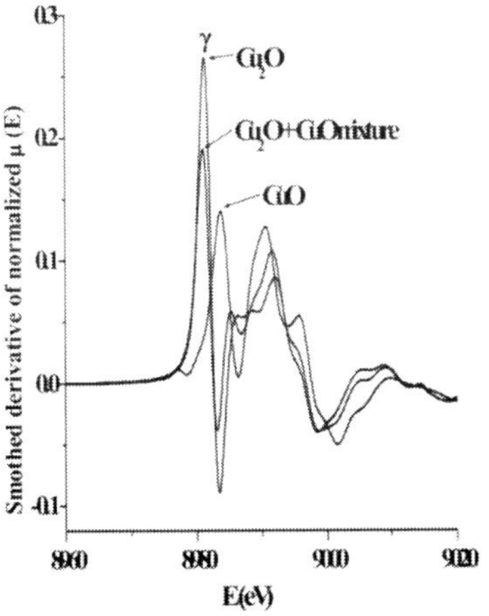

Figure 8. First derivatives of XANES region for Cu_2O, CuO and their mixture. Feature γ corresponds to the Cu (I) component. The height of this feature has been used to determine the percentage of Cu (I) in the mixture.

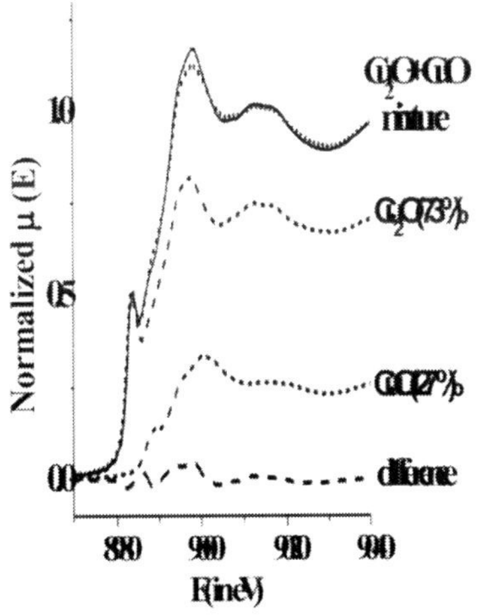

Figure 9. Results of the LCF for copper K-edge XANES spectra of mixture of Cu_2O and CuO.

6.2 Target transformation (TT): In target transformation, which is also known as transformation factor analysis (TFA), the spectra from reference compounds are individually compared with the collection of mathematical principal components (PCs). The experimental reference spectra are transformed to fit the number of PCs selected from PCA. The degree to which the reference must be changed to fit the PCs is measured by SPOIL value. A high SPOIL value indicates a poor fit, suggesting that this standard is not well represented in the collection of sample spectra (represented by the PCs) [4]. Target transformation has been performed in both XANES and EXAFS regions. However, the results of TT have been presented in figure 7 only for the EXAFS region in the range 2 to 12 $Å^{-1}$. One of the standards, i.e., copper metal has been rejected as a potential target because its SPOIL value has come out to be > 6. Cu_2O has proved to be a very good target, with a SPOIL value < 3 and CuO has come out as marginal candidate in the SPOIL test, with SPOIL value between 3 and 6.

6.3 Method based on derivative spectra: The derivative spectra of the mixture and the standards Cu_2O and CuO are shown in figure 8. The derivative spectrum of a mixture has a distinct feature: 'γ' at 8981 eV corresponding to the Cu(I) component. It is the most intense feature of the derivative XANES spectra of Cu(I) species and can be used to determine the percentage of a species in the mixture. Also, it lies in a region (from 8980 to 8984 eV) where the Cu(II) species have a null derivative. The intensity of this peak varies as the amount of Cu(I) species is varied. If I_m and I_{Cu2O} are the intensities of the first maximum in the derivative spectra at 8981 eV in the mixture and Cu_2O respectively, then the ratio of I_m to I_{Cu2O} gives the fraction of Cu(I) in the mixture. The percentages of Cu_2O and CuO in the mixtures as obtained from this method are 74% and 26% respectively with an error of ± 2%.

6.4 Method based on position of the edge: It is well known that position of the absorption edge varies with the oxidation state of central metal atom. Also, significant difference between the positions of the Cu K-edges of Cu(I) and Cu(II) has been observed. In a mixture the position of the Cu K-edge is between the positions of the edges of Cu(I) and Cu(II). This shift in the value of edge energy of the mixture w.r.t. Cu(I), can be used to measure percentage of Cu(II) present in the mixture.

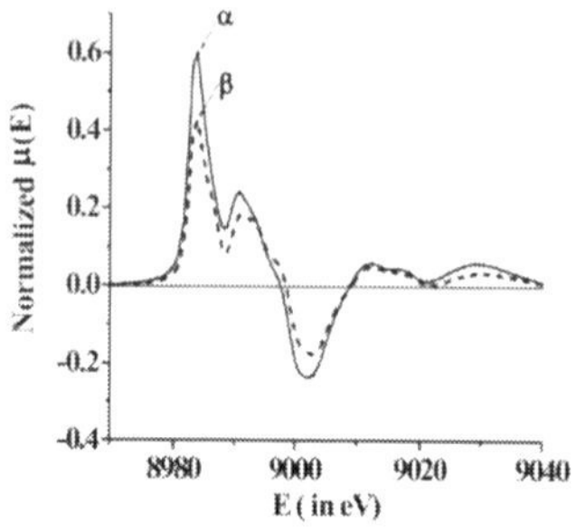

Figure 10. Normalized difference absorption edge spectra for copper K-edge XANES. The ratio of the height of positive peak β to that of peak α gives the proportion of Cu(I) in the mixture.

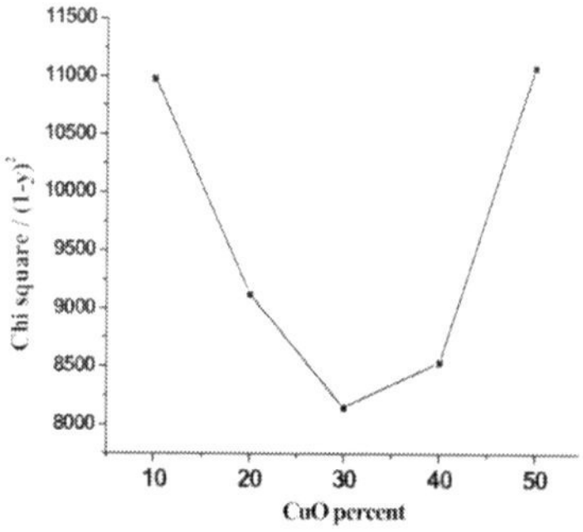

Figure 11. Reduced statistical chi square obtained by varying the percentage of CuO using RPA performed in EXAFS region.

For this method, the position of the edge is taken as the point corresponding to $\mu x = 0.5$ in the normalized spectrum. Writing E_{Cu2O}, E_{CuO} and E_m as the energies of the position of the edge in Cu_2O, CuO and the mixture respectively, the fraction x_m of Cu(II) in the mixture can be determined from relation: $x_m = \Delta E_m/\Delta E_{CuO}$, where $\Delta Em = (E_m\text{-}E_{Cu2O})$ and $\Delta E_{CuO} = (E_{CuO}\text{-}E_{Cu2O})$ are the edge energy shifts w.r.t. Cu_2O. The percentages of Cu_2O and CuO in the mixtures as determined from this method are 79% and 21% with an error of ± 5%.

6.5 Linear combination fitting (LCF): In least-squares LCF, one determines the proportions of the spectra for selected standards that, when summed, yield the least fit to the spectrum for an unknown sample. Also, least-squares LCF can be done across a wide energy range of the spectrum, i.e., even in the EXAFS region and thus enhances the discrimination of different species having identical spectral features at a given energy but distinct features at other energies.

The results of LCF are shown in figure 7. The fractions of the Cu_2O and CuO making up the fitted spectra are also shown in this figure. The percentages of Cu_2O and CuO in the mixture, as determined from LCF method are 73% and 27% respectively with an error of ± 2%. In the case of LCF method, R-factor and chi-square values are 0.00016 and 0.02309 respectively. The little difference between the experimental data and LC fitted data can be seen from the difference curve which is also shown in figure 9.

6.6 Normalized difference absorption edge analysis (NDAES): In this method, to quantitate copper content, the normalized difference edge absorption spectra have been calculated by subtracting normalized edge of Cu(II) from that of normalized edge of the Cu(I) which produces characteristic derivative shaped spectra. The difference spectrum has a positive peak at 8983-8984 eV and a broad negative feature at 8990- 9000 eV. The amount of cuprous ion present can be determined from the height of the positive peak. In other elements, similar characteristic features are present which can be used for the purpose of speciation.

The results of NDAES are shown in figure 10. In this figure, α is the positive peak obtained when CuO spectrum is subtracted from Cu_2O spectrum and β is obtained when the spectrum of mixture is subtracted from the Cu_2O spectrum. The heights of the peaks β and α have been measured very precisely. The ratio of the height of positive peak β to that of peak α gives the proportion of Cu(I) in the mixture. The percentages of Cu_2O and CuO in the mixture, as determined from this method are 71% and 29% respectively with an error of ± 5%.

6.7 Residual phase analysis (RPA): RPA utilizes one of the known components as a ''starting phase.'' The ''starting phase'' is then fractionated and iteratively subtracted from the total EXAFS signal to produce corresponding residual spectra. The individual residual spectra are analyzed to obtain the best fit as follows.

Taking pure Cu_2O EXAFS data as the known phase and the EXAFS data of mixture as total EXAFS signal, the residual phase data has been constructed, for all y (adjustable mixing fraction) between 0 and 1, with an increment of 0.1. The statistical chi-square values obtained in each fit were then plotted as the function of y, and the minimum has been obtained as shown figure 11. The corresponding mixing fraction is found from the position of the minimum. The percentages of Cu_2O and CuO in the mixture have thus been found to be 70% and 30% respectively with an error of ± 5%.

The two studies presented here were done at two different beamlines at two synchrotron facilities. As already pointed out earlier, in India, the work on EXAFS has been done on recently developed Indus-2 synchrotron facility which was not earlier available to Indian workers. For the last forty years research workers in Indian laboratories were doing experiments on laboratory XAFS set-ups. Now, it is hoped that future research workers from Indian laboratories will utilize to the maximum the XAFS facilities at Indus-2.

References

[1] Bunker G 2010 *Introduction to XAFS - A Practical Guide to X-ray Absorption Fine Structure Spectroscopy* (Cambridge University Press, NewYork).
[2] Joshi S K, Gaur A, Johari A and Shrivastava B D 2009 *J. Phys. Conf. Ser.* **190** 012044
[3] Johari A, Gaur A, Shrivastava B D, Hinge V K, Joshi S K and Gaur D C 2011 *Indian J. Pure & Appl. Phys.* **49** 308
[4] Kelly S D, Hesterberg D and Ravel B 2008 *Chapter 14 in Methods of Soil Analysis, Part 5, Mineralogical Methods* (Medison, Soil Science Society of America) 387
[5] Gaur A, Johari A, Shrivastava B D, Gaur D C, Jha S N, Bhattacharyya D, A Poswal and S K Deb 2011 *Sadhana* **36** 339
[6] Gaur A, Shrivastava B D, Gaur D C, Prasad J, Srivastava K, Jha S N, Bhattacharyya D, Poswal A and Deb S K 2011 *J. Coord. Chem.* **64** 1265
[7] Haddad M S, Wilson S R, Hodgson D J and Hendrickson D N 1981 *J. Am. Chem. Soc.* **103** 384
[8] Casey A T, Hoskins B F and Whillans F D 1970 *Chem. Comm.* **708** 904
[9] Gaur A, Shrivastava B D and Joshi S K 2009 *J. Phys. Conf. Ser.* **190** 012084
[10] Gaur A and Shrivastava B D 2012 *Acta Phys. Polon. A* **121** 1111

Soft magnetic thin films: influence of annealing on magnetic properties

Marco Coïsson, Gabriele Barrera, Federica Celegato, Paola Tiberto, Franco Vinai
INRIM, Electromagnetism Division, strada delle Cacce 91, 10135 Torino (TO), Italy

m.coisson@inrim.it

Abstract. Soft magnetic materials are currently used in a variety of applications in electrical machines, sensors and elements of devices. If prepared in thin film form, they can be applied to micro- and nano-patterned devices. However, with respect to ribbons and bulk materials, thin films of the same composition usually display worse soft magnetic properties, thus requiring suitable solutions to restore high permeability and low coercivity. In this review, the magnetic properties of thin films prepared by sputtering are presented. Several compositions of soft magnetic materials are investigated, including Co-Fe-Si-B, Fe-Cu-Nb-Si-B, Fe-Si-B, Fe-Co-Nb-Si-P-B, Fe-Zr-Nb-Cu-B. Their amorphous-to-crystallization processes are studied by means of furnace annealing and Joule heating. Hysteresis loops and magnetic domain imaging, together with structural techniques, are used to follow the effects of annealing, which include stress relaxation and crystallization. The effects of magnetic field annealing are investigated and discussed.

1. Introduction

Soft magnetic materials have been widely studied in the last decades because of their interesting fundamental properties and their important applications in electrical machines, sensors and energy saving technologies [1,2]. The integration with the semiconductors industry and the progresses in the preparation of micro- and nano-devices are stimulating the research on soft magnetic thin films [3,4], which should be prepared with conventional sputtering techniques, coupled with other magnetic materials or non magnetic layers, properly annealed to improve their properties or to optimize interfaces, and ideally preserve the excellent soft magnetic properties of ribbons, microwires or bulks. Within this framework, this review paper discusses some recent results on soft magnetic thin films, prepared by sputtering in an amorphous condition, and subsequently annealed to induce (nano)crystallization and to tailor their properties. Results on selected compositions such as Co-Fe-Si-B, Fe-Cu-Nb-Si-B, Fe-Si-B, Fe-Co-Nb-Si-P-B, Fe-Zr-Nb-Cu-B will be analyzed.

2. Experimental techniques

Soft magnetic thin films can be prepared by means of sputtering on different substrates (e.g. glass, Si-O, Si_3N_4). Compositions close to those commonly used in amorphous ribbons can be obtained starting from targets made of the desired alloys. It is often possible to prepare these targets directly with amorphous ribbons prepared by planar flow casting. The resulting thin films are usually amorphous to X-ray diffraction, and preserve the same composition as the originating target (as detected e.g. with energy dispersive X-ray spectrography, EDS-EDX). The evolution of their microstructure (amorphous-to-crystalline transformation) can be studied by means of structural techniques (e.g. differential

scanning calorimetry, DSC, X-ray diffraction, XRD) and magnetic measurements (hysteresis loops, magnetization as a function of temperature) on samples which are submitted to thermal treatments. These include annealing in furnace (in vacuum, or in an inert gas), at temperatures at which diffusion of the atoms of the film in the substrate is negligible, or current annealing (Joule heating). Magnetic measurements are usually performed by means of high-sensitivity magnetometry (e.g. vibrating sample, VSM, or alternating gradient field, AGFM, magnetometry) or by optical means (magneto-optic Kerr effect, MOKE). Magnetic domain patterns can be studied by means of MOKE and magnetic force microscopy, MFM.
M

3. Results

The techniques usually employed for studying amorphous ribbons can be exploited also for thin films. In this case, however, several difficulties arise which need to be addressed. As an example, the amorphicity of a specimen can be investigated through DSC measurements, as reported in Figure 1 [5].

But since in a thin film very little material is available, the contribution of the substrate has to be taken into account. For this reason, it is sometimes preferrable to deposit the film on a resist, which can then be chemically dissolved: in this way, small flakes of the film can be put in the DSC crucible, without the substrate. The film thickness has to be of a few hundreds of nanometers to have a reasonable signal, which can then be compared with that of the amorphous ribbon used as target. As evidenced in Figure 1, even when the composition of the film is close to that of the ribbon (as evidenced e.g. by EDX), DSC curves show different features: in thin films, the crystallization peaks are often broader and may occur at slightly different temperatures. Additionally, broad halos at lower temperatures often indicate relaxation, oxidation or early relaxation effects, which are not present in the ribbons.

A technique complementary to DSC and applicable to amorphous magnetic thin films consists in measuring their magnetization at constant field, as a function of temperature, for example by means of a VSM equipped with a furnace. Figure 2 shows an example [6], whith a comparison to the corresponding DSC curve.

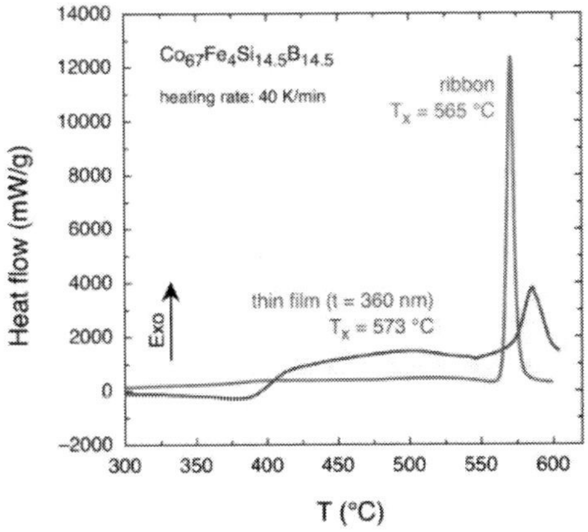

Figure 1. Differential scanning calorimetry on a $Co_{67}Fe_4Si_{14.5}B_{14.5}$ thin film (blue curve) with a thickness of 360 nm, compared to that of an amorphous ribbon with the same composition (red curve).

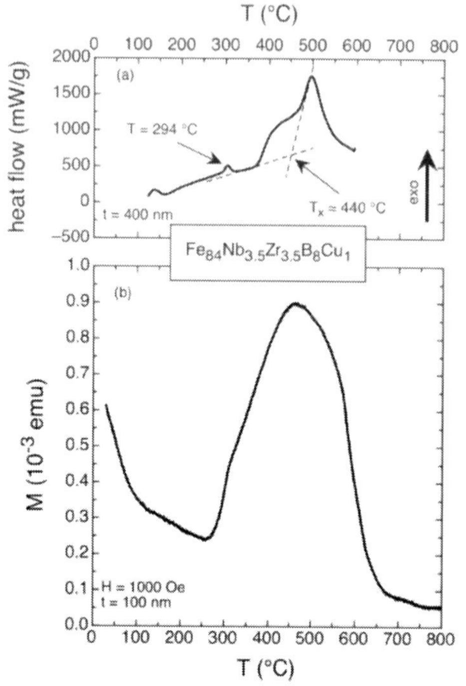

Figure 2. (a) Differential scanning calorimetry on a $Fe_{84}Nb_{3.5}Zr_{3.5}B_8Cu_1$ thin film, 400 nm thick. (b) Magnetization vs. temperature curve on a film with the same composition and a thickness of 100 nm. The measurement is done under an applied field equal to 1000 Oe.

M vs. *T* measurements are more sensitive than DSC in detecting the Curie temperature of the amorphous phase. The other features of the two curves are easily comparable: a very early crystallization process occurs slightly below 300 °C according to DSC, and indeed the increase of the magnetization, indicating the precipitation of a crystalline phase from the amorphous matrix, begins at comparable temperature values. Similarily, the onset of crystallizaion at 440 °C on the DSC curves corresponds precisely with the peak of the magnetization as a function of temperature. Additionally, *M* vs. *T* measurements allow the detection of the Curie temperatures of the crystalline precipitates, thus helping in their identification. In fact, in order to identify crystalline phases, XRD spectra are usually analyzed. However, on thin films the results are often hardly decisive, as the contribution of the substrate tends to be dominant, especially for samples thinner than a few hundreds of nanometers. In most cases, only the first diffraction peak can be detected, which can only be used to infer the presence of crystalline precipitates in annealed samples, as shown in Figure 3 [6] where the spectrum of an as-prepared sample is shown for comparison.

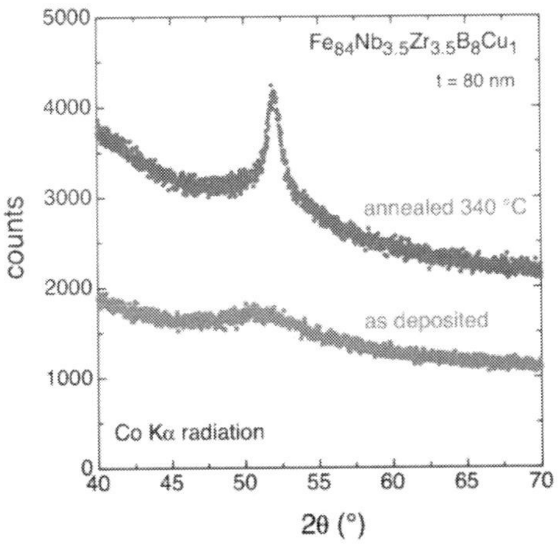

Figure 3. X-ray spectra of $Fe_{84}Nb_{3.5}Zr_{3.5}B_8Cu_1$ thin films, 80 nm thick, in the as-prepared state (red curve) and after annealing at 340 °C (blue curve).

Different from the usual practice with amorphous or crystalline magnetic ribbons, measuring hysteresis loops on thin films with an inductive technique is often very difficult, because of the very low amount of material and the fact that most of the cross section is attributed to the substrate. On the contrary, magnetometric techniques making use of a VSM or an AGFM are usually viable alternatives, especially because, given the very low thickness of the samples with respect to their lateral dimensions, demagnetizing coefficients are usually negligible when the field is applied in the sample plane. For very thin films, proper actions need to be applied to compensate the diamagnetic contribution of the sample holder and of the substrate. Figure 4 [5] shows the remarkable effects of annealing on a Co-based alloy in thin films form. The crystallization, induced by means of furnace annealing in vacuum at selected temperatures, causes a significant increase of the coercive field: after annealing at 600 °C, the soft magnetic properties are lost and a very hard loop is instead measured.

However, the opposite may also be true, e.g. with alloys that develop a nanocrystalline structure upon annealing. An example is shown in Figure 5: with respect to the as-prepared state, the annealed samples display a coercive field which is more than halved: this is the combined result of stress relaxation and nanocrystallization, which act to promote the soft magnetic properties of this alloy.

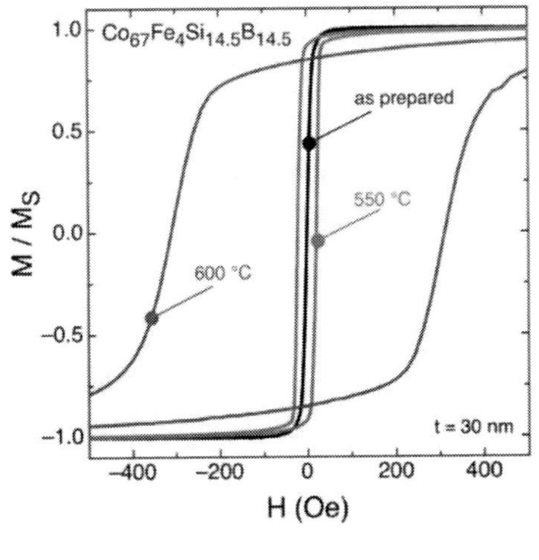

Figure 4. Hysteresis loops of $Co_{67}Fe_4Si_{14.5}B_{14.5}$ thin films, 30 nm thick, in the as-prepared condition (black curve), and after annealing in furnace at 550 °C (red curve) and 600 °C (blue curve).

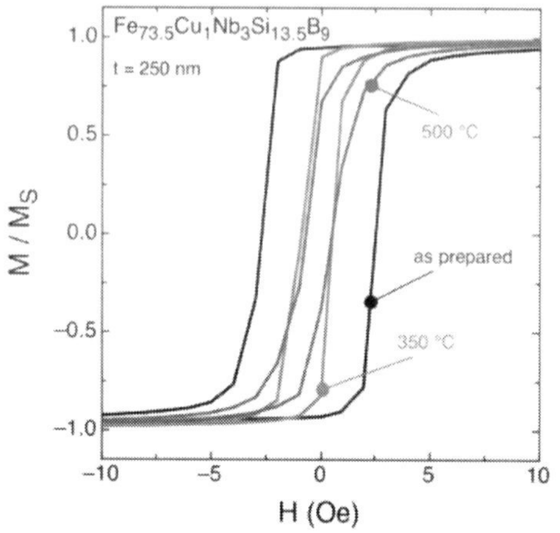

Figure 5. Hysteresis loops of $Fe_{73.5}Cu_1Nb_3Si_{13.5}B_9$ thin films, 250 nm thick, in the as-prepared condition (black curve), and after annealing in furnace at 350 °C (red curve) and 500 °C (blue curve).

In any case, with respect to nanocrystalline ribbons of the same composition, thin films are usually characterized by worse soft magnetic properties, possibly due to the significant role played by the interface between the substrate and the film. The relaxation of quenched-in stresses by means of low temperature annealing is clearly evidenced in Figure 6 [7]: these amorphous thin films are characterized by a spin reorientation transition effect [8], which is responsible for the development of a perpendicular anisotropy when the sample thickness is within a certain range. The domain patterns detected by means of a MFM demonstrate that at low and high thickness values the component of the magnetization perpendicular to the film plane is negligible. On the contrary, at intermediate thickness values, a dense stripe domain structure develops, which has been demonstrated to be due to the presence of large stresses in the as-prepared materials. These stresses can be attributed to implantation effects during the sputtering process [9], and in materials with a positive magnetostriction constant result in the development of a perpendicular magnetic anisotropy.

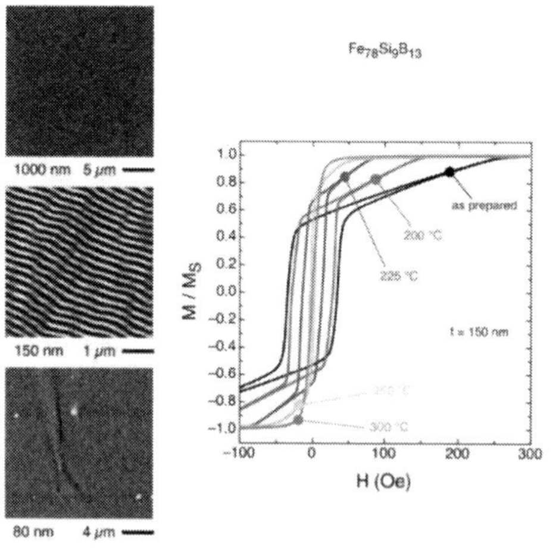

Figure 6. (Left) MFM images of $Fe_{78}Si_9B_{13}$ thin films with thicknesses of 80, 150 and 1000 nm. (Right) For the sample 150 nm thick, hysteresis loops in the as-prepared condition and after annealing at selected temperatures.

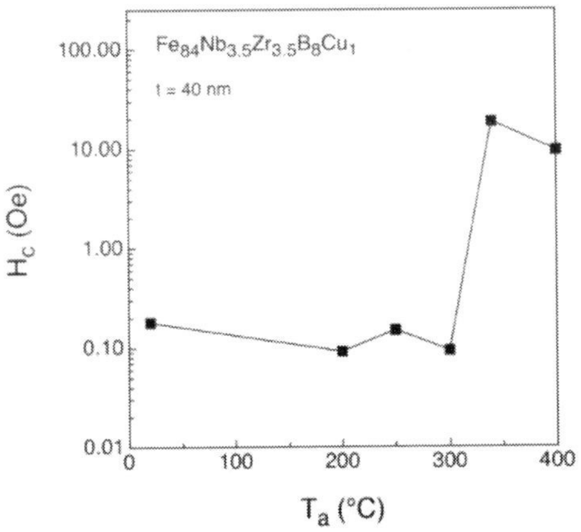

Figure 7. Coercive field as a function of annealing temperature for $Fe_{84}Nb_{3.5}Zr_{3.5}B_8Cu_1$ thin films with a thickness of 40 nm.

This effect is quite common in amorphous magnetic thin films, and can also be observed in other crystalline soft materials such as $Ni_{80}Fe_{20}$ for specific thickness values [10]. In presence of this large perpendicular anisotropy, the hysteresis loops display a "transcritical" shape [11] (Figure 6). When submitted to low temperature annealing, which are not sufficient to induce crystallization, the samples are progressively relieved from the stresses, and the perpendicular anisotropy progressively reduces, eventually restoring the soft magnetic properties of the samples. The stress relaxation effects are also shown in Figure 7 [6] for a different composition: the coercive field slightly decreases with respect to the as-prepared state when low temperature annealing is performed.

At temperatures greater than the onset of crystallization, the coercive field suddenly increases, marking a significant change in the sample microstructure. Low temperature annealing can also be performed below the Curie temperature of the amorphous phase in presence of a conditioning magnetic field, which can be used to induce a uniaxial anisotropy in the sample. An example is shown in Figure 8 [5], where the presence of uniaxial anisotropy is clearly shown by the loop shapes along the easy and hard magnetization axes.

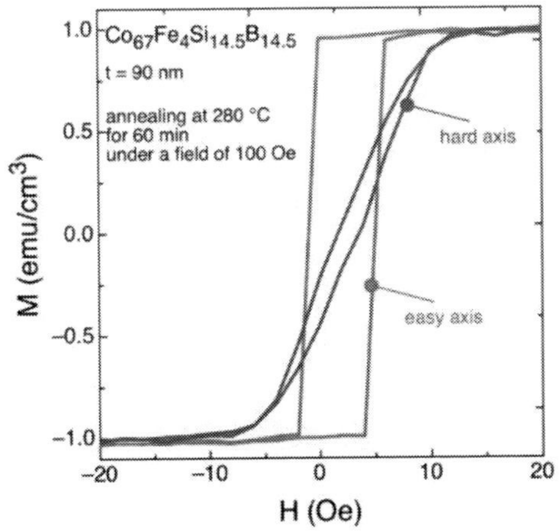

Figure 8. Hysteresis loops of $Co_{67}Fe_4Si_{14.5}B_{14.5}$ thin films, 90 nm thick, submitted to furnace annealing at 280 °C under an applied field of 100 Oe. Red curve: measurement along the direction of the field applied during annealing. Blue curve: measurement at 90° with respect to it.

Although widely used in ribbons [12], current annealing (Joule heating) techniques have not yet been widespread adopted to tailor the soft magnetic properties of thin films. This kind of annealing is less straightforward in films than in ribbons, because of the difficulty in making electrical contacts and of the presence of the substrate, which strongly affects the dissipation of the heat generated by the Joule effect. Figure 9 shows early data on Joule heating on thin films. In Figure 9a the evolution of the electrical resistance variation as a function of time is reported for selected annealing current intensities. At low current values, a reduction of the resistance is observed, which can be ascribed to stress relaxation effects. At itermediate current values (5 and 7 mA), a subsequent slight increase of the resistance marks the first changes in the sample microstructure, which still need to be investigated. Higher current intensities (10 mA) result in a peak of the resistance, which can be attributed to the crystallization of the sample. The hysteresis loops reported in Figure 9b support this picture, as the samples annealed at 5 and 7 mA display the softest magnetic properties, whereas the sample annealed

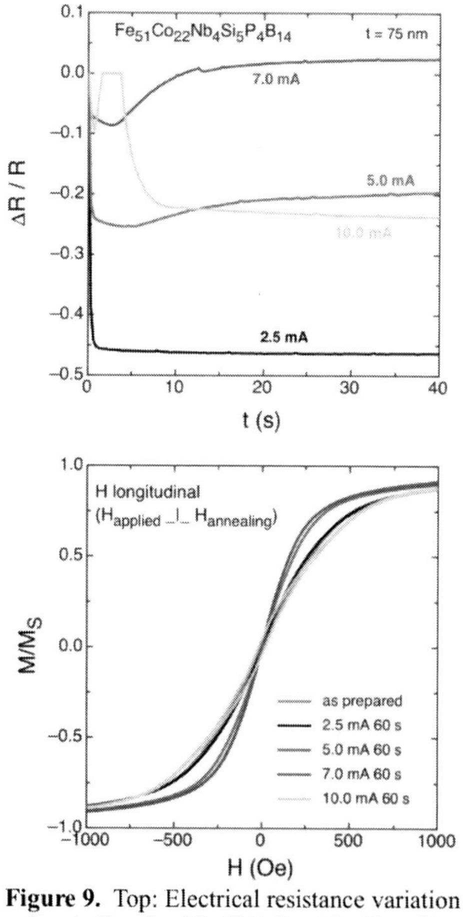

Figure 9. Top: Electrical resistance variation vs. time in $Fe_{51}Co_{22}Nb_4Si_5P_4B_{14}$ thin films 75 nm thick submitted to Joule heating at selected current intensities. Bottom: Hysteresis loops on as-prepared and Joule heated samples with the measurement field perpendicular to the annealing field.

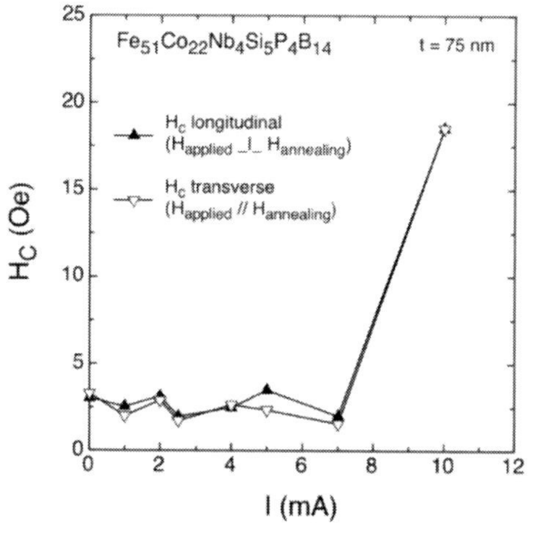

Figure 10. Coercive field as a function of annealing current intensity in Joule heated $Fe_{51}Co_{22}Nb_4Si_5P_4B_{14}$ thin films, 75 nm thick. Full symbols: measurement field perpendicular to annealing field. Open symbols: measurement field parallel to annealing field.

at 10 mA is characterized by a much larger coercive field. Its dependence on the annealing current intensity is shown in Figure 10 for the same samples. A slight anisotropy is induced, as expected, by the annealing technique, which exploits a self generated field, and is evidenced by the slightly lower coercivity value when the measurement field is parallel to the annealing field.

4. Conclusion

In conclusion, amorphous soft magnetic thin films can be prepared by sputtering with the same compositions used for ribbons. Even if thin films usually display worse soft magnetic properties with respect to ribbons with the same composition, thermal treatments can be used to tailor them by relieving the quenched-in stresses, and by inducing (nano) crystallization or a uniaxial anisotropy in the case of field annealing.

References

[1] Knipling K E, Daniil M and Willard M A 2009 *Appl. Phys. Lett.* **95(22)** 22516
[2] Makino A, Hatanai T, . Naitoh, . Bitoh, Inoue A., and Masumoto T., 1997 *IEEE Trans. Magn.* **33(5)** 3793
[3] Arnalds U B, Papaioannou E Th, Hase T P A, Raanaei H , Andersson G, Charlton T R, Langridge S, Hjörvarsson B 2010 *Phys. Rev. B*, **82(14)** 144434
[4] Cho J U, Kim D K, Tan R P, Isogami S, Tsunoda M, Takahashi M and Kim Y K 2009 *IEEE Trans. Magn.* **45(6)** 2364
[5] Celegato F, Coïsson M, Olivetti E, Tiberto P and Vinai F 2008 *Phys. Stat. Sol. A*, **205(5)** 1047
[6] Coïsson M, Celegato F, Olivetti E S, Tiberto P, Vinai F, Kane S N, Gan'shina E A, Novikov A I and Perov N S 2011 *J. All. Comp.* **509** 4688
[7] Coïsson M, Celegato F, Olivetti E, Tiberto P, Vinai F and Baricco M 2008 *J. Appl. Phys.* **104** 033902
[8] Sharma P, Kimura H, Inoue A, Arenholz E and Guo J H 2006 *Phys. Rev. B* **73(5)** 052401
[9] Windishmann H 1992 *Crit. Rev. Sol. St. Mater. Sci.*, **17**, 547
[10] Svalov A V, Aseguinolaza I R, Garcia-Arribas A, Orue I, Barandiaran J M, Alonso J, Fdez-Gubieda M L and Kurlyandskaya G V 2010 "*IEEE Trans. Magn.* **46(2)** 333
[11] Alvarez-Prado L M and Alameda J M 2004 *Phys. B* **343** 241
[12] Allia P, Tiberto P, Baricco M and Vinai F 1993 *Rev. Sci. Instr.* **64**(4) 1053

Towards a Wearable Non-invasive Blood Glucose Monitoring Device

Joseph Thomas Andrews,
J. Solanki, Om P Choudhary, S. Chouksey, N. Malvia, P. Chaturvedi and P. Sen.
National MEMS Design Centre, Applied Photonics Laboratory, Department of Applied Physics, Shri G S Institute of Technology & Science, Indore 452003 India.
E-mail: jtandrews@sgsits.ac.in

Abstract. Every day, about 150 Million people worldwide face the problem of diabetic metabolic control. Both the hypo- and hyper- glycaemic conditions of patients have fatal consequences and warrant blood glucose monitoring at regular interval. Existing blood glucose monitors can be widely classified into three classes viz., invasive, minimally invasive, and noninvasive. Invasive monitoring requires small volume of blood and are inappropriate for continuous monitoring of blood glucose. Minimally invasive monitors analyze tissue fluid or extract few micro litre of blood only. Also the skin injury is minimal. On the other hand, noninvasive devices are painless and void of any skin injury. We use an indigenously developed polarization sensitive Optical Coherence Tomography to measure the blood glucose levels. Current trends and recent results with the device are discussed.

1. Introduction

Diabetes mellitus is a medical condition in which the body does not adequately produce the quantity or quality of insulin needed to maintain normal circulating glucose level in the blood. Insulin is a hormone that enables glucose (sugar) to enter the body's cells to be converted into energy. Two types of diabetes are common. Type I is also known as Insulin Dependent Diabetes Mellitus (IDDM or T1DM) and is found to be 5-10% of the generalized cases of diabetic patients. Type II or Non-Insulin Dependent Diabetes Mellitus (NIDDM or T2DM) occurs in the rest of the diabetic population. In case of IDDM, the disease occurs in childhood and requires healthy eating, regular exercise and insulin doses to maintain a reasonably healthy life. On the other hand, NIDDM occurs usually at the human age around 40 years and require an externally supplied insulin dosage in addition to regular exercise and controlled diet.

T2DM is a non-autoimmune, complex, heterogeneous and polygenic metabolic disease condition in which the body fails to produce enough insulin, characterized by abnormal glucose homeostasis [1]. Its pathogenesis appears to involve complex interactions between genetic and environmental factors. T2DM occurs when impaired insulin effectiveness (insulin resistance) is accompanied by the failure to produce sufficient beta-cell insulin [2].

The recent survey given by World Health Organization is shocking. India tops the list with the number of T2DM patients as much as 31.7Million. By the year 2030, the estimated number of affected by T2DM in India will be 79.4Million. Actual interpolation of data put the value much larger than this. This data is alarming [3]. A data of global scenario of T2DM is shown in Figure 1. The diabetes mellitus is associated with the symptoms such as Polyuria (frequent urination), Polydipsia

Published under licence by IOP Publishing Ltd

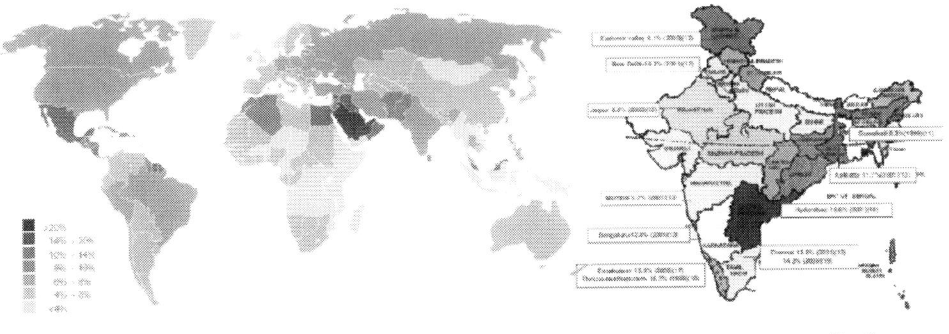

Figure 1. The global and Indian trend of number of cases having T2DM [2-4].

(increased thirst), Polyphagia (increased hunger). These symptoms are rapid in T1DM and slow with T2DM. The Prolonged DM may also cause vision problems.

On the other hand, the control and treatment of T2DM requires regular monitoring, sometimes very frequently. The diagnosis methods are invasive. The pain associated with the diagnostic methods of blood glucose make the patients traumatic. Also the existing diagnostic tools are bulky, time consuming, requires blood and is available only at clincs. The authors group are aiming at a device, which could overcome these problems and is fabricating a device which is non-invasive and wearable or handheld. The design, fabrication and characterization of the device are discussed in the following Sections.

2. Non-Invasive Diagnosis Methods

Non-invasive glucose monitoring techniques can be grouped as subcutaneous, dermal, epidermal and combined dermal and epidermal glucose measurements. Matrices other than blood under investigation include interstitial fluid, ocular fluids and sweat. Test sites being explored include finger tips, cuticle, finger web, forearm and ear lobe. Subcutaneous measurements include microdialysis, wick extraction, and implanted electrochemical or competitive fluorescence sensors. Microdialysis is also an investigational dermal and epidermal glucose measurement technique. Epidermal measurements can be obtained via infrared spectroscopy as well. Combined dermal and epidermal fluid glucose measurements include extraction fluid techniques (iontophoresis, skin suction and suction effusion techniques) and optical techniques. A summary of possible methods for the non-invasive measurement of blood glucose is given in Figure 2.

The range of measurement techniques usually based upon optical properties of the sample is wide that includes some of the sophisticated methods like near infrared spectroscopy [5], infrared spectroscopy [6], Raman spectroscopy [7], photoacoustic spectroscopy [8], scatter and polarization change measurements, etc. Non-invasive optical measurement of glucose is performed by focusing a beam of light onto the body. The light is modified by the tissue after transmission through the target area. An optical signature or fingerprint of the tissue content is produced by the diffuse light that escapes the tissue has penetrates. The absorbance of light by the skin is due to its chemical components (i.e., water, hemoglobin, melanin, fat and glucose). The transmission of light at each wavelength is a function of thickness, color and structure of the skin, bone, blood and other material through which the light passes [9-11].

The glucose concentration can be determined by analyzing the optical signal changes in wavelength, polarization or intensity. The sample volume measured by these methods depends on the measurement site. The correlation with blood glucose is based on the percent of fluid sample that is interstitial, intracellular or capillary blood. Not only is the optical measurement dependent on concentration changes in all body compartments measured, but also the changes in the ratio of tissue fluids (as altered by activity level, diet or hormone fluctuations) and this, in turn, affects the glucose measurement. Problems also occur due to changes in the tissue after the original calibration and the lack of transferability of calibration from one part of the body to another. Tissue changes include:

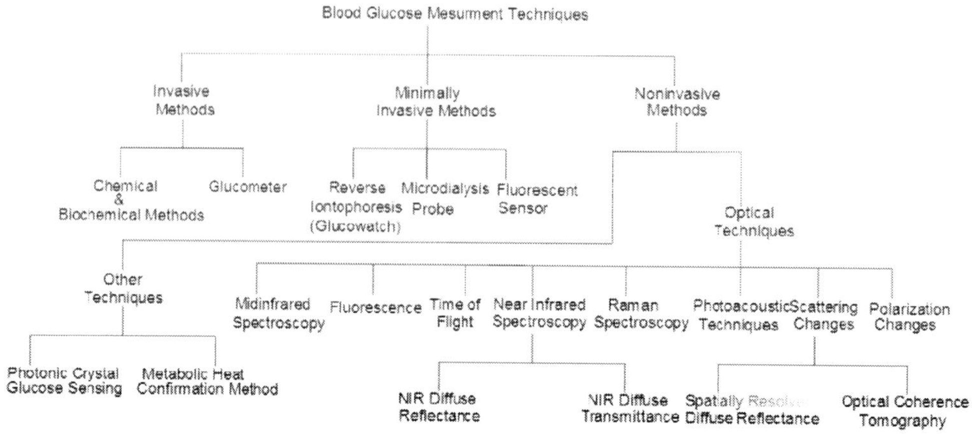

Figure 2. A summary of glucose measurement techniques used. All non-invasive methods are use optical models with associated mathematical equations.

body fluid source of the blood supply for the body fluid being measured, medications that affect the ratio of tissue fluids, day-to-day changes in the vasculature, the aging process, diseases and the metabolic activity in the human body [12].

3. Design and Characterization

Of late, the advances in the area of Micro-Opto-Electro-Mechanical Systems (MOEMS) based [13], devices are utilized widely for various applications in the area of biomedical applications. We explore the possibility of using a MOEMS devices as a optical coherence tomography for non-invasive measurement of blood glucose. We use nanophotonic Silicon on Insulator (SOI) as the platform for the fabrication of the structures. SOI waveguides normally have a silicon core (refractive index n_0=3.45) surrounded by cladding layers of air or silica (SiO_2), with typical refractive indices n_1 between 1 and 2. An SOI wafer consists of a thin top Si layer sitting on silica layer, which is carried on a thick Si substrate. Photonic components are realized by etching the top Si layer, resulting in high

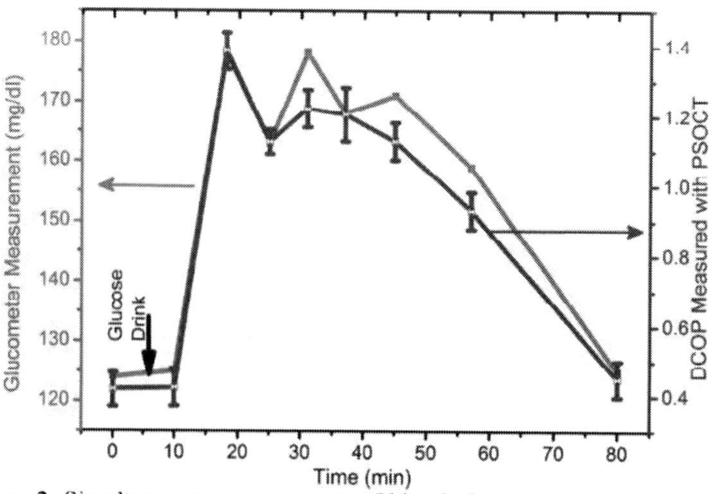

Figure 3. Simultaneous measurements of blood glucose using the current device as well as with commercial grade glucometer. The PS-OCT monitors the degree of circular polarization of backscattered light, which is a non-invasive method, while the glucometer measurements are minimally invasive method.

refractive index contrast in all directions between waveguide core and cladding, allowing good confinement of optical modes and reduction of device dimensions. Using wafer scale CMOS compatible processes, low loss waveguides with core cross-sectional area of 0.1 μm^2 and bend radii of 5 μm can be realized [14].

Figure 3, demonstrates some of the results obtained with various human volunteer subjects. The PS-OCT device developed by the authors measures the degree of circular polarization of the back-scattered signal. Simultaneously, the glucose concentration is measured with a commercial grade glucometer. The measurements were carried out for every ten minutes. The results reported in Figure 3, establishes a linear correlation between the measurements.

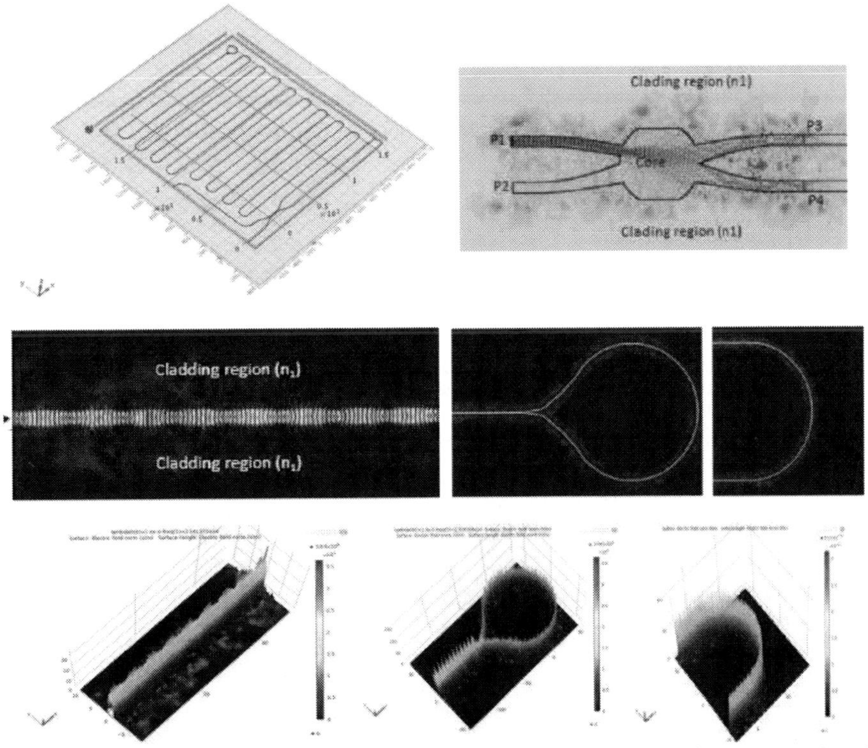

Figure 4. The complete MOEMS based optical coherence tomography setup and various components integrated to complete the setup. The first row shows the MOEMS-OCT after integration. The next image shows the simulated results of bi-directional coupler. The second and third rows shows the simulated results of a straight waveguide, waveguide radiation in a circular path and a MOEMS mirror. The second row shows the radiation pattern after finite element analysis. The 3d radiation patterns are shwon in third row.

4. Simulation model and results

Design and simulation of whole MOEMS OCT system done using Finite element method (FEM) based analysis. For that, we employed COMSOL Multiphysics as the frontend to simulate and report results. We optimize all parameter for active and passive waveguides through it, which is required for MOEMS-OCT. The MOEMS based TD-OCT has various components. We study and optimize each components systematically. At the end all the components are integrated to bring all of them to a single chip. The various components such as the following are analyzed independently (i) Waveguide (ii) Directional coupler (equivalent of a beam splitter), (iii) Reference arm, (iv) mirror and (v) light coupling and decoupling. Since the light coupling and decoupling is a complicated mechanism and needs more understanding, we discuss them in a separate paper. In the following sections, we discuss each component separately.

5. Conclusions

To conclude, we have designed and studied MOEMS based optical coherence tomography setup. The results show that the successful demonstration of MOEMS-OCT is possible with Silicon on Insulator type structures.

Acknowledgements

The authors thank Prof. P. K. Sen, for fruitful discussions and support. They also thank NPMASS, Bangalore for the facilities provided at the National MEMS Design Center, MPCST- Bhopal and DBT, New Delhi for financial support.

REFERENCES

[1] Gupta V, Khadgawat R, Saraswathy KN, Sachdeva MP and Kalla AK. 2008 *Int J Hum Genet,* **8(1-2)**: 199

[2] Permutt MA, Wasson J, Cox N Genetic Epidemiology of Diabetes. 2005 *Journal of clinical Investigation,* **115**:1431

[3] Wild, S; Roglic, G; Green, A; Sicree, R; King, H (2004). "Global prevalence of diabetes: Estimates for the year 2000 and projections for 2030". *Diabetes care* **27** (5): 1047

[4] Gupta V, Type – 2 Diabetes Mellitus in India, Report submitted to South Asia Network for chronic Disease (www.sancd.org).

[5] Maruo K, T O (2006) New Methodology to Obtain a Calibration Model for Noninvasive Near-Infrared Blood Glucose Monitoring. *Applied Spectroscopy* **60** 441

[6] Heise HM, R M J. 1989 *Analytical Chemistry* **31** 2009

[7] Barman I, CR K 2010 *Analytical Chemistry* **82** 6104

[8] Kulkarni O C, Mandal P, Das S S and Banerjee S 2010 *4th International Conference on Bioinformatics and Biomedical Engineering (iCBBE),* 1

[9] Andrews J T, Patel H S and Gupta P K 2002 *J. Optics,* **30** 151. Poddar R, Sharma S R, Bose K, Sen P and Andrews J T 2006 *American Journal of Physics,* **74** 569

[10] Poddar R, Sharma S R, Sen P and Andrews J T 2006 *J. Biochem. & Biotech.,* **10** 312. Sharma S R, Poddar R, Sen Pand Andrews J T 2008 *Afr. J. Biotechnology* 7 2049

[11] Andrews J T, Poddar R, Sen P and Sharma S R 2008 *OCTNews,* 480761; Poddar R, Sharma S R, Andrews J T and Sen P 2008 *Current Science,* **95** 340

[12] Solnaki J S, Andrews J T, Sen P, 2012 *J. Mod. Physics,* **3(1)** 64

[13] Qiang Wu, H P Chan, Pak L Chu, D P Hand and Chongxiu Yu 2008 *J. Opt.Comm.* **281** . Saleh B E A and Teich M C, Fundamentals of Photonics, Second Ed., (John Wiley, New Jersey, 2007). 290

[14] Bogaerts W, Baets R, Dumon P, Wiaux V, Beckx S, Taillaert D, Luyssaert B, Van Campenhout J, Bienstman P, Van Thourhout D 2005 *Journal of Lightwave Technology* **23(1)** 401

Analytical and numerical investigation of high-harmonic and attosecond pulse-train generation : the sliding mirror model

Vineeta Jain[1], H K Malav[2], K P Maheshwari[2] and N K Jaiman[1]

[1]Department of Pure and Applied Physics, University of Kota, Kota
[2]DST-Project, Vardhaman Mahaveer Open University, Kota, India

k_p_maheshwari@rediffmail.com

Abstract. Pulse-shape dependence of high harmonic generation and attosecond pulse generation as a result of the interaction of a short ultra relativistic laser pulse with a thin layer of overdense plasma for normal and oblique incidence and different polarizations is studied analytically and numerically. The temporal profile effects in the relativistic regime on HHG and optimization of the parameters for most effective attosecond pulse train generation in the transmitted pulse are also investigated.

1. Introduction :

Much attention has been paid to aspects of the interaction of intense laser light with solid-density targets in which the plasma is overdense in terms of the critical density[1-2]. In the relativistic range of amplitude of the laser radiation, when intensity is as high as 10^{18}-10^{20}W/cm^2 the ratio $a = \dfrac{eE}{mc\omega_0}$ is greater than 1. At these ultrahigh intensities, electron quiver velocities are close to the velocity of light and the motion of even free electrons becomes highly nonlinear. High-order harmonic generation by reflection of an intense laser beam from an overdense plasma has been observed with very different types of lasers, from nano-second CO_2 laser to picoseconds Nd-glass and femtosecond Ti-Sa laser. High order harmonics attract a great attention due to a wide range of their applications for the diagnostics, the UV and coherent x-ray sources, lithography etc [3-5]. In this paper we investigate analytically and numerically the effect of the temporal profile of the laser pulse when it interacts with a ultra-thin, overdense plasma layer leading to the generation of high-order harmonics.

2. Reduction of the oblique incidence to the normal incidence

Making use of Lorentz transformation equations and following the basic equations and figure (1) of reference [6] one can reduce the oblique incidence of the laser pulse to that of normal incidence. The laser pulse is incident on the vacuum-plasma thin layer at an angle θ with respect to the normal to the thin layer. In the boosted reference frame the plasma moves in the opposite direction.

The electric field of the incident pulse in the laboratory frame has the form:

$$\vec{E}_0 = E_p(t_r)\left[-\sin\theta\,\hat{e}_x + \cos\theta\,\hat{e}_y\right] + E_s(t_r)\hat{e}_z, \qquad (1)$$

$$E_p(t_r) = E_{p0}P(t_r)\cos(\omega_0 t_r + \psi_p), \qquad (2)$$

$$E_s(t_r) = E_{s0} P(t_r)\cos(\omega_0 t_r + \psi_s). \tag{3}$$

The pulse-envelopes $P(t_r)$ for a Gaussian, Lorentzian, and hyperbolic secant pulses are taken to be [7]

$$P_g(t_r) = \exp\left[-\left(\frac{1.17t_r}{\tau}\right)^2\right],\ P_l(t_r) = \left[1+\left(\frac{1.29t_r}{\tau}\right)^2\right]^{-1},\ \text{and}\ P_h(t_r) = sech\left(\frac{1.76t_r}{\tau}\right)$$

respectively, where τ is full width at half maximum (FWHM) of pulse intensity. Here E_{p0} and E_{s0} are the electric-field amplitudes of the p- and s- polarized components of the pulse E_p and E_s, $t_r = t - \vec{n}.\vec{r}/c$, $n = \cos\theta\,\hat{e}_x + \sin\theta\,\hat{e}_y$, is the unit vector along propagation direction, $\vec{r} = x\hat{e}_x + y\hat{e}_y + z\hat{e}_z$, and ψ_p, ψ_s are the carrier-envelope phases. The electric field of the incident pulse in the boosted reference frame is

$$E_0' = \cos\theta\left[E_p(t_r'\cos\theta)\hat{e}_y + E_s(t_r'\cos\theta)\hat{e}_z\right], \tag{4}$$

here $t_r' = t' - x/c$. In the boosted frame the central frequency becomes $\omega_0' = \omega_0\cos\theta$ and the duration becomes $\tau' = \tau/\cos\theta$. The number of cycle in a pulse ($\sim \omega_0\tau$) and carrier-envelope phase are Lorentz invariants. In the boosted frame the plasma density becomes $n' = n/\cos\theta$; the electrons and ions move with the speed $\vec{v}_0' = -c\sin\theta\,\hat{e}_y$ along the plasma layer, which corresponds to the electron momentum $\vec{p}_0' = -mc\tan\theta\,\hat{e}_y$. A prime denotes quantities in the boosted reference frame. The pulse in the boosted frame has a form of a plane wave propagating along the x direction i.e. normal to the plasma layer. The relativistic factor is equal to $\sec\theta$.

3. Sliding-mirror model

We consider the limit of dense plasma, in which the displacement of the reflecting electron layer in the direction perpendicular to the target surface is negligibly small. In this limit the plasma density is so high that the charge separation electric field suppresses the electron-layer motion inside and outside the target. The electron layer can move only along the target surface. Under this assumption of the so called "sliding mirror" model we investigate numerically the electromagnetic field evolution inside the plasma. The crucial parameter of the laser-thin-plasma slab interaction is the normalized plasma density:

$$\varepsilon_p = \frac{\int \omega_p^2(x)dx}{2\omega_0 c} = \frac{\int n(x)dx}{n_{cr}\lambda_0}, \tag{5}$$

here ω_p is the plasma frequency, n is the electron density, n_{cr} is the critical electron density. For the constant-density plasma of thickness l, $\varepsilon_p = (\pi n l)/(n_{cr}\lambda_0) \approx a_0$. The parameter $\varepsilon_p \sim a_0$ or larger is important from the point of view of attosecond pulse generation.

4. High harmonic generation & attosecond pulse train generation

The electric fields of the reflected and transmitted pulses are expressed in boosted frame of reference as [8]

$$E_r'(x,t') = -\frac{\varepsilon_p}{\cos\theta}\left\{\frac{\vec{p}_0'}{\sqrt{1+|\vec{p}_0'|^2}} - \frac{\vec{p}_0' + \vec{A}'(0,t'+x)}{\sqrt{1+|\vec{p}_0' + \vec{A}'(0,t'+x)|^2}}\right\},\ x<0, \tag{6}$$

$$E_t^{'}(x,t') = E_0^{'}(x,t') - \frac{\varepsilon_p}{\cos\theta}\left\{\frac{\vec{p}_0^{'}}{\sqrt{1+\left|\vec{p}_0^{'}\right|^2}} - \frac{\vec{p}_0^{'} + \vec{A}'(0,t'-x)}{\sqrt{1+\left|\vec{p}_0^{'} + \vec{A}'(0,t'-x)\right|^2}},\ x.>0\right\}, \tag{7}$$

where $\vec{p}_0^{'}$ is normalized by mc and the normalized vector potential inside the plasma layer $\vec{A}'(0,t'+x)$ satisfy the following ordinary differential equation

$$\frac{d\vec{A}'(0,t')}{dt'} - \frac{\varepsilon_p}{\cos\theta}\left\{\frac{\vec{p}_0^{'}}{\sqrt{1+\left|\vec{p}_0^{'}\right|^2}} - \frac{\vec{p}_0^{'} + \vec{A}'(0,t')}{\sqrt{1+\left|\vec{p}_0^{'} + \vec{A}'(0,t')\right|^2}}\right\} = -\vec{E}_0^{'}(0,t') \tag{8}$$

5. Results and discussions :

We solve the system of coupled Maxwell-Lorentz equations of the incoming and scattered radiation numerically by using GEAR method. Numerical results for HHG and attosecond pulse train are obtained for Gaussian, Lorentzian and hyperbolic secant laser pulses with identical full width at half maxima of intensity.

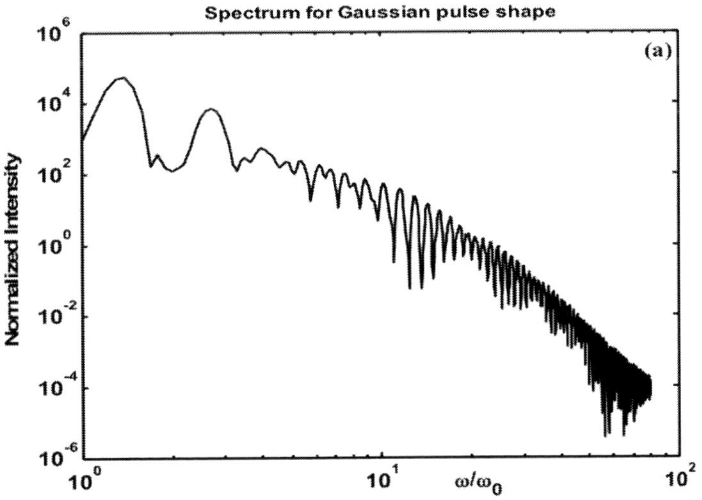

Figure 1. Reflected pulse spectra for p-polarized incident double cycle Gaussian pulse ($\lambda = 800\,nm$) for parameters $a_0 = 10$, $\varepsilon_p = 5$, $\omega_0\tau = 4\pi$, $\psi_p = 0$ and $\theta = 30^0$

The left-hand side of equation (8) represents the current density of the sliding mirror and is responsible for HHG. The right hand side is the driving force. We have chosen the driving force to be pulse shape dependent. We consider the case of p-polarized few-cycle intense laser pulse incident on a thin overdense plasma slab. The transmitted electric field resembles the trains of attosecond pulses. Besides the dependence of the pulse train on the angle of incidence, carrier envelope phase, plasma density, number of cycles in a pulse, intensity of the incident pulse, it also depends on the pulse-shape of the incident laser-pulse. In order to qualify an attosecond pulse train with high peak intensity and shorter pulse duration we define a merit function $S = I_{max} / \tau^2_{FWHM}$, where I_{max} is the maximum intensity and τ_{FWHM} is full width at half maxima of intensity of the main peak of the pulse train. The parameters that govern the nonlinear interaction of the laser pulse with the thin plasma are a_0, density parameter ε_p, carrier envelope phase (CEP), angle of incidence, and number of cycles in the pulse . Figure 1 shows the reflected pulse spectra showing harmonics of p-polarized incident double cycle Gaussian pulse ($\lambda = 800\,nm$). Figure 2 shows the variation of intensity of the attosecond pulse train generated by the p-polarized single cycle incident pulse for optimized parameters as a function of normalized time. Results of attosecond pulse train generation are found to depend on the temporal profile of the pulse.

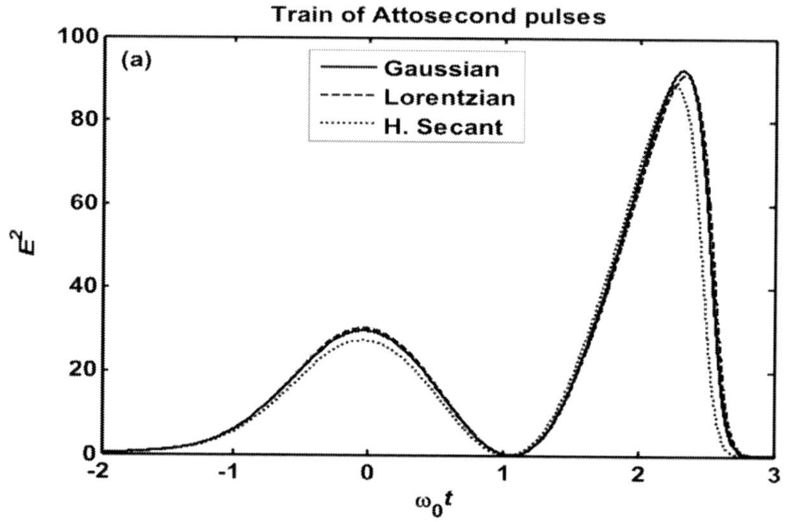

Figure 2. Variation of intensity in the attosecond pulse train generated by the optimized parameters of the p-polarized single cycle incident pulse as a function of normalized time. The optimized parameters are $a_0 = 10$, $\psi_p = 0.0$, and $\theta = 40^0$. $\varepsilon_p = 7.7$ for Gaussian, $\varepsilon_p = 7.6$ for Lorentzian, and $\varepsilon_p = 8.1$ for hyperbolic secant pulse shape

References

[1] Linde D V 1999 *Applied physics B* **68**, 315
[2] Gibbon P 2007 *Short pulse laser interactions with matter(an introduction)*(Imperial college press, L.)
[3] Zhou J, Peatross J, Murnane M M, Kapteny H C and Christov 1996 *Phys. Rev. Lett.* **76** 752
[4] Mourou G, Chang Z, Maksimhuk, Nees J, Bulanov S V, Bychenkov V Y, Esirkepov T Z, Naumova N M, Pegoraro F, and Ruhl H 2002 Plasma Phys. Rep. **28** 12
[5] Bulanov S V, Esirkepov T Z, Naumova N M, Sokolov I V 2003. *Phys. Rev. E.* **67** 016405
[6] Pirozhkov A S, Bulanov S V, Esirkepov T Z, Mori A S and Daido H 2006 *Physics of Plasmas.* **13** 013107
[7] Brabec T, & Krausz F 2000 *Rev. Mod. Phys.* **72** 545
[8] Vshivko V A, Naumova N M, Pegoraro F, and Bulanov S V 1998 *Physics of Plasmas* **5** 2727

Journal of Physics: Conference Series **365** (2012) 012006 doi:10.1088/1742-6596/365/1/012006

Photocatalytic and photoelectro-chemical investigations of Fe/ Sn/ Nb containing oxides for energy application: Comparative study

Pramod H. Borse

International Advanced Research Centre for Powder Metallurgy and New Materials (ARC International), Balapur PO, Hyderabad, AP, 500 005, India

Email : phborse@arci.res.in, pborse@physics.org

Abstract. The physico-chemical property of any bulk material is mainly determined by its lattice structure. To improve the understanding of opto-electronic response of any photocatalyst system, it is thus necessary to investigate the underlying physics of these materials. The world is exploring various metal-oxides to identify an efficient energy material, specifically with the interest towards solar energy conversion. The work discusses typical important example of ferrite, stannate and niobate containing photocatalyst systems; and compares their photocatalytic aspects for solar hydrogen production. Though ferrites and stannate are expected to be abundant eco-friendly systems, in contrast the niobates are highly efficient photocatalytic systems. Experimental results validate niobates as efficient system as compared to the ferrites/stannates. Computation using density functional theory demonstrates that in Fe-based oxides, d-orbital play a vital role to render visible light character, which is not the case in niobate like $PbBi_2Nb_2O_9$. The best suited band energetics w.r.t. redox levels of water splitting, yields high efficiency photocatalyst. The present work displays a direct correlation of experimental observations with computed results for the photocatalytic system those are used for solar energy conversion

1. Introduction

Energy crisis faced by mankind is demanding exploration of efficient energy generating materials by making use of the present day advanced tools of science. The physico-chemical property of any material system is mainly determined by its lattice structure, which in turns determines their optical, electrical and magnetic properties [1, 2]. The photocatalysts require certain stringent requirements with respect to the optical and electrical behavior of the constituent material. On one hand, it demands low band-gap material to absorb larger fraction of solar spectrum rather than just 4% that is absorbed by conventional UV-active materials as TiO2 [3]. On the other hand these photocatalyst demands that their band-edges straddle the redox-potential of water, to enable it to be used as visible-active photocatalysts [4]. In order to identify or engineer such challenging photocatalysts, the knowledge of their electronic and optical structure is of immense utilization. We discuss here that how a correlation of theoretically computed electronic-structure and experimentally obtained physico-chemical properties can be a useful tool specifically in case of certain metal-oxide photocatalysts.

Unlike, UV-active photocatalyst, visible-active photocatalysts gained lot of attention during recent times [5, 6]. Iron containing oxide and tin containing metal oxides are known to occur naturally on earth. Thus, Ferrites *viz.* Fe_2O_3, MFe_2O_4 (M- metal ion) and stannate *viz.* SnO_2, $BaSnO_3$ have been investigated in view producing the eco-friendly visible light active photocatalysts. On the other hand, $PbBi_2Nb_2O_9$ type niobate system, one of efficient visible active material photocatalyst is also presented in the study. It is interesting to understand the effect of electronic structure on the photocatalytic or photoelectrochemical performance of the material. We show here the results of the theoretically computed properties, and experimentally observed properties of three different *viz.* $ZnFe_2O_4$ (ZFO), $BaSnO_3$:Pb (BSPO) and $PbBi_2Nb_2O_9$ (PBNO). The summarized results were further used to conclude the efficacy of visible light active photocatalyst for photo-splitting of water to generate the renewable energy.

2. Theoretical and Experimental Methods

2.1. Computational details

In order to simulate the electronic or optical structure of metal oxides following methodology was followed. Accordingly, theoretical calculations were performed with the full potential linearized augmented plane-wave (FP-LAPW) method as implemented in the Wien97 code with density functional theory (DFT) in generalized gradient approximation (GGA) [7]. In the LAPW method the unit cell is divided in two types of regions, one with the atomic spheres centred upon nuclear sites and other being the interstitial region between non-overlapping spheres. Inside the atomic spheres, the wave functions are replaced by linear combination of atom like wave functions, while in the intersphere region the wave functions are replaced by plane wave- like expansion. The muffin-tin radii for various atoms were chosen in the range of 2.0-1.6 a.u. in these calculations. The convergence parameter RK_{max} was set to 6.0. The calculation was iterated with the charge convergence criteria of 0.0001. Depending on the case the self-consistency was carried out on a 10/ 18 *k*-point mesh in the irreducible Brillouin zone for peticular system. In addition for a typical example of $BaSnO_3$:Pb, a fine mesh of 100 **k** points in the irreducible Brillouin Zone (BZ) of cubic unit cell was used to calculate the frequency dependent absorption coefficient $I(\omega)$ from real and imaginary parts of dielectric function $\varepsilon(\omega) = \varepsilon_1(\omega)+i\varepsilon_2(\omega)$. A procedure similar to Gupta et al [8, 9] was adopted for these calculation of $I(\omega)$ which is given as $I(\omega)=\sqrt{2}(\omega)\left(\sqrt{\varepsilon_1(\omega)^2+\varepsilon_2(\omega)^2}-\varepsilon_1(\omega)\right)^{1/2}$. The atomic coordinates were directly adopted from literature as reported earlier in our past work [9-11].

2.2. Preparation of photocatalysts

ZFO and BSPO

The powder photocatalysts of ZFO and $BaSnO_3$:Pb (BSPO) were prepared by chemical as well solid state reaction method as described below. The samples were synthesized by modified citrate complex method [12-14] in which typically stoichiometric concentration of Zinc nitrate hexahydrate ($Zn(NO_3)_2$): iron nitrate nona hydrate ($Fe(NO_3)_2.9H_2O$) were mixed in 0.4 mol ethylene glycol (Mallinckrodt Chemicals AR grade) to which 0.1 mol of citric acid (Fischer Chemicals, 99.5 %) was added. Thus obtained mixture was stirred at 60 °C to obtain a thick brown (yellow) colored viscous solution, which was further heated for more than 10 hrs at 130 – 140 °C to finally obtain brown or dark brown color gel. The thick gelatinous liquid was further heated in alumina boat to 350 °C for 8 hr in an electric furnace. Thus resulting brownish black powder was crushed and calcined at various temperatures in the range of 600-1200 °C for 10 hrs. This yielded the desired spinel (cubic perovskite) metal oxides. $BaPb_xSn_{1-x}O_3$ samples were prepared by following similar protocol. Accordingly, Barium nitrate, stannous chloride and lead acetate were used as respective precursor. The lead precursor concentration was varied from 0.1mol to 0.8mol to increase the Pb doping concentration. Thus, this enabled us to obtain pale white, yellow, red, brown, black colored powders.

PBNO

In order to synthesize the powder photocatalyst of PBNO by solid state reaction method, a stoichiometric mixture of PbO, Bi2O3, and Nb2O5 was ground in a mortar and dried in an oven. Pelletized powders were calcined at 1273 K for 24 h in static air and sintered at 1200 oC for 24 h. Thus obtained poweder was used for photocatalytic experiment. TiO2-xNx was prepared by the reported procedure [15]. The photocatalysts were loaded with 0.1 wt% Pt by impregnation with PtCl2, and further used for the photo-splitting experiments.

2.3. *Characterization of photocatalyst systems*

The photocatalysts were comprehensively characterized before being subjected to photochemical experiments. Structural phase confirmation being the first important objective, X-ray diffraction (XRD) studies was carried out for all photocatalysts. An advanced XRD equipment (Bruker AXS D8) that is equipped with a Lynx eye detector and CuK_{α}(1.542Å) radiation was used to ascertain the sample phase compositions, within the detection limit of this technique. The accelerating potential and the filament current were fixed to 40kV and 40mA respectively. The XRD scan was recorded in the range of 20-80^0 2θ angles with a step size of 0.01^0. The results were compared with the respective standard JCPDS cards. The optical properties were studied by UV-Visible Diffuse reflectance spectroscopy (DRS) using Perkin-Elmer Lambda-750 spectrometer in the wavelength range of 200-800nm. The raw data obtained was further analyzed in detail. For the sake of normality other morphological studies were also carried out but not described here.

The photocatalytic behavior of the photocatalysts was studied by carrying out the photocatalytic hydrogen and oxygen evolution studies. Accordingly, the photocatalytic hydrogen/oxygen (over ZFO, PBNO/ BSPO) evolution was studied under visible light irradiation (λ≥420nm). A typical schematic of photocatalytic experimental setup is shown in figure 1. As seen, the Hg-arc lamp (S) was connected to collimator and infra-red (IR) water filter to avoid heating of the reaction mixture. A cut-off filter was used to select the visible light (λ>420nm) for carrying out the reaction at the desirable wavelength. For hydrogen evolution studies known fixed weight of the photocatalyst was taken in a pyrex/ quartz cell containing water-methanol mixture, the solution was purged with nitrogen prior to irradiation. The rate of H_2 evolution was detected using a gas chromatograph (Shimadzu 2010AFT) equipped with a thermal conductivity detector. For hydrogen evolution study, typically light was made "ON" for certain duration of time evolved hydrogen gas was estimated. Under Light "OFF" there was no evolution of gas. Similarly, for oxygen evolution studies and aqueous $AgNO_3$ solution was used during the experiments.

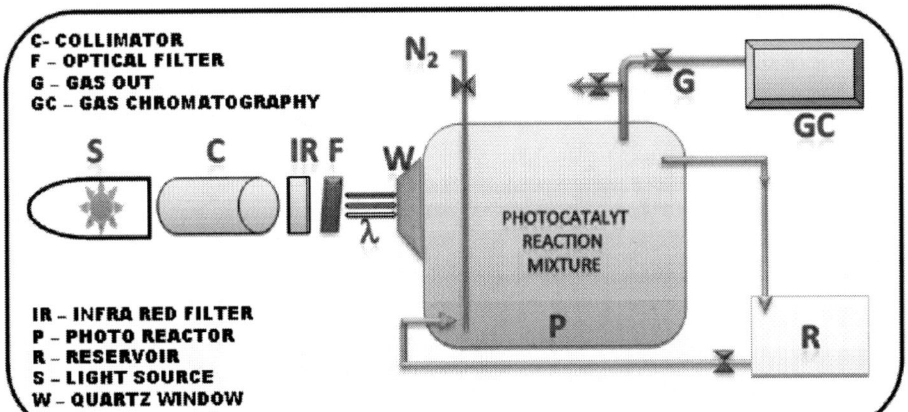

Figure 1. A typical schematic of experimental setup used to carryout the photocatalytic reactions and estimate the quantitative gas evolution.

3. Results and Discussions

3.1. $ZnFe_2O_4$ photocatalyst system

Structural characterization of the photocatalysts was carried out to study the calcination temperature dependent crystal phase evolution. Figure 2 displays the XRD pattern of the samples calcined at

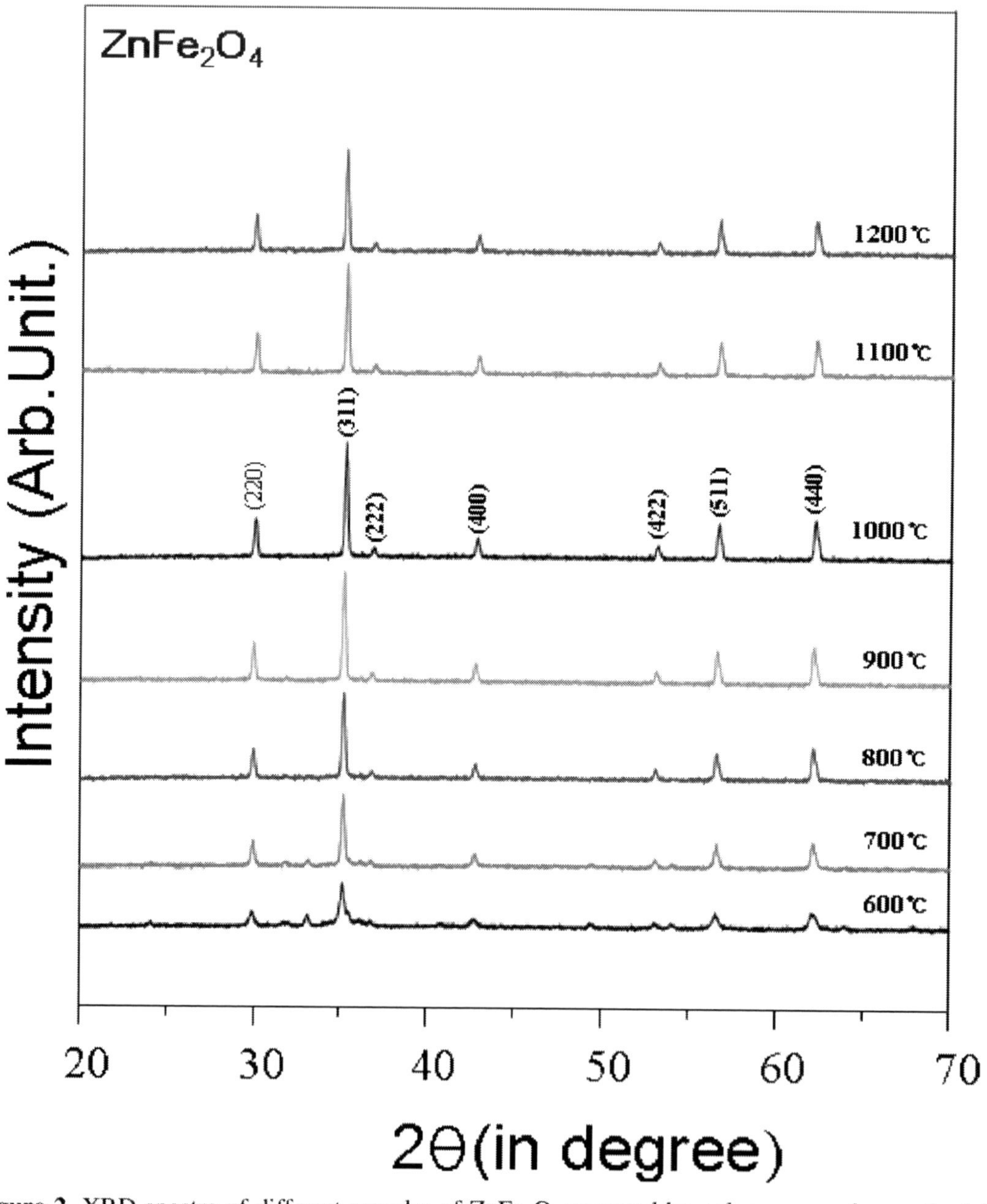

Figure 2. XRD spectra of different samples of $ZnFe_2O_4$ prepared by polymer complex methodology and calcined at various temperatures in the range of 500 ̄ 1200°C, for 4h.

various temperatures, showing the peaks related to the $ZnFe_2O_4$ crystal structure. The formation of the spinel $ZnFe_2O_4$ structure, along with the other impurity phases, was observed in the all the samples calcined in range of 500 - 700°C. The sample calcined in the range of 800 - 1200°C exhibited a single $ZnFe_2O_4$ phase without any impurity phase. This indicates that by making use of the chemical method, the $ZnFe_2O_4$ phase can be formed at a low temperature, yielding a pure phase at relatively lower temperatures [16, 17].

Figure 3 shows the UV-Vis diffuse reflectance spectra (DRS) for the ZFO sample calcined at 1000 °C and the $TiO_{2-x}N_x$ sample, indicating that ZFO efficiently absorbs the visible light photons. The ZFO sample showed a sharp edge without a shoulder indicating that absorption can be attributed to an intrinsic band transition. It is known that, in spinel $ZnFe_2O_4$, the band structure is defined by the O2*p* valence band (VB) and the Fe3*d* conduction band (CB). Thus, the $ZnFe_2O_4$ absorption can be attributed to the photo-excited electron transition of O2*p* → Fe3*d* [18]. On the contrary, $TiO_{2-x}N_x$ showed two absorption edges; a main edge due to the oxide at 390 nm and a shoulder due to the nitride at 451 nm [15]. Estimated band-gaps were found to be 1.90 eV and 2.76 eV, respectively, for ZFO and $TiO_{2-x}N_x$. The dark brown and yellow colors of these materials also indicate that they absorb visible light photons, as desired for a visible-light photocatalyst.

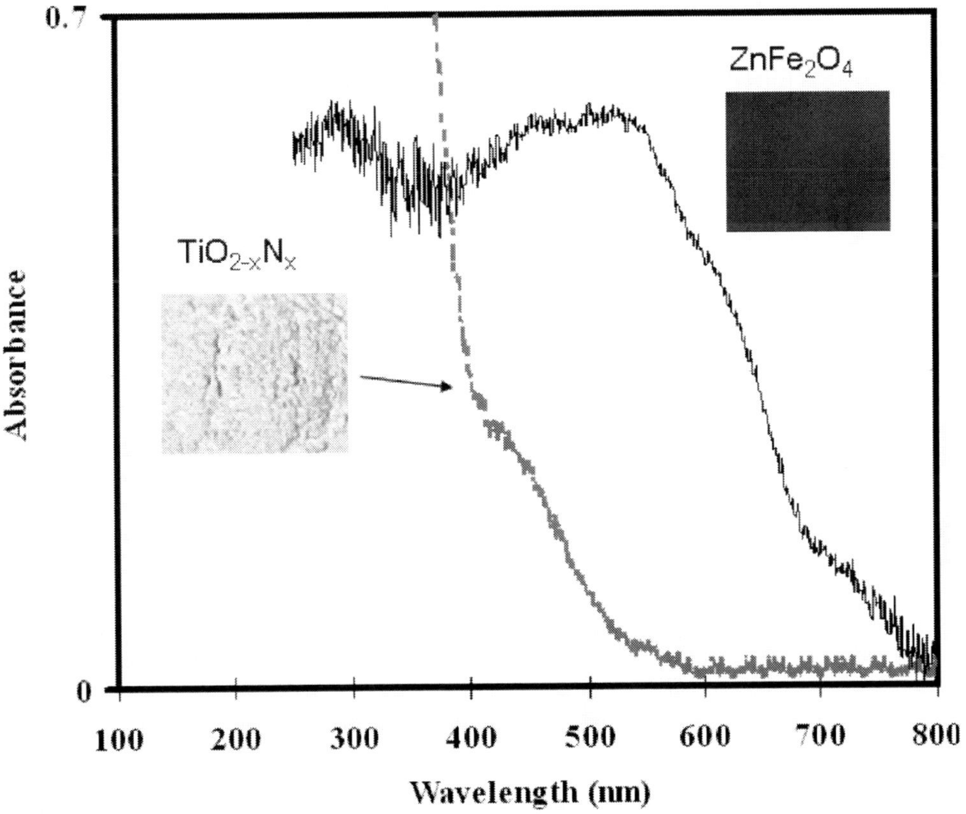

Figure 3. UV-Vis diffuse reflectance spectra of $TiO_{2-x}N_x$, and $ZnFe_2O_4$ (calcined at 1000 °C/ 4h); Inset shows the color of respective as-synthesized sample.

Figure 4 shows the computed total and the projected partial density of states (DOS) of $ZnFe_2O_4$ spinel lattice structure. The inset of the figure shows the schematic of spinel lattice used in the present computation studies. The DOS clearly reveals that Fe-*d* orbital contributes to the states in the inter-band forbidden gap of ZFO. The observation of DOS indicates that VB is constituted of dominant O2*p*

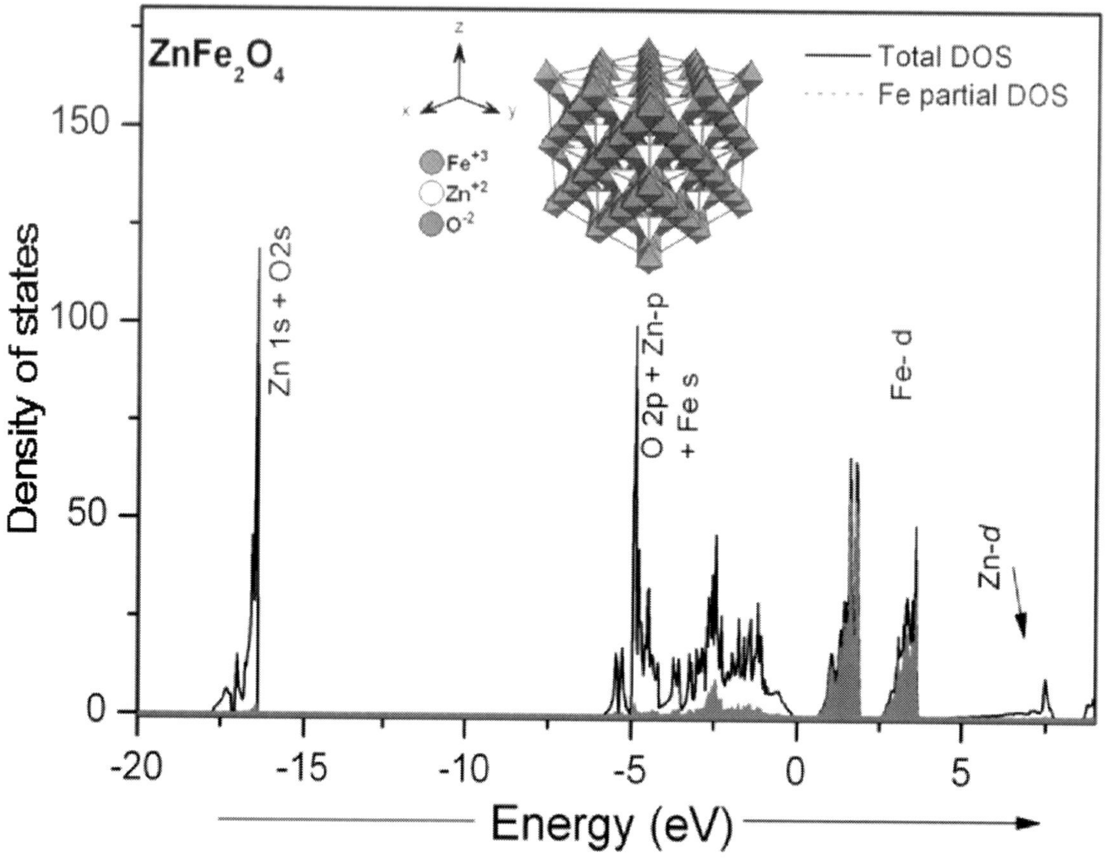

Figure 4. Calculated total and projected partial density of states (DOS) of spinel $ZnFe_2O_4$ crystal structures using FLAPW. The Fe-contribution to DOS is distinctly shown (red shaded area) revealing its importance in the formation of d-band in forbidden region. The unit of each vertical axis is states/ (eV/unit). Fermi level is adjusted to 0eV. Inset shows the schematic of spinel unit cell.

rbital, whereas CB is composed of Zn-*d* states, thus indicating the forbidden gap of ZFO. However, the dominant effect of *d*- orbitals is clearly revealed from the DOS studies. The Fe-PDOS shows a dominant contribution in the forbidden gap making ZFO suitable to absorb low energy photons. The calculated DOS shows a displaced valence band signifying it's the *n*-type conductivity. However, in present case the Fermi-level is adjusted to 0eV as a general convention utilized during reporting the DOS studies. The Fe-PDOS shown by shaded area (red colour) indicates that they promote the inter-band transition in the band gap of ferrite systems, thereby making them absorb the low energy photons. This is in accordance with the observed optical absorption spectra of Fig. 3(a) that shows that ZFO strongly absorb in the range of 400-700nm wavelength range.

3.2. $BaSnO_3$: Pb photocatalyst system

Figure 5 shows the effect of Pb substitution on the XRD pattern of different $BaPb_xSn_{1-x}O_3$ (x ~ 0 to 1) samples. The spectra reveal that as the Pb concentration increase from x~0 to 1, the 110 reflection of cubic $BaSnO_3$ structure shifts towards the lower 2θ values, finally yielding a structure close to the cubic $BaPbO_3$ in case of highest value of *x*. This clearly indicates that the degree of Pb incorporation in the $BaSnO_3$ cubic structure increases with the increase in Pb doping concentration. The existence of Pb as confirmed by energy dispersive spectroscopy (not shown here) is in line with the above fact. The additional impurity peaks are due to the possible existence of trace amounts of BaO, SnO_2 or PbO phases. A detail analysis shows that there is concentration dependent increase in the lattice constant from $BaSnO_3$ (a=4.1163Å) [19] to $BaPbO_3$ (a=4.262 Å) [20]. The increase in lattice constant (a) is expected due to the difference in the ionic radii of Sn^{+4} (8.3Å) and Pb^{+4} (9.1 Å).

The photon absorption (UV-Vis) experiments are necessary to understand the optical response of

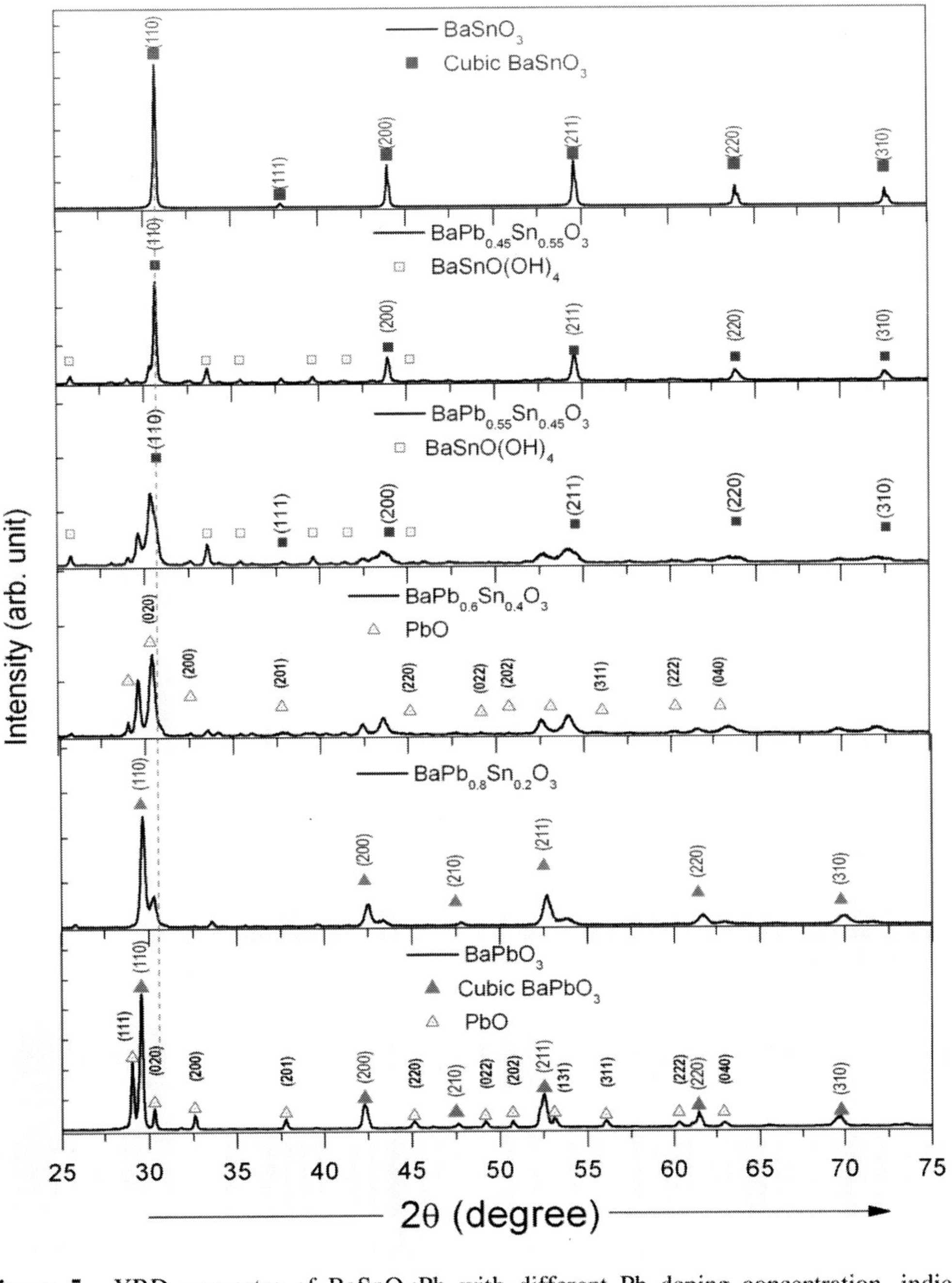

Figure 5. XRD svgpectra of $BaSnO_3$:Pb with different Pb doping concentration, indicating the concentration dependent transformation from $BaSnO_3$ to $BaPbO_3$. The red dotted line indicates the concentration dependent change in 2θ, induced due to Pb-doping.

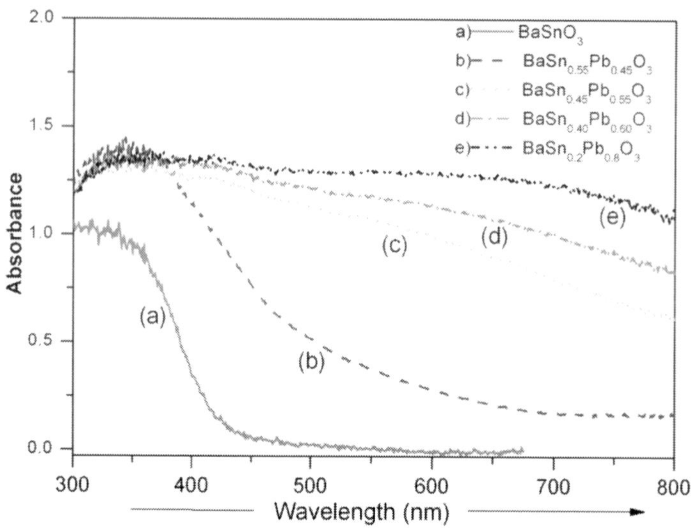

Figure 6. UV-Vis diffuse reflectance spectra for $BaSnO_3$:Pb structures.

any material. The UV Visible DRS for the BSPO samples is displayed in Fig. 6. It can be noted that the absorption of $BaSnO_3$ clearly extended to the whole visible region (maximum to λ>750nm) with the increase in the Pb doping level. This is an important aspect desired for solar active materials. With the Pb doping level of x=0.45, the absorption significantly enhanced for the photons around wavelength of λ>450nm. For the higher Pb doping levels, the absorption totally extended to the visible region of electromagnetic spectrum, indicating that BSPO structures (for x>0) absorb the photons necessary for a visible light active photocatalyst. It was observed that for 0>x>1 in BSPO, the white color of $BaSnO_3$ changes to yellow, reddish brown, dark brown and finally to black. This clearly demonstrates the tunability of the spectral response in the BSPO structures and there by their potential usefulness in the visible light optical applications

Figure 7 shows the computed total and the projected partial DOS $BaPb_{0.5}Sn_{0.5}O_3$, along with calculated absorption coefficient I (ω) of the $BaPb_{0.5}Sn_{0.5}O_3$ structure, as a function of photon energy. The comparison is made with the computed absorption of $BaSnO_3$ structure, indicating a clear increase in the visible photon absorption for BSPO (see magnified inset of Figure 7(b)). It is clearly evident from Pb-PDOS (Figure 7(a)) that Pb-6s orbital controllably affects the bottom of CB of $BaSnO_3$. It further reveals that, in forbidden gap, the dominant contribution to the top of VB comes from the O 2p orbitals, while the states at the CB minimum are predominantly due to Sn 5s and Pb 6s states. Thus, the Pb substitution induces the Pb 6s states, which are located close to the lower edge of the CB to form a lowest unoccupied band. This kind of Pb doping renders the solar photo activity to photocatalytically inactive $BaSnO_3$. This is validated by the experimental confirmation as seen in Figure 6. Additionally, the doped system was found to photo-oxidize the water molecule under visible light irradiation as described in later section.

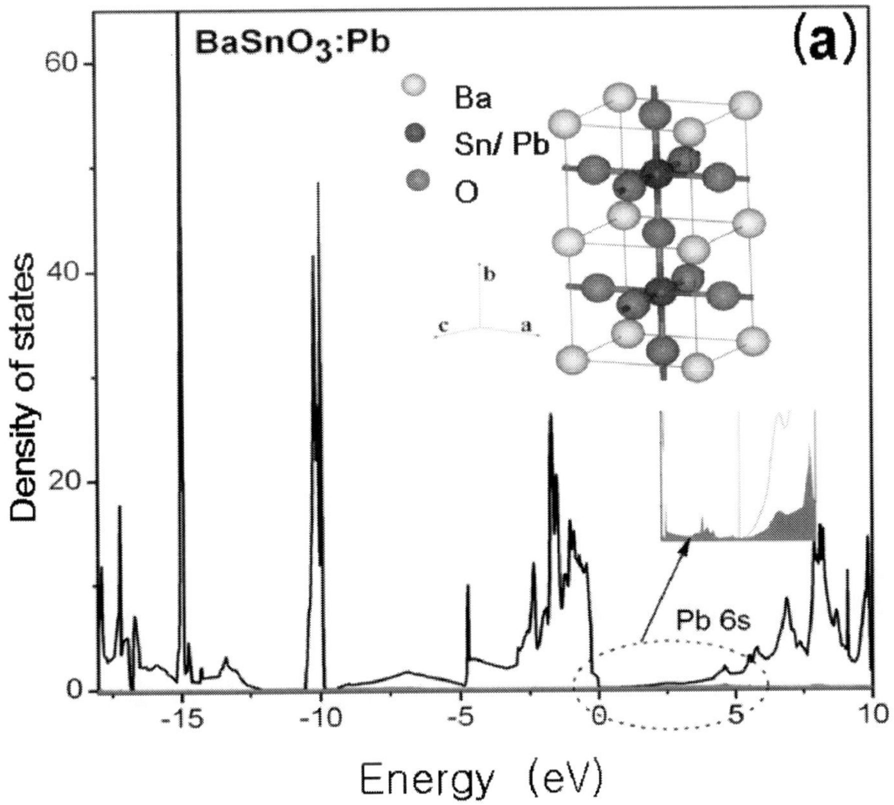

Figure 7.(a) Theoretically calculated total and partial DOS for $BaSnO_3$:Pb indicating the role of Pb-6s orbital (shaded in red) in the forbidden gap. The Fermi level is set at 0 eV. Inset shows the magnified view of DOS from -0.5 to +5 eV.

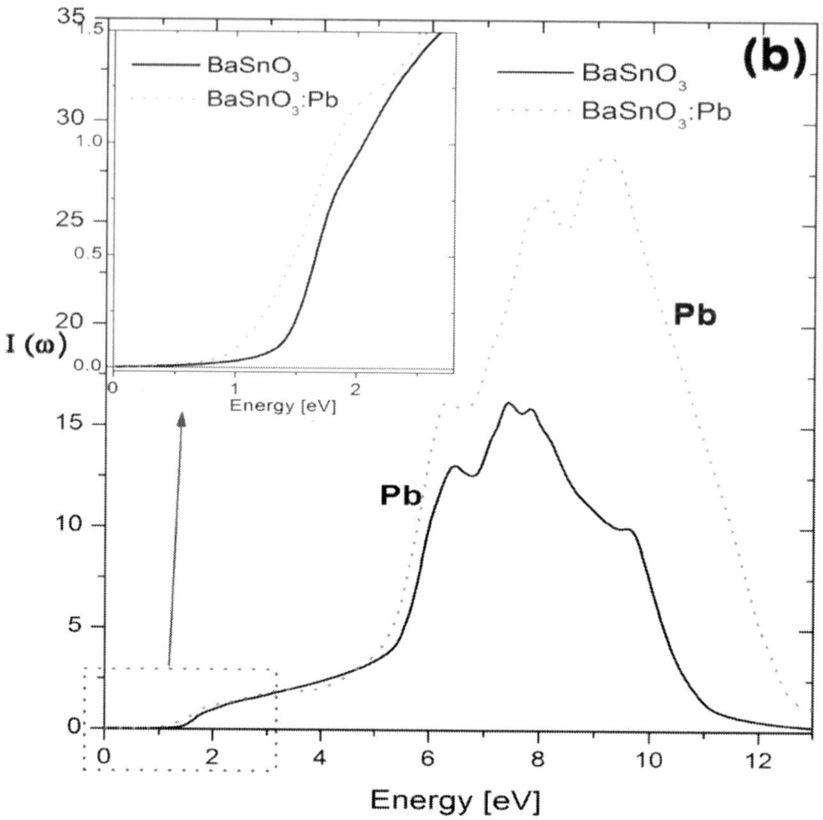

Figure 7(b). Computed absorption coefficient I (ω) of the $BaSnO_3$ and Pb doped structure as a function of photon energy. The inset clearly shows the enhanced absorption in doped sample.

3.3. $PbBi_2Nb_2O_9$ (PBNO) photocatalyst system

Figure 8 displays the results of structural and optical characterisation of most important photocatalytic system under study. The XRD pattern in Figure 8 (a) exhibits the formation of highly crystalline $PbBi_2Nb_2O_9$ aurivillius perovskite phase [11] identified with a space group of *A21am*, a = b = 5.496Å, c = 25.55Å. It is known to posses a layered structure as shown in the right corner structure model. Figure 8(b) showing the UV-Visible diffuse reflectance spectrum over the range of 200 – 800nm is compared along with the spectrum of $TiO_{2-x}N_x$, indicating larger absorption by PBNO in the visible region of electromagnetic spectrum. It shows a conspicuous absorption edge indicating its low-band gap (2.88eV); in contrast, as mentioned earlier section, $TiO_{2-x}N_x$ displays an absorption alongwith a weak shoulder at around 451nm. This clearly indicates the superior optical property of PBNO system.

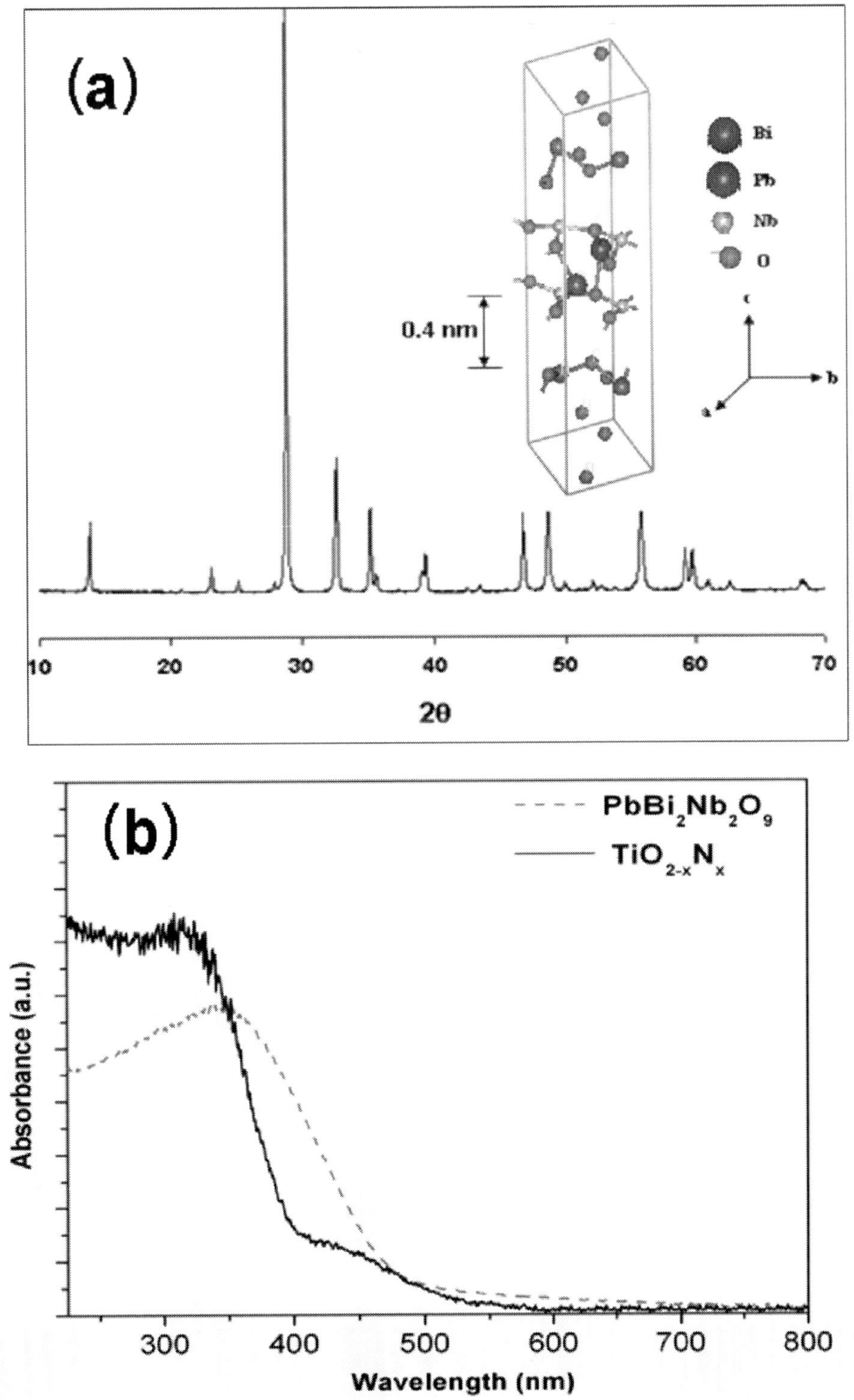

Figure 8. XRD (a) and UV-Vis DRS spectra (b) for $PbBi_2Nb_2O_9$ Inset in (a) shows the unit cell for the corresponding orthorhombic phase. $TiO_{2-x}N_x$ spectrum is used as reference.

Figure 9 shows the computed total and the projected partial DOS of PBNO lattice structure. According to the DFT calculation, we found that the band gap energy of PBNO was less than 2.88eV. The result illustrated in Figure 9 ascertains that in such compounds, the empty Nb4d and occupied O2p orbital are contributing to the formation of CB and VB, respectively. Importantly, the hybridization of the occupied Pb6s and O2p orbital seem to result in a narrower band gap in PBNO. This hybridization would push up the position of the valence band giving the smaller band gap compared to their respective lead-free counterparts. The Bi6s orbital is also located nearby and seems to form hybridization with O2p orbital (O2, O3 and O5 in Figure 9), although this hybridized band of Bi6s and O2p is located below the top of the valence band. The study clearly indicates that the Pb-6s orbital contributes to the top of VB in PBNO.

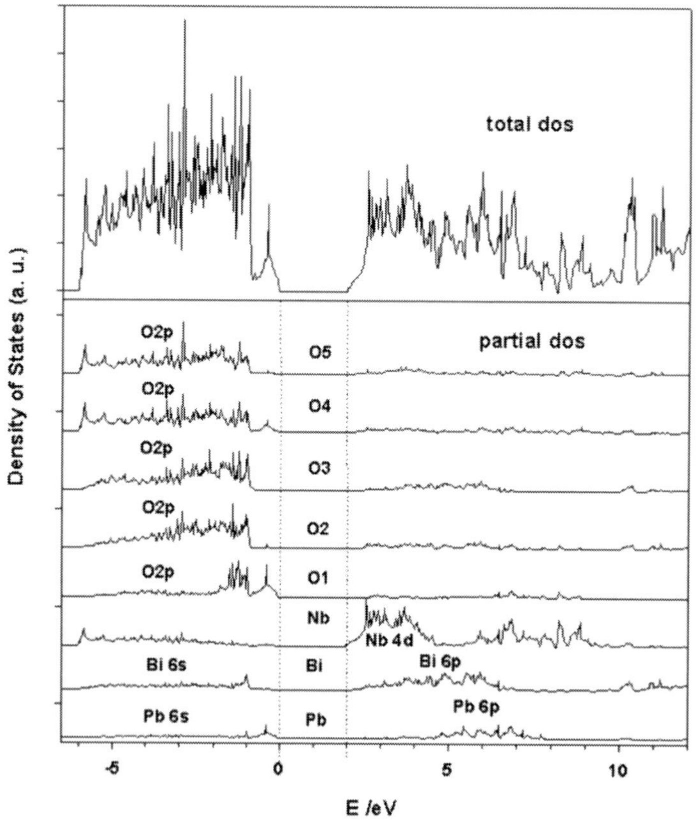

Figure 9. The computed total and the projected partial DOS of PBNO lattice structure indicating the role of different orbital towards the total DOS. Fermi energy was set to 0eV.

Table 1. Photocatalytic activities for respective H_2 evolution with sacrificial agent and O_2 evolution in presence of electron scavenger.

Catalyst	**Band Gap Energy** (eV)	**Hydrogen Evolution** (μmol/gcat.hr)	**Oxygen Evolution** (μmol/gcat.hr)
[a]$ZnFe_2O_4$	1.9	1.3	NR
[b]$BaSnO_3$:Pb	VL	NF	32
[a]$PbBi_2Nb_2O_9$	2.88	7.6	520
[a]$TiO_{2-x}N_x$	2.73	Trace	221

[a] Catalyst loaded with 0.1wt% Pt, 0.3g; light source, 450W Hg-Arc lamp(Oriel) with UV cut-off filter($\lambda \geq 420$nm). Reaction was performed in aqueous methanol solution (methanol 30ml + distilled water 170ml) for the half-reaction of water-splitting.

[b] Photocatalytic oxidation was under $\lambda \geq 420$nm, was tested for 0.1g sample in presence of 0.01M $AgNO_3$ as electron scavenger with 300W Xe arc lamp (Oriel). ;

VL-full visible range; NR-Not reported for this half reaction; NF-Not favourable band conduction band.

4. Photocatalytic perception: Band energetics for ferrites, stannates and niobates

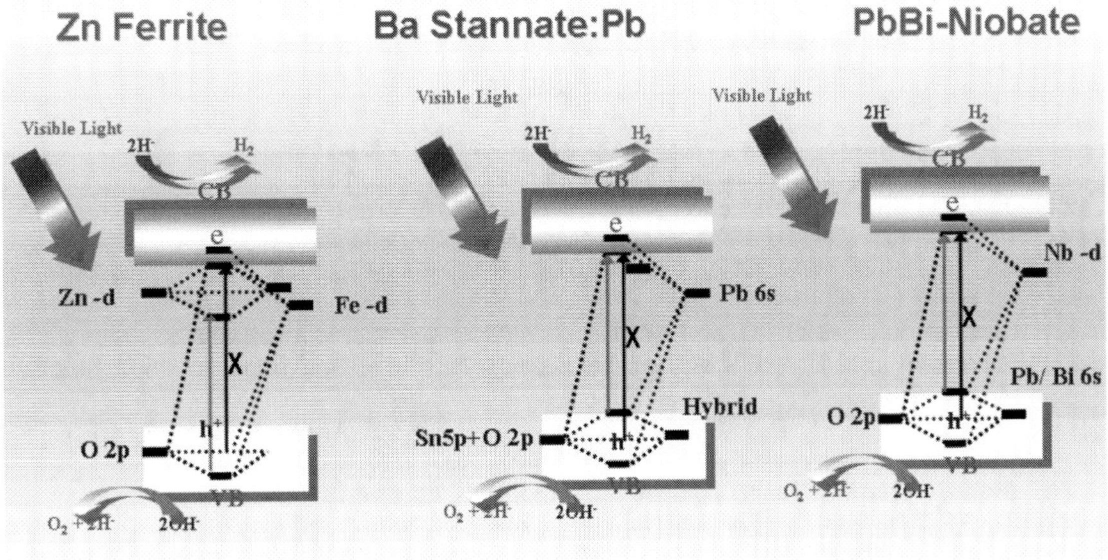

Figure 10. The schematic band diagram for ferrite, stannate and niobate indicating the role of Fe/ Pb orbital towards photocatalytic behaviour.

As described in above section, we have systematically presented the experimental and computed results for respective important system of Ferrite, stannate and niobate. Table – I shows the results of the photocatalytic experimentation on these systems. It shows the band gap energies, and respective hydrogen or oxygen evolution over the photocatalyst under consideration. These results are compared with those of known visible light active reference photocatalyst of $TiO_{2-x}N_x$. All are found to be photocatalytically active systems, for hydrogen or oxygen evolution. However, in each low band gap system, the role of orbital around VB and CB, are decisive in controlling the photocatalytic mechanism. This can be easily understood from the schematic given in Figure 10 as deduced from respective DOS studies. It clearly reveals the role of Fe and Pb orbital in ZFO and BSPO/PBNO respectively. It is noteworthy that Fe-*d* orbital controllably induce the mid band gap states, such *d-d* transitions enable the absorption of visible light photons, thus making it visible-active photocatalyst that can generate photocatalytic hydrogen. The role of Pb-6*s* orbital is interestingly different in case of BSPO and PBNO type systems. The Pb-6*s* orbital is found to contribute to the bottom of CB in case of BSPO; in contrast, it contributes to the top of VB in case of PBNO. In either cases, the role of Pb 6*s* orbital is highly important. Accordingly, in case of BSPO, it converts photocatalytically inactive $BaSnO_3$ (>3.2eV) system into visible active photocatalyst. In case of PBNO type materials, it not only narrows the band gap of corresponding lead-free counterpart [11] of $(Bi_2O_2)^{2+}(A_{m-1}B_mO_{3m+1})^{2-}$ [say here A=Pb, B=Nb; *m*=2] but also yields an efficient visible light active photocatalyst.

The above discussion demonstrates that the correlation of experimental observations with computed electronic structure is very useful in improvement of understanding of the photocatalyst materials. Such correlation is very desirable to discover new solar active materials for solar energy related power generation.

5. Acknowledgement
Permission granted by Director, ARCI to publish the work is gratefully acknowledged.

6. References

[1] Kudo A 2003 *Catalysis Surveys from Asia* 7 31
[2] Borse PH, Lee JS, and Kim HG 2006 *J.App.Phys.* **100** 124915
[3] Fujishima A and Honda K 1972 *Nature* (London) **238** 37
[4] Gratzel M 2001 *Nature* (London) **414** 338
[5] Sivula K, Formal F L, Gratzel M 2009 *Chem. Mater.* **21** 2862
[6] Hoffman M R, Martin S T, Choi W, Bahnemann D W 1995 *Chem. Rev.* **95** 69
[7] Blaha P, Schwarz K and Luitz J, WIEN97, Vienna University of Technology, 1997. Improved and updated UNIX version of the original copyrighted WIEN code.
[8] Gupta G, Nautiyal T and Auluck S 2004 *Phys. Rev. B* **69** 052101
[9] Borse PH, Kim HG and Lee JS 2005 *J.App.Phys.* **98** 043706
[10] Dom R, Subasri R, Radha K, Borse PH 2011 *Solid State Commun.* 151 470
[11] Kim H G, Becker O S, Jang J S, Ji S M, Borse P H and Lee J S 2006, *J. Solid State Chem.* **179** 1211
[12] Hwang DW, Kim HG, Lee JS, Li W, Oh SH 2005 *J. Phys. Chem. B* **109** 2093
[13] Borse P H, Joshi U A, Ji S M, Jang J S, Jeong E D, Kim H G and Lee J S 2007 *Appl. Phys. Lett.* **90** 034103
[14] Borse P H, Jang J S, Hong S J, Lee J S, Jung JH, Hong T E, Ahn C W, Jeong E D, Hong K S, Yoon J H and Kim H G 2009 *J. Korean Phys. Soc.* **55** 1472
[15] Asahi R, Morikawa T, Ohwaki T, Aoki K, Tao Y 2001 *Science* **293** 269
[16] Kim H G, Hwang D W, Bae S W, Jung J H and Lee J S 2003 *Catal. Lett.* **91** 193
[17] Jung E D, Borse P H, Jang J S, Lee J S, Cho C R, Bae J S, Park S, Jung O S, Ryu S M and Kim H G 2009 *J. Nanosci. Nanotechnol.* **9** 3568
[18] Jang J S, Borse P H, Lee J S, Jung OS, Cho CR, Jeong E D, Ha M G, Won M S and Kim H G, 2009 *Bull. Korean Chem. Soc.* **30**1738
[19] Smith A J and Welch AJ E 1960 *Acta Crystllogr.* **13** 653
[20] Bente K, Blum W, Gaebel R, Pebler J and Schmid-Beurmann P 1992 *Physica C* **202** 379

Journal of Physics: Conference Series **365** (2012) 012007 doi:10.1088/1742-6596/365/1/012007

Optical tweezers: light for manipulating microscopic world

Raktim Dasgupta
Laser Biomedical Applications and Instrumentation Division, Raja Ramanna Center for Advanced Technology, Indore, M.P., India

Email: raktim@rrcat.gov.in

Abstract: Optical tweezers make use of a tightly focused laser beam to trap, move, guide, rotate and even sort microscopic objects solely with light. Although the basic laser tweezers, making use of a TEM00 laser beam to create a single trap point, have proved to be useful for any applications in areas ranging from physics to biology, a major breakthrough in this field came as the use of computer generated holograms enabled researchers to create multiple trap sites from single laser source (holographic optical tweezers). Coupled with microfluidic techniques, holographic optical tweezers have promised development of optical techniques for high throughput sorting of different cell types under a single micro-chip platform. The holographic methods have also helped the use of specialized laser beams like Laguerre-Gaussian beams instead of the conventional laser beam for interesting applications like orienting/rotating the trapped objects or trapping cells with minimum photodamage. Further, combining optical tweezers with Raman spectroscopy is becoming increasingly popular for studying single cell biochemistry as use of optical forces to immobilize the cells under investigations not only avoids the negative effects of fixing the cells onto substrate but also improve the quality of the recorded spectra. These advanced optical trapping techniques as outlined above along with some illustrative biophotonics applications have been explored.

1. Introduction

During the past two decades after its invention, single-beam optical gradient force trap, commonly known as optical tweezers [1], has become an indispensable tool for biophotonics research [2]. When a microscopic particle is illuminated by a laser beam, it experiences two types of forces: the scattering force acts along the direction of the laser beam; and the gradient force, proportional to the gradient of light intensity (as exists in a Gaussian profiled beam), pulls the particle to the beam axis. Near the focus of a laser beam, there also exists an axial gradient force pulling the particle towards the focus, and for a sufficiently tightly focused laser beam, this force overcomes the scattering force resulting trapping of the particle at beam focus. Unlike mechanical micro tools, the optical trap is gentle and absolutely sterile and can be used to capture, move and position single cells or sub cellular particles without direct contact or significant damage.

2. Optical sorting with holographic optical tweezers

In the last decade the field of optical trapping has seen several new developments. One very significant advancement in optical trapping technique is the holographic optical tweezers [3], that uses computer generated holograms to create three-dimensional array of optical traps out of a single laser beam. The computer-generated holograms splits the incident laser beam into any desired fan-out of beams, each of which is relayed to a strongly converging objective lens and focused into a distinct optical trap. This approach can project hundreds of simultaneous optical traps in arbitrary three-dimensional configurations. By offering trap array comprising of number of traps placed with micrometer-scale spatial resolution and real-time reconfigurability, holographic optical tweezers provide unsurpassed access to the microscopic world. For example the holographic trapping technique has proven to be useful for a wide range of filtering and sorting methods that are important for lab-on-a-chip applications. One intersting approach is to exploit the difference in degree of interactions of flowing colloidal particles (having varying size, shape or composition) with the array of traps to selectively deflect a particular set of particles away from the fluid flow direction. This approach has been successfully used for sorting of objects from a co-flowing mixture of two types of colloids using either a linear array pattern [4] or a three dimensional optical lattice [5]. We recently studied the influence of trap separation, within an array of holographic traps, on the force experienced by flowing colloids [6]. The force required by a microsphere to move from a trapping position to the next was observed to strongly depend on the ratio of the trap separation and the microsphere size as the inter traps separation was reduced. For some particular ratios the force reaches minima and a linear array of traps behaves like a potential trench consisting of small potential barriers which can be surmounted when a small external force is applied and thus particle can be guided along the array. Such size selective response of particles in an array could form a strong basis for sorting of particles of different size. In a two dimensional trap array, the trap separations in two orthogonal directions could be tailored to simultaneously lower the potential energy barriers for two different sizes of particles in two orthogonal directions. When a mixed suspension of 3 and 5 μm particles was flown diagonally to a suitably designed two dimensional array, the 3 and 5 μm particles moved along vertical and horizontal directions respectively in the array.

3. Optical manipulation with Laguerre-Gaussian beams

Another important improvement in optical trapping technique is the use of special laser beams like Laguerre-Gaussian (LG) beams, Bessel beams etc that can be generated from conventional TEM00 laser beam by inserting a suitable hologram [7]. Like for example, we have recently shown that use of LG beams provides a more efficient and gentler means for trapping biological cells [8]. Due to its annular light profile, LG01 beam is known to have better trapping efficiency as confirmed with spherical particles. Further, LG modes have lower peak intensities compared to the usual Gaussian mode for identical beam power. We realized that these properties of LG beams can offer important advantage for manipulating biological cells by reducing the level of photodamage. Since nowadays spermatozoa are being routinely manipulated by laser tweezers for assessing quality of sperm samples and assisting the in vitro fertilization method, we chose spermatozoa as the model biological cells. The results of our studies.

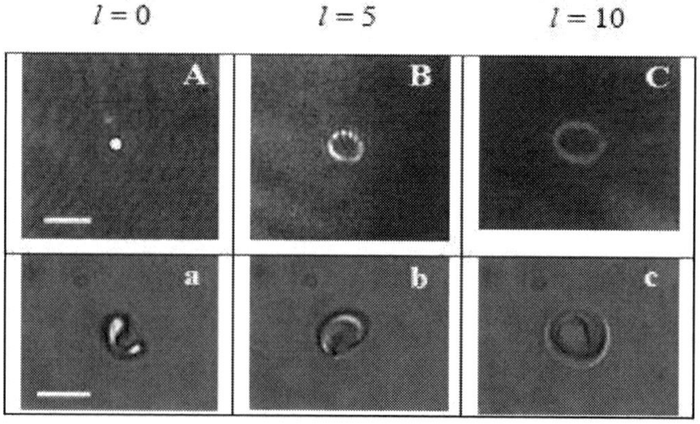

Figure 1. Trapping LG beam profiles for topological charges 0, 5 and 10 (A-C) and the corresponding orientation of a trapped RBC (a-c) respectively. Scale bar, 2.5 μm. (A-C) and 6 μm (a-c).

revealed two important advantages: firstly, LG01 mode offers better trapping efficiency than TEM00 mode for highly motile spermatozoa and second, the light damage on the cells is significantly reduced with LG beams. The reason for reduced light damage on the cells is believed to be due to lowered degree of nonlinear absorption of trapping light and experiments performed with acridine orange stained cells supported this conjecture. Considering high level of concern over the genetic purity of spermatozoa the results should motivate evaluation of the special laser beams for manipulating the cells

Another very interesting application of LG trapping beams was their use for orientation and rotation of trapped red blood cells (RBCs) [9]. It was observed that since the size of the intensity annulas of LG modes increases with the topological charge the cells get trapped at larger tilt angle with respect to the beam axis and thus provided additional control on the orientation of the cells under trap. Further, the RBCs could also be driven as micro-rotors by a transfer of orbital angular momentum from the LG trapping beam having large topological charge or by rotating the profile of LG mode having fractional topological charge. The technique for controlling the orientation of the trapped RBC may find useful applications while studying polarized Raman spectra of the cells.

4. Studying red blood cells with Raman optical tweezers

Optical tweezers are also being used for spectroscopic studies on single cells, which helps account for the problem of heterogeneity present in bulk cell samples. In particular Raman spectroscopy is receiving considerable current interest for studies of the chemical composition and conformation of macromolecules in individual cells since this technique avoids the necessity of any exogenous stain. However, due to the inherent weak nature of the Raman signal, a long acquisition time, often tens of seconds to few minutes, is required to acquire spectra with a good signal to noise ratio. The cell should therefore be immobilized. But the physical or chemical methods used for immobilization of cells in micro-Raman technique often lead to undesirable surface-induced effects on the cells or lead to strong background spectra originating from the substrate medium. The use of optical tweezers to immobilize cell without direct contact helps to avoid these problems and therefore Raman optical tweezers or a setup facilitating acquisition of Raman spectra from an optically trapped cell, are receiving much attention [10]. In particular the use of near infrared (NIR) radiation for Raman studies is gaining rapid interest due to much reduced fluorescence background that often obscures the small but important Raman bands. Raman optical tweezers have already been utilized for several interesting studies such as monitoring the real-time heat denaturation of yeast cells [11], the transition from the oxygenated to deoxygenated condition of a RBC on application of mechanical stress [12], sorting and identification of microorganisms [13] etc. Raman optical tweezers are being extensively used for studying RBCs since Raman spectroscopy is a powerful technique to monitor the oxygen carrying capacity of RBCs beacause the binding or the dissociation of oxygen with heme leads to significant conformational changes of hemoglobin that can be sensitively monitored by this technique [14]. For recording Raman spectra from a single optically trapped RBC, we used a Raman tweezers incorporating a 785 nm cw beam from a Ti:Sapphire laser for both trapping the cells and exciting the Raman spectra. In Fig 2 we show the mean Raman spectrum of RBCs collected from blood samples of five healthy volunteers as well as from five patients suffering from malaria (Plasmodium vivax infection). For the malaria samples investigated, the number of parasitized

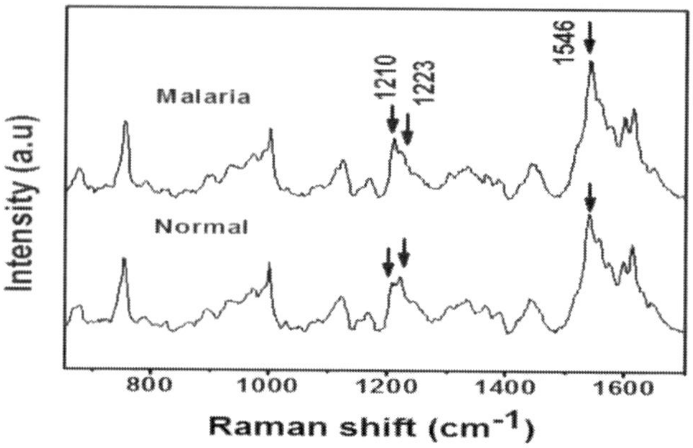

Figure. 2. Mean Raman spectra of RBCs collected from healthy donors and malaria patients. The arrows indicate Raman bands where significant changes are observed.

cells was ~ 1-2%, as confirmed by acridine orange staining. For this study no distinction was made between parasitized and non-parasitized cells. As compared to RBCs from healthy donors, in RBCs taken from malaria patients (IRBCs), a significant decrease in the intensity of the low spin (oxygenated-haemoglobin) marker Raman band at 1223 cm-1 (ν13 or ν42) along with a concomitant increase in the high spin (deoxygenated-hemoglobin) marker bands at 1210 cm-1 (ν5+ν18) and 1546 cm-1 (ν11) was observed. These changes suggest a reduced hemoglobin-oxygen affinity for the IRBCs [15]. For this study all the spectra were acquired with trap laser beam power of ~ 2 mW and an integration time of 30 s since it was observed that at higher trap beam power the intensity of Raman bands at 975 cm-1, 1244 cm-1 and 1366 cm-1, increases with exposure time, first slowly until a critical exposure time, beyond which it increase rapidly and then again levels off. The time interval at which the steep increase in intensity occurred was 90-100 s, for excitation power of ~ 5 mW and reduced to 30-40 s, at ~ 9 mW. These changes in the Raman spectra suggest photoinduced aggregation of intracellular heme [16]. Further the simultaneous decrease in intensity of the Raman band at 1544 cm-1 (ν11) suggests photo induced hemichrome formation in the cells [17].

5. Conclusion

The important recent developments in the field of optical tweezers have been briefly discussed alongwith some useful applications of these. It is pertinent to note here that being fairly new techniques the full potentials of holographic optical trapping and Raman optical tweezers are yet to be explored.

References

[1] Ashkin A, Dziedzic J M, Bjorkholm J E and Chu S 1986 *Opt. Lett.* **11** 288
[2] Svoboda K and Block S M 1994 *Ann. Rev. Biophys. Biomol. Struct.* **23** 147
[3] Curtis J E, Koss B A and Grier D G 2002 *Opt. Comm.* **207** 169
[4] Ladavac K, Kasza K and Grier D G 2004 *Phys. Rev. E* **70** 010901
[5] MacDonald M P, Spalding G C, Dholakia K *Nature* **426** 421
[6] Ahlawat S, Dasgupta R, Verma R S, Kumar N V and Gupta P K 2012 *National Laser Symposium-20, Anna University,* Chennai, India, January 9-12,
[7] Garbin V, Cojoc D, Ferrari E, Proietti R Z, Cabrini S and Fabrizio E D 2005 *J Appl. Phys.* **44** 5773
[8] Dasgupta R, Ahlawat S , Verma R S,Shukla S and Gupta P K 2010 *J. Biomed. Opt.* **15** 065010
[9] Dasgupta R, Ahlawat S, Verma R S and Gupta P K 2011 *Opt. Exp.* **19** 7680

[10] Petrov D V 2007 *J. Opt. A: Pure Appl. Opt.* **9** S139
[11] Xie C, Li Y, Tang W and Newton R J 2003 *J. Appl. Phys.* **94** 6138
[12] Rao S, Bálint S, Cossins B, Guallar V and Petrov D 2009 *Biophys. J.* **96** 209
[13] Xie C, Chen D and Li Y Q 2005 *Opt. Lett.* **30** 1800
[14] Wood B R, Tait B and McNaughton D 2001 *Biochem. Biophys. Acta* **1539** 58
[15] Dasgupta R, Verma R S, Ahlawat S, Uppal A and Gupta P K 2011 *J. Biomed. Opt.* **16** 077009
[16] Wood B R, Hammer L, Davis L and McNaughton D 2005 *J. Biomed. Opt.* **10** 014005
[17] Dasgupta R, Ahlawat S, Verma R S, Uppal A and Gupta P K 2010 *J. Biomed. Opt.* **15** 055009

Journal of Physics: Conference Series **365** (2012) 012008 doi:10.1088/1742-6596/365/1/012008

Role of Plasmons in Improved Photovoltaic Devices : Nano Clusters of Sn and SnO_x in SnO_2 Films

Shailendra Kumar[1] and C Mukherjee[2]
[1]Indus Synchrotron Utilization Division
[2]Mechanical and Optical Support Section
Raja Ramanna Centre for Advanced Technology, Indore-452013, India

Email : Shailendra Kumar : shail@rrcat.gov.in

Abstract: Recently, plasmonic concepts have been used to increase absorption length of incident light and seperation of excess carriers in photovoltaic devices. Doped SnO2 films are used as transparent conducting oxide (TCO) in photovoltaic devices. Sputter etched surfaces of SnO2 films have nano clusters of Sn and SnOx. Valence bulk plasmons (VBPs) of nano clusters have been studied using ultraviolet photoelectron spectroscopy (UPS) in in-situ conditions. Transmission and atomic force microscopy (AFM) are used in ex-situ conditions to study surface morphology and optical properties of sputtered SnO2 thin films.

1.Introduction

Applications of photovoltaic devices for various uses are increasing as cost of production is decreasing. For strategic applications, such as in space and defense, higher efficiency (η) is more important than cost of solar cells (SCs). On the other hand for mass consumptions, low cost and high η are required. Use of concepts of plasmonics to improve η of SCs is an active research topic of last decade and presently many groups are active in this field. Review and research articles are available on the use of plasmonics to increase η of SCs [1-6]. Maximum η of about 30% have been reported of SCs based three junctions of III-V and II-VI semiconductors [7-9]. Maximum η of single crystal Si SCs is about 27%, of polycrystalline Si SCs is about 20% and of thin film SCs is about 15% [9]. Maximum η of SCs based on polymer semiconductors is about 5% [9]. These maximum η are without incorporating concepts of plasmonics. Role of plasmonics is more important in increasing η of low cost thin film SCs [2,3,10].

Books and review articles related to bulk and surface plasmons and their applications in various fields are available [11-12]. There is little confusion about bulk plasmons in metals, semiconductors and insulators. The free carrier density in semiconductors is much less than in metals and it is zero in insulators at room temperature (≈300K), then expected question is that what is the origin of bulk plasmons in insulators? For example, bulk plasmons energy of Au, Al (metals), Si (semiconductor) and SiO2 (insulator) are 9.1eV, 15.3eV, 16.5eV and 24eV respectively [11, 13]. In metals, intraband transitions of electrons generate bulk plasmons (transitions within a band having the Fermi level). In semiconductors and insulators, interband transitions of electrons (from the valence band to the conduction band) generate excitons and bulk plasmons [13,14]. Electrons of the valence band of semiconductors and insulators participate in excitation of bulk plasmons, hence let us denote them as valence bulk plasmons (VBPs). Further, let us denote bulk plasmons of metals as free-carriers bulk plasmons (FBPs) to differenciate these with respect to VBPs of semiconductors and insulators. Free carrier density in F doped SnO2 transparent conducting oxide (TCO) is about 9x1020 cm-3 and corresponding energy of FBPs is about 0.43eV. In comparison, energy of VBPs in SnO2 is about 18eV as reported in literature [15, 16]. The energy of FBPs in metallic Sn is about 14eV [16,17]. Causes of energies of FBPs and VBPs in SnO2 are good examples to differenciate between bulk plasmons due to free electrons and valence electrons in semiconductors and insulators.

TCO is part of many opto-electronic devices. These are used in large area panels of SCs. Schemes had been proposed to increase the effective absorption length near the band gap energy and in the sub band gap region using concepts of plasmonics [1-6]. Many research groups are working to increase η of SCs, without increasing cost of SCs. It is required to investigate low cost processing steps, which can use benefits of plasmonics, to enhance η of SCs. It had been observed that sputtering creates nano metallic clusters due to difference in the sputtering rates of different constituent elements of semiconductors and TCOs. For example, In clusters in InP, Ga clusters in GaAs, Sn clusters in SnO2 etc. had been observed on sputtered surfaces [18-20]. Many groups had used nano clusters of costly metals such as Au and Ag, to increase η of SCs by incorporating concepts of plasmonics [1-6]. One can use nano clusters produced during sputtering of semiconductors and TCOs in geometries of SCs, in place of nano clusters of costly metals, to incorporate schemes of plasmonics. Sputtering had been used to produce textured surfaces of TCOs, used in SCs [10]. However, a systematic research is needed to study sizes of nano clusters as a function of sputtering parameters, substrate temperature and ambient conditions to maximize η, using concepts of plasmonics. Oxide formation of nano metallic clusters in ambient conditions, hampers the study of actual size, such as formation of GaOx on GaAs, InOx on InP etc. Hence, study of nano clusters under in-situ conditions is required. In the present work, formation of nano clusters of Sn and SnOx on SnO2 as a function of sputtering time is studied using ultraviolet photoelectron spectroscopy (UPS) under in-situ conditions. Atomic force microscopy (AFM) and optical transmission measurements are used in ambient conditions. Results of these studies are given below.

2.Experimental:

Commercial F doped SnO2 thin films deposited on glass have been used for present study. SnO2 films, cleaned with acetone were loaded in the preparation chamber. UPS spectra had been measured on angle integrated photoelectron spectroscopy (AIPES) beam line of INDUS-1 synchrotron source. Vacuum in the AIPES experimental chamber was better than 2x10-9 Torr. Other details of AIPES experimental chambers are given in previously published paper [21]. Ar+ ions sputtering was done at 2x10-5 Torr partial pressure of argon, at room temperature, in the preparation chamber separated by the gate valve from the experimental chamber. Energy of Ar+ ions was 500 eV and current was 5μA. After sputtering, samples were transferred to experimental chamber and UPS spectra were recorded within half an hour after sputtering. Photoelectron counts are collected in steps of 20meV with pass energy of 20eV. Polycrystalline Au foil was also loaded along with SnO2 samples. UPS spectrum from the Au foil is used as a reference to characterize the monochromatic synchrotron light and the Fermi energy level (EF) of the metallic sample holder.

AFM studies were done to measure sizes of clusters on the surface of sputtered samples after taking out samples to ambient conditions. Surface topographies of 10 minutes sputtered SnO2 samples were imaged using a multimode scanning probe microscope (NT-MDT, SOLVER-PRO).

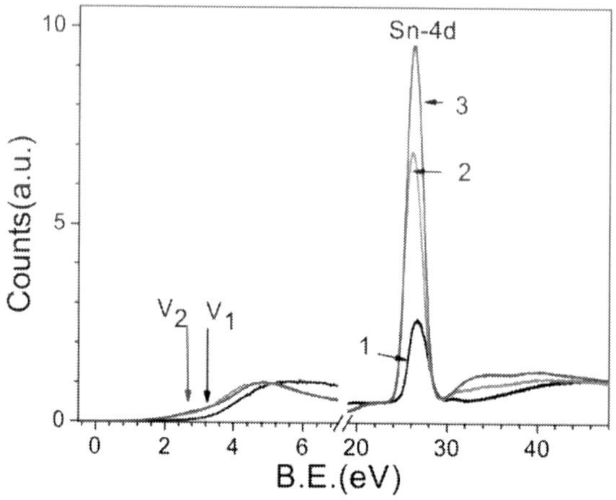

Figure 1. Photoelectron spectra of SnO_2 film measured with incident energy of 100eV . Curves 1(black), 2(red) and 3(blue) show spectra of un-sputtered, 3 minutes sputtered and 15 minutes sputtered surfaces of SnO_2 film. Arrows V_1 and V_2 show valence band onset of un-sputtered and sputtered surfaces.

AFM measurements were carried out in a non-contact mode using silicon cantilever tips having radius of curvature ~ 20nm and a spring constant 5.5 N/m with a resonant frequency of 190 KHz under ambient conditions. Top surface was electrical grounded during measurements. Mean length and width of individual nano clusters were estimated using the imaging software (SOLVER-PRO). Analysis was carried out at various places on the surface of samples. Transmission spectra of un-sputtered and sputtered SnO2 films were measured in the range 200nm-1000nm (VARIAN, CARY5000).

3. Results and Discussion

3.1 Photoelectron Spectroscopy:

SCs deposited on transparent substrates such as ordinary glass and polymers have one electrode usually of thin film of TCO. For example, incident light enters through TCO in p-i-n solar cells of amorphous silicon hydrogen alloy (a-Si:H) and micro-crystalline silicon hydrogen (μc Si:H) [10]. Patterned Ag layer and nano clusters of noble metals (Ag, Au etc) had been used, at the interface of bottom electrode and semiconductor layer, to enhance the absorption of incident light [2,3]. Can nano clusters of constituent elements of TCO play similar role to that of noble metals? This is a research topic and worth exploring. UPS spectra of sputtered surfaces of SnO2 had been measured to study the presence of nano clusters of Sn and SnOx. UPS spectra of SnO2 film measured with incident energy of 100eV are shown in figure 1. Curves 1(black), 2(red) and 3(blue) show spectra of un-sputtered, 3 minutes sputtered and 15 minutes sputtered surfaces of SnO2 film respectively. Arrows V1 and V2 show valence band onsets of un-sputtered and sputtered surfaces.

The difference between the Fermi level (EF) and the valence band onset (Vb) of un-sputtered surface (= EF -Vb) is 3.3±0.05eV. EF - Vb of 3 minutes and 15 minutes sputtered surfaces are 2.7±0.05eV and they are same within error levels. There are two possible causes for this difference between un-sputtered and sputtered surfaces.

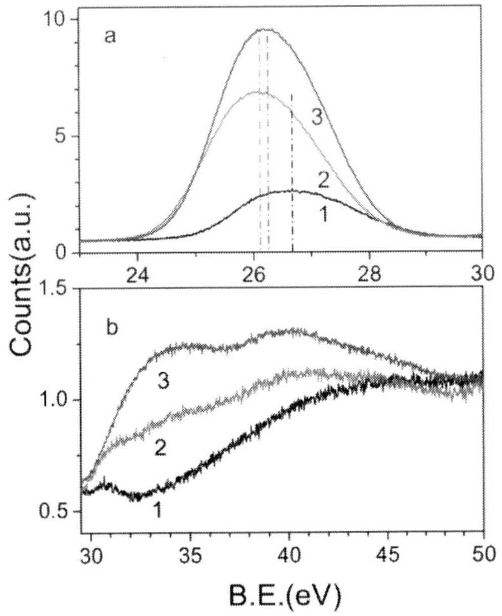

Figure 2. Curves 1,2 and 3 are same as in fig.1. (a) Sn-4d core spectra of un-sputtered and sputtered surfaces, (b) Valence Band Plasmons peaks of un-sputtered and sputtered surfaces.

First cause is that there is more upward band bending at the sputtered surfaces resulting in the reduction of (EF – Vb). Second cause is that the optical band gap (Eg) of top layers after sputtering is less than Eg of un-sputtered surface. Any one of these or both will reduce (EF-Vb). This information is important in theoretical calculation of η of SCs and it is discussed after comparing with transmission spectra of un-sputtered and sputtered SnO2 thin films.

Figure 2 (a) shows expanded UPS spectra of Sn-4d core level and (b) shows expanded UPS spectra due to bulk plasmons of un-sputtered and sputtered surfaces. Curves 1, 2 and 3 in figure 2 are same as in figure 1. The ratio of intensity of Sn-4d peak to intensity of O-2p is increasing as a function of sputtering time. This shows that the concentration of Sn is increasing after sputtering. The origin of bulk plasmons peaks at higher binding energies (B.E.) in UPS spectra (figure 2) is due to reduction of kinetic energy (K.E.) of core photoelectrons by scattering with plasmons [15]. The difference between the B.E. of scattered photoelectrons by plasmons and B.E. of core elastically scattered photoelectrons is equal to the energy of plasmons. A comparison of present UPS spectra with published UPS and electron energy loss spectra (EELS) of Sn, SnO and SnO2 helps in assigning the origin of various peaks in figure 2(b) [15, 20]. The broad peak at 18eV in the UPS spectrum of un-sputtered surface is due to VBPs of SnO2. Broad plasmon peaks near 10 and 14eV in UPS spectra of sputtered surfaces, (curves 2 and 3 in figure 2 (b)), can have contributions from clusters of both SnO and metallic Sn. Reported plasmon peaks at 13.5 and 9 eV are due to bulk and surface components of VBPs of SnO. Plasmon peaks at 14 and 10 eV in UPS spectra of sputtered surfaces are due to bulk and surface components of FBPs of metallic Sn. The intensities of peaks near 14 and 10 eV are increasing as

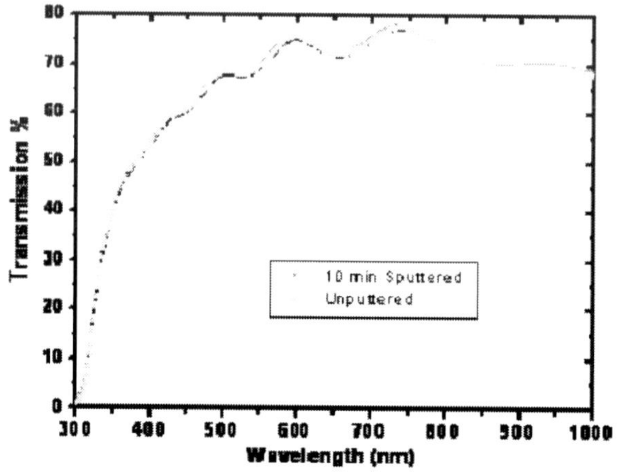

Figure 3. Transmission spectra of un-sputtered and 10 minutes sputtered SnO_2 thin film. There is very little change in overall transmission spectra after 10 minutes sputtering.

sputtering time is increased. Clearly, UPS spectra of sputtered surfaces show presence of both Sn and SnO.

3.2 Transmission and AFM studies:

Transmission spectra of un-sputtered and sputtered SnO2 thin films are shown in figure 3. There is no noticeable difference in the transmission spectra of un-sputtered and sputtered SnO2 films. Transmission spectra were measured within half an hour after taking out sputtered film in ambient room conditions. Oxidation of nano clusters of metallic Sn into oxides in ambient conditions makes it difficult to differentiate between sputtered and un-sputtered thin films by transmission measurements in ex-situ conditions [17]. Hence, in-situ measured UPS spectra are more useful to study the effect of sputtering. The origin of the observed decrease in the difference (EF – Vb) in UPS spectra after sputtering, figure 1, can not be answered definitely as the difference on transmission spectra of un-sputtered and sputtered SnO2 thin films is very small. Increase in the upward band bending at the sputtered surface looks more probable, but more studies carried out in in-situ conditions after sputtering will resolve this issue.

AFM 2D pictures of un-sputtered and 10 min. sputtered surfaces of SnO2 thin films are shown in figure 4(a) and (b) respectively. The un-sputtered surface shows grains of size ~ 200nm. Few smaller size grains are also visible. RMS roughness is ~ 28nm. The main difference between the 10min. sputtered and the un-sputtered sample is the appearance of large number of small size (~ 30 – 50nm) clusters sitting over the underneath large size grains. The rms roughness of the sputtered surface remains nearly same. It appears that these nano clusters are formed due to preferential oxygen sputtering from the surface.

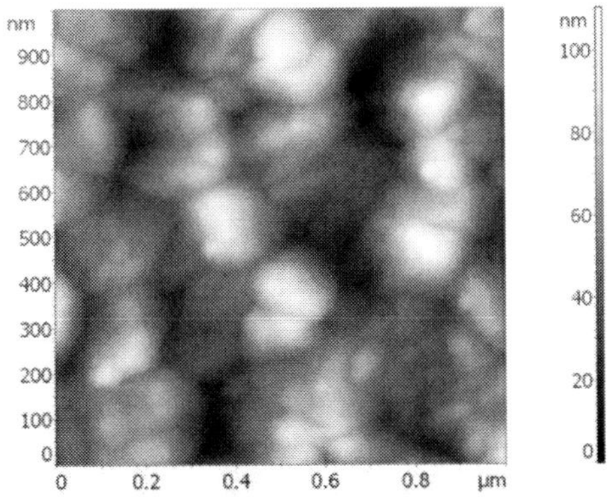

Figure 4a. AFM 2D picture of un-sputtered surface of SnO_2 thin film.

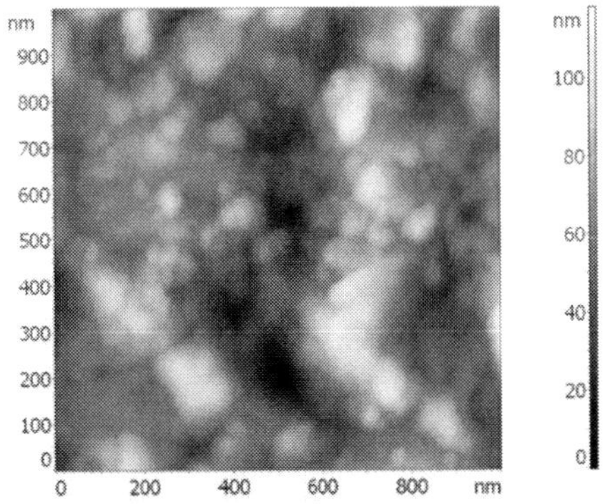

Figure 4b. AFM 2D picture of 10 minutes sputtered surface of SnO_2 thin film.

Textured surfaces of TCOs, prepared by sputtering without breaking vacuum in fabrication line had been used in SCs [10]. The role of nano clusters had not been considered in the performance of SCs. Present results indicate that nano clusters of Sn, SnOx are present at the sputtered surface. Distribution of sizes of nano clusters of Sn, SnOx and SnO2 depends upon the sputtering conditions and duration of sputtering. If one can deposit nano clusters of SnO2 of appropriate sizes such that their surface components of FBPs are in the energy range 0.5 - 1.5eV, then these will be useful in enhancing the absorption of incident radiation in the sub band gap region of semiconductors with band gap less than 1.5eV such as c-Si, a:Si:H, µc:SiH thin films, GaAs etc. Energy of plasmons of nano clusters of diameters (d) more than about 50nm is independent of size, however it increases as 1/d2 for sizes less than 50nm [22]. It is a topic of further research to measure energy of FBPs of nano clusters of TCO as a function of size, so that they will be useful in enhancing η of SCs. This will be a cost effective process as sputtering equipments are already in use in the fabrication line.

4. Conclusions

UPS spectra of sputtered SnO2 thin films show presence of nano clusters of Sn and SnOx at the surface. There is very little change in transmission spectra of un-sputtered and sputtered SnO2 thin films, measured in ambient conditions. AFM studies show grains of average size 200nm on un-sputtered surface and formation of nano clusters of average size 30-50nm sitting over the underneath large size grains after sputtering. It is proposed that plasmons due to free carriers in nano clusters of SnOx of sizes less than 10nm can be used in SCs to increase absorption of incident light in the energy range 0.5-1.5eV.

Acknowlwdgement

Authors are thankful to Dr D M Phase and Mr Wadikar for help in UPS measurements, to Mr K Rajiv for help in optical measurements and to Drs S K Deb and L Abhinandan for their interest in this work. Authors are also thankful to members of program committee-ICRTP-2012 for inviting S Kumar to give talk on this topic.

References

[1] Atwater H A and Polman A 2010 *Nature Materials* **9** 205
[2] Curtin B, Biswas R and Dalal V 2009 *Appl. Phys. Lett.* **95** 231102
[3] Ferry V E, Verschuuren M A, Hongbo B T Li, Scropp Rudd E I, Atwater Harry A and Polman A 2009 *Appl. Phys. Lett.* **95** 183503
[4] Nakayama K, Tanabe K and Atwater Harry A 2008 *Appl. Phys. Lett.* **93** 121904
[5] Matheu P, Lim S H, Derkacs D, Mcpheeters C, and Yu E T 2008 *Appl. Phys. Lett* .**93** 113108
[6] Derkacs D, Chen W V, Matheu P M, Lim S H, Yu P K L and Yu E T 2008 *Appl.Phys. Lett.* **93** 091107
[7] Farrell R M, Neufeld C J, Cruz S C, Lang J R and Iza M 2011 *Appl. Phys. Lett* **98** 201107
[8] Guter W, Schone J, Phillips S P, Steiner M, Siefer G, Wekkeli A, Welser E, Oliva E, Bett A W and Dimroth F 2009 *Appl. Phys. Lett.* **94** 223504
[9] Green M A, Emery K, Hishikawa,Y and Warta W 2009 *Prog. Photovolt: Res. Appl.***17** 85 and references therein
[10] Müller J, Kluth O, Wieder S, Siekmann H, Schöpe G, Reetz W, Vetterl O, Lundszien D, Lamberts A, Finger F, Rech B and Wagner H 2001 *Solar Energy Materials and Solar Cells,* **66**, 275
[11] Kreibig U and Vellmore M , Optical Properties of Metal Clusters, Springer Verlag, Berlin, Germany
[12] Hutter E and Fendler J H 2004 *Adv. Mater.* **16** 1685
[13] Egerton R F 2009 *Rep. Prog. Phys.***72** 016502
[14] Egri I 1985 *Phys. Reports* **119** 363
[15] Köver L, Moretti G, Kovacs Zs, Sanjines R, Cserny I, Margaritondo G, Palikas J and Adachi H, 1995 *J. Vac. Sci. Technol.* **A13** 1382
[16] Hoflund Gar B and Corallo G R 1992 *I Phys. Rev. B* **46** 7110
[17] Bevolo A J, Verhoeven J D and Noack M 1983 *Surf. Sci.* **134** 499
[18] Kumar S, Mukherjee C, Dixit V K, Singh S D and Jha S N 2011 *Appl. Surf. Sci.* **258** 143
[19] Kang, H J, Moon Y M, Kang T W, Leem J Y, Lee J J and Ma D S 1989 *J. Vac. Sci. Technol.* **A7** 325
[20] Egdell R G, Erikson S and Flavell W R 1987 *Surf. Sci.* **192** 265
[21] Dixit V K, Kumar S, Mukherjee C, Tripathi S, Sharma A, Brijpuriya R and Chaudhari S M 2006 *Appl. Phys. Lett.* **88** 083115
[22] Wang Y W, Kim J S, Kim G H and Kim Kwang S 2006 *Appl. Phys. Lett.* **88** 143106

Influence of Copper doping on physical properties of Barium Titanate

A Mishra and Niyati Mishra
School of Physics, Devi Ahilya University, Khandwa Road, Indore 452001, India
nmishra.sop@gmail.com

Abstract. A finest possible sample of Copper (Cu) doped micro particles of $BaTiO_3$ (BTO) with possible tetragonal structure via a solid-state route was prepared. Prepared samples of $BaTi_{1-x}Cu_xO_3$ (x=0.05) were structural characterized by X-ray diffraction (XRD). Fourier Transform Infrared spectroscopy (FTIR) technique used for determines the Ti-O bond shifting as an influence of Cu in BTO. Finally Dielectric Constant measurements of the sample above and below the Curie temperature were carried out.

1. Introduction

Barium titanate ($BaTiO_3$) is one of the best known Perovskite ferroelectric compounds ($A^{2+}B^{4+}O^{-3}$) that have been extensively studied [1, 2] due to the simplicity of its crystal structure, which can accommodate different types of dopant. This has led to the possibility of tailoring the properties [3] of doped $BaTiO_3$ for specific technological applications, such as capacitors, sensors with positive temperature coefficients of resistivity, piezoelectric transducers and ferroelectric thin-film memories. Because of the intrinsic capability of the Perovskite structure to host ions of different size, a large number of different dopants can be accommodated in the lattice [4]. The mechanism of dopant incorporation into $BaTiO_3$ has been extensively investigated and the behaviour of some transition metal ions has been well elucidated. The ionic radius is the main parameter that determines the substitution site [5]. Ferroelectrics are materials which have the outstanding property of possessing spontaneous electric polarization, the reversibility of the permanent polarization by an electric field. Due to their possessing large piezoelectric values, make them more attractive in different piezoelectric application. $BaTiO_3$ has been a ferroelectric and piezoelectric with an extensive application. It can be used as a capacitor, thermistor, transducer or accelerometer [6]. $BaTiO_3$ chemically and mechanically more stable, which exhibits ferroelectrics properties at and above room temperature and it can be prepared and used in the form of ceramic polycrystalline samples [7]. Modified BaTiO3 is suitable for pyroelectric applications [8] as their infrared response can be adjusted over a wide range of operating temperatures. [9] Our present focus is to develop Cu modified BaTiO3 materials to get a maximum IR response near room temperature, as room temperature IR sensors are attractive for a wide range of civilian and military applications.[10, 11] The sample of Cu doped $BaTiO_3$ (Cu-BTO) by solid-state route has been synthesized. This sample was then characterized by X-ray diffraction. Also, different techniques as FTIR and dielectric constant measurement were carried out. The purpose of this paper is to throw some light on the influence of Copper impurity on the structural and Dielectric properties of $BaTiO_3$.

2. Experimental

2.1. Preparation of Copper doped Barium Titanate (Cu-BTO).
Sample of Copper doped Barium Titanate were prepared via Solid State Route. The starting materials were analytical reagent grade $BaCO_3$, TiO_2 and CuO. Weighed mixtures of the three components were homogenized by grinding and preheated at 900^{0}C for 12 h. They were then grounded, pressed into pellets and heated at 1250^{0}C for 12 hours in air followed by a furnace cooling.

2.2. Physical Measurement

2.1.1. *X-Ray Diffraction.* XRD measurements were carried out using Bruker D8 Advance X-ray diffractometer. The x-rays were produced using a sealed tube and the wavelength of x-ray was 0.154 nm (Cu K-alpha). X-rays were detected using a fast counting detector based on Silicon strip technology (Bruker Lynx Eye detector).

2.1.2. Fourier Transform Infrared Spectroscopy. Infrared spectra (KBr disc) of the sample were obtained using Perkin-Elmer FTIR spectrometer in the 400-4000 cm^{-1} spectral range.

2.1.3. Dielectric Constant Measurement. In Dielectric constant measurement, Dielectric-constant of the sample was recorded on hp- Hewlett Packard 4192 A, LF Impedance 5 Hz – 13 Hz Analyzer.

3. Result and Discussion

3.1. X-ray Diffraction:
X-ray diffraction (XRD) is a versatile, non-destructive technique that reveals the detailed information about the chemical composition and crystallographic structure of natural and manufactured materials. Figure1 shows that the XRD patterns reports the single phase tetragonal crystal system of the space group P4mm (99) and pattern matched with the standard pattern JCPDS no. 79-2265. Diffraction data of Cu doped $BaTiO_3$ has concluded that all major peaks of Cu- BTO are matching with reported XRD data of pure BTO and sample is in single phase. Which shows doping is perfect. The value of 2θ is in between 20 to 80°. There is a comparison of XRD peaks of samples sintered at 900°C and 1250°C.

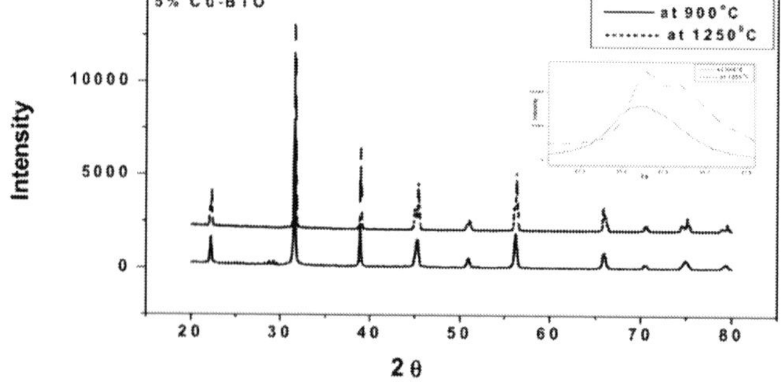

FIGURE1. XRD data of Cu doped $BaTiO_3$ at 900°C & 1250°C. Insert shows the peak width broadening after sintering.

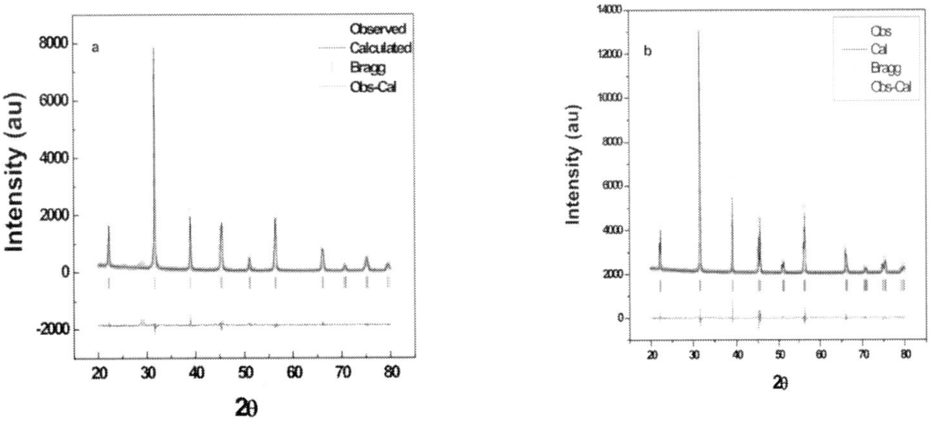

FIGURE2. Rietveld fit of 5% Cu doped Barium Titanate sintered at (A) 900 ^{0}C and (B) 1200

3.2. Fourier Transform Infrared Spectroscopy:
Figure3 indicates FTIR spectra of the 5% Copper doped Barium Titanate (Cu-BTO) particles, synthesized at 1250 °C for 12 Hours. A broad absorption band in the range of 400-4000 cm^{-1} suggests the presence of considerable amount of H_2O and OH in the particles [12, 13]. KBr has been used as a binder for background fit, Hydroxyl group occurs because of moisture in KBr. In particular the sharp absorption band at 3600 cm^{-1} was assigned to the stretching vibration of the Hydroxyl group with an intermolecular hydrogen bond, which shows the presence of hydroxyl group in Barium Titanate lattice. Also the sharp absorption band observed at around 532 cm^{-1} was attributed to the bending vibration of O_6 octahedra deformation mode which was assigned by Slater [14] and confirmed by Spitzer [15].
3.3. Dielectric Measurement:
Barium titanate is a cubic paraelectric above 120^{0}C. Below this temperature, it is ferroelectric, with a tetragonal structure down to 5^{0}C. Considerable effort has been devoted to understanding the behaviour of the dielectric constants of Cu-BTO. In Dielectric-constant measurement, the transition temperature Cu-BTO is occurring at 125^{0}C. The reported value of transition temperature (Curie- temperature) of pure BTO is 120^{0}C.
Table1. Comparison of the lattice parameters of un-doped $BaTiO_3$ and Cu doped $BaTiO_3$

Direct cell parameters	a	b	c	Direct cell volume
$BaTiO_3$	3.9900	3.9900	4.0302	64.2800
$BaTi_{0.95}Cu_{0.05}O_3$ (At 900 ^{0}C)	3.9895	3.9895	4.0077	63.7848
$BaTi_{0.95}Cu_{0.05}O_3$ (At 1250 ^{0}C)	3.9830	3.9839	4.0139	63.6791

In FIGURE4 one can see that after doping Cu in pure BTO, on increasing the temperature on higher side, the transition temperature was shifted from 120^{0}C to 125^{0}C. While on cooling the sample, the transition temperature was shifted to 110^{0}C.

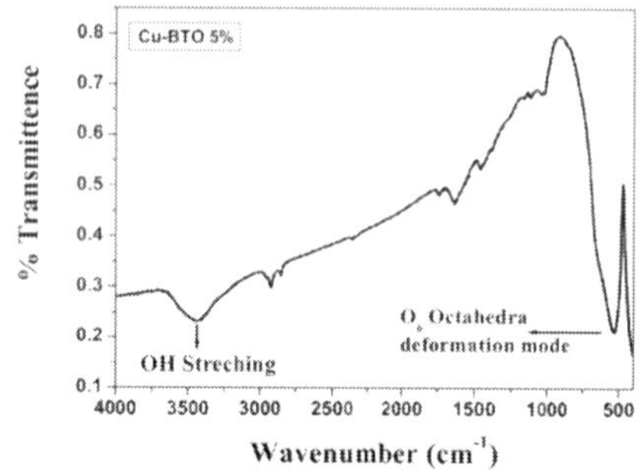

FIGURE3. FTIR spectra of 5% Cu doped $BaTiO_3$

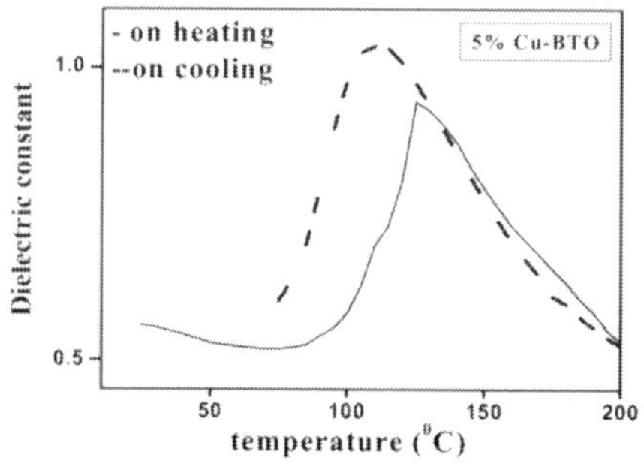

FIGURE4. Dielectric Constant Measurement plot of Cu Doped $BaTiO_3$

4. Conclusion

4.1. The $BaTi_{0.95}Cu_{0.05}O_3$ ceramics was prepared by the solid state reaction method. XRD patterns reports the single phase tetragonal crystal system of the space group P4mm (99) and pattern matched with the standard pattern JCPDS no. 79-2265.

4.2. FTIR Spectra shows that bending vibration of O_6 octahedra deformation mode, which confirms its ferroelectric behaviour.

4.3. Finally the dielectric constant measurements of the sample above and below the Curie temperature were carried out at 1 MHz, the Transition temperature is found shifted towards higher side from that of pure $BaTiO_3$. The Thermal Hysteresis is indicative of the fact that the transition is indeed of First Order.

5. Acknowledgment

The authors are thankful to Dr. Ajay Gupta, Dr. Alok Banerjee, Dr. T. Shripathi, Mr. Uday Deshpandey and Dr. Mukul Gupta UGC-DAE-CSR, Indore for providing the experimental facilities and for the precious guidance. One of the author (Niyati Mishra) is thankful to UGC-DAE-CSR Indore, India for providing the CRS Fellowship.

References

[1] Makovec D, Samadmija Z and Drofenik M 2004 *J. Am. Ceram. Soc.* **87**, 1324
[2] Maga D, Igor P and Sergei M 2000 *J. Mater. Chem.* **10**, 941
[3] Jana A, Kundu T K, Pradhan S K and Chakravorty D 2005 *J. Appl. Phys.* **97 (4)**, 44311
[4] Jin Z, Ang C and Yu Z 1999 *J. Am. Ceram. Soc.* **82(5)** 1345
[5] Yongping P, Wenhu Y and Shoutian C 2005 *J. Rare Earths.* **25**, 154
[6] Chen-Feng Kao & Wein-Duo Yang 1999 *Appl Organomet. Chem*, **13** ,383
[7] Stojanovic B D, mastelaro V R, Paiva Santos C O & Varela J A 2004 *Science of Sintering*, **36** 179
[8] Herbert J M 1980 Ferroelectric Transducer and Sensors (Gordon and Breach, New York)
[9] Mohammed M S, Auner G W, Naik R, Mantese J V, Schubring N W and Micheli A L 1998 *J. Appl. Phys.* **84**, 3322
[10] Scribner D A, Kruer M K and Killiany J M 1991 *Proc. IEEE* **79**, 66
[11] Dereniak E L and Crowe D G 1984 Optical Radiation Detectors (Wiley, New York)
[12] Vivekanandan R, Philip S and Kutty T N 1986 *Mater.Res.Bull.*, **22**, 99
[13] Hennings D and Schreinemacher S 1992 *J.Eur.Ceram.Soc.*, **9**, 41
[14] Slater J C 1950 *Phys.Rev.*, **78**, 448
[15] Spitzer W G, Miller R C, Kleiman D A and Howarth L E 1962 *Phys.Rev.*, **126**, 1710.

Journal of Physics: Conference Series **365** (2012) 012010 doi:10.1088/1742-6596/365/1/012010

Spectroscopic Studies of Some Transition Metal Copper and Iron Complexes

A Mishra[1], R Sharma[2], B D Shrivastava[3] and N Mishra[1]
[1]School of Physics, Devi Ahilya University, Khandwa Road, Indore 452001, India
[2]Physics Department, M.B. Khalsa College, Indore, India
[3]Physics Department, Govt. P.G. College, Dhar, India
rsharma.sop@gmail.com

Abstract. The present paper deals with the spectroscopic studies of transition metal complexes of copper and iron. Metal complexes are characterized by XRD, FTIR, Mössbauer and VSM (Vibrating sample magnetometer) measurements. The XRD analysis reveals the crystalline nature of both compounds. Mössbauer study interprets paramagnetic nature of iron complex at room temperature. VSM measurements also reveal the paramagnetic nature of iron complex at higher temperature range, while it follows magnetic phase transition from paramagnetic to antiferromagnetic nature at low temperature.

1. Introduction

The metal complexes of Schiff bases have been playing an important role in the development of science and technology. These have been widely studied for industrial, antifungal and biological applications[1,2]. It is found that the complexation of iron, copper, cobalt and other metals with Schiff bases has enhanced their activity[3,4]. These complexes show their antifungal potential against Alternaria alternata, Fusarium oxysporum and Myrothecium roridum by inhibiting spore germination[5]. In this communication, the ligand 2- [N- (2-hydroxy phenyl)ethanimidoyl] phenol has been synthesized and its complexation with iron metal salts has been reported. Mentioned properties of metal complexes are due to varied geometrical arrangements of ligands around the metal ion. Due to biological and industrial applications associated with the Schiff base complexes, these have been chosen for spectroscopic investigations. Therefore, the resulting complexes were studied by XRD, FTIR, Mössbauer and VSM (Vibrating sample magnetometer) measurements in order to elucidate their structures, geometry and magnetic nature.

2. Experimental

2.1. Synthesis of Complexes.

Ligand 2- [N- (2-hydroxy phenyl)ethanimidoyl]phenol was synthesized by condensing ortho hydroxy acetophenone ortho-aminophenol in equimolar proportion. Now one third mole of anhydrous $FeCl_3$ was added to this ligand in ethanol and the mixture was refluxed for 4-6 hours on water bath to get (2N2HPEPhenol) Fe complex. Copper complex is prepared by using same method as iron complex. In case of copper complex one half mole of copper chloride was added to ligand instead of $FeCl_3$. After refluxing we get (2N2HPEPhenol) Cu complex.

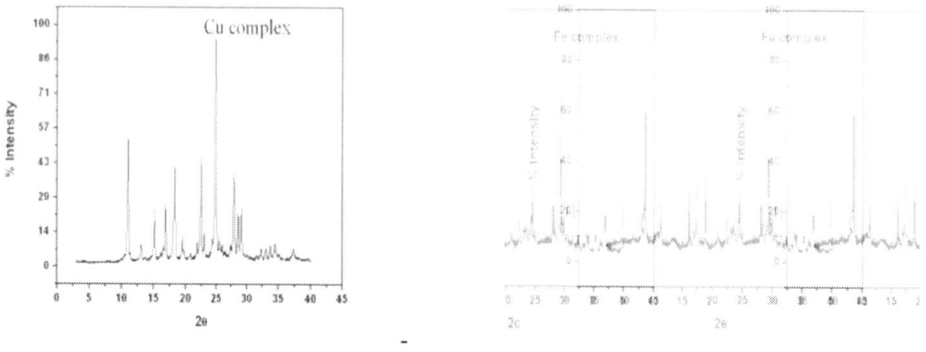

Figure1. XRD Pattern of Cu(II) and Fe(III) Complexes

2.2. Experimental Analysis by Different Methods

All the spectroscopic investigations have been performed in powdered form of the samples. XRD patterns were recorded by a RigakuRINT-2000 diffractometer equipped with a rotating anode with a tube voltage of 40 kV and current of 100 mA. Cu target was used as a source of X-rays having wavelength 1.54 Å. FTIR spectra of the ligand and their complexes were recorded in the range 4000 to 400 cm^{-1} on a Perkin-Elmer 16 FPC FT-IR using KBr pellet technique. Mössbauer measurements of iron complexes were carried out using standard PC- based spectrometer equipped with Weissel velocity drive operating in the constant acceleration mode. The data was fitted with the NORMOS-SITE program and the obtained parameters are with respect to natural iron. Measurements of DC magnetization have been recorded on commercial 14 T vibrating sample magnetometer from quantum design.

3. Result and Discussion

3.1. X-ray Diffraction:

The XRD pattern of the complexes shows sharp intense peaks throughout the spectrum. It is indicating crystalline nature of the samples. The particle size has been calculated using Scherrer's formula given by $t=0.9\lambda /B \cos \theta$[6], where t is the particle size (same unit as λ), B is half width (in radians) of diffraction line, θ is the Bragg angle and λ is the wavelength. The particle size of Cu(II) and Fe(III) complexes was found to be 57.85nm and 81.07nm respectively. The XRD pattern of complexes are shown in Figure 1.

3.2. Fourier Transform Infrared Spectroscopy:

Important IR bands of the Schiff base ligands and complexes support their structure. A broad and weak band occurring around 3059-3080 cm^{-1} may be attributed to the phenolic group due to intra molecular hydrogen bond between phenolic hydrogen and azomethinic nitrogen. This band disappeared in the IR spectra of complexes indicating the deprotonation of phenolic group suggesting that -OH group of the ligand has been utilized in the formation of metal-oxygen bond[7].

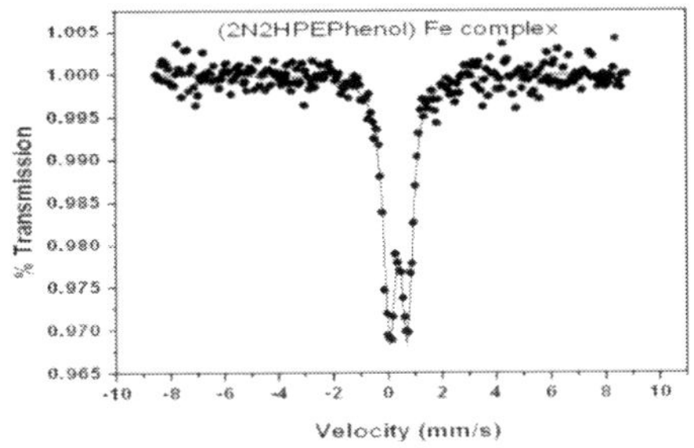

Figure 2. Mössbauer Spectra of Iron Complex

3.3. Mössbauer Spectral Studies:

The Mössbauer effect provides a powerful probe to chemical state and the environment of iron atoms. The Mössbauer spectrum for Fe(III) complex at room temperature is shown in Figure 2. The least square fitting and lorenzian line shape were used for the spectral plots which indicate the presence of quadrupole splitting doublets. This is consistent with the presence of an identical oxidation state of the Fe^{3+} nuclei in the crystal lattice of the present complexes.

The spectra of quadrupole splitting at room temperature suggest that the presence of electric field gradient at the metal ion is due to the presence of ligand around the ion. This may be due to the distorted octahedral geometry[8]. The spectra did not show any magnetic splitting on applying external magnetic field which is indicative of absence of effective internal magnetic field at the Mössbauer nuclei in each case and the state remains doubly degenerate, showing the Kramer's degeneracy[9]. Mössbauer parameters are reported in Table 1.

3.4. Magnetic studies on VSM

The DC magnetization of Fe(III) complex was measured from 10K to 300K. Figure 3 shows the graph between inverse of magnetic moment and temperature, follows Curie Weiss law[10] χ=C/T+θ, where θ is Curie Weiss temperature and C is Curie's constant.

Table 1. Mössbauer parameters of iron complex

Compound	Isomer shift (mm/sec)	Quadrupole splitting (mm/sec)	Linewidth (mm/sec)	% Area	Oxidation state
(2N2HPEPhenol) Fe(III) complex	0.368±0.005	0.563±0.008	0.331±0.039	100 %	Fe^{3+}

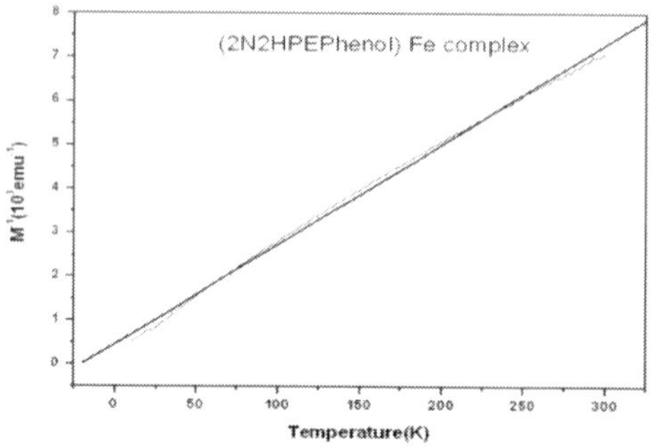

Figure 3. Temperature versus (1/magnetic moment) for Fe(III) complex

The Curie Weiss Temperature is found to be negative in reported complex indicating anti ferromagnetic interaction at low temperature. Hence the interaction between Fe(III) ion in the iron complex is antiferromagnetic coupling[11].Thus iron complex follow magnetic phase transition from paramagnetic to antiferromagnetic nature at low temperature.

4. Conclusion

X-ray diffraction studies of complexes indicate crystalline nature of the samples. FTIR Spectra confirms the structure of the synthesized ligands and their complexes. Mössbauer studies show paramagnetic nature of Fe(III) complex. It also shows oxidation state of iron to be Fe^{3+}.VSM supports Mössbauer studies at room temperature. At low temperature VSM shows magnetic phase transition.

Acknowledgment
Authors are thankful to Dr. R. J. Chaudhary for XRD measurements, Dr. V.R. Reddy for Mössbauer study and Dr. Alok Banerjee for VSM measurement at UGC-DAE Consortium for Scientific Research, Indore-452001, India.

References

[1] Herzfeld R and Nagy P 1999 *Spectrosc. Lett.* **32**, 57
[2] Singh K, Barwa M S and Tyagi P 2007 *Eur. J. Med. Chem.* **42**, 394
[3] Mandlik P R, More M B and Aswar A S 2003 *Indian J. Chem.* **42A**, 1064
[4] Kumar S, Dhar D N and Saxena P N 2009 *J Sci. Ind. Research* **68**,181
[5] Hothi H S, Makkar A, Sharma J R and Manrao M R 2006 *Eur. J. Med. Chem.* **41**, 253
[6] Cullity B D 1978 Elements of X-Ray Diffraction (Addison-Wesley, Philippines)
[7] Chohan Z H and Pervez H H 1993 *Synth. React. Inorg. Met. -Org. Chem.* **23**, 1061
[8] Ali S, Alimuddin and Reddy V R 2005 *Pramana-J. Phys.* **65**,1121
[9] Siddiqi Z A, Kumar S, Khalid M. and Shahid M 2009 *Spectrochimica acta A.* **72**, 616
[10] Dekker A J Solid State Physics 1991 (Macmillan India Limited, Madras)
[11] Yang J H Zhao L Y Zhang Y J Wang Y X and Liu H L 2007 *Solid State Commun.* **143**, 566

Correlation of X-Ray Absorption Parameters with Schultz index

Pravin A. Kekre[1] , Sunil Khatri[2] , A. Mishra[3] and K.P. Joshi[4]

[1] Department of Physics, P.M.B. Gujarati Science College, Indore.
[2] Shri R.G.P. Gujarati Professional Institute, Indore.
[3], [4] School of Physics, D.A.V.V. Indore.

Abstract. A novel application of topological Index in estimating some X-ray parameters is described. X-Ray Absorption parameters (Chemical Shift and Effective Charge) are correlated with Schultz index. Some Cobalt (II) complexes were used to establish this type of correlation. The result have indicated that the chemical shift and effective charge are sensitive to the topological structure of coordinating ligand moieties. The topological understanding of molecular properties can lead to the development of new areas of present and future interest. i.e. designing of new drugs, tracking the effects of pollutants in environment and the prediction of carcinogenicity of a molecule.

1 Introduction

Earlier, we have established a between topological Index (Winer and Randic) and X-ray parameters[1,2] . In continuation of our earlier work, we now establish a correlation between Schultz index and some X-ray absorption parameters:

2 Generation of Schultz index

The Schultz index[3] is also known as molecular topological index (MTI) and is defined as

$$\mathrm{MTI} = \sum_{i=1}^{n} e_i \qquad 1$$

Where, $e_i's(i, 2, \ldots\ldots\ldots\ldots\ldots n)$ represent the elements of the following row matrix of order n.

$$V(G)[A(G) + D(G)] = [e_1, e_2, e_3, \ldots\ldots\ldots\ldots\ldots\ldots, e_n] \qquad 2$$

Where V(G) is the valency row matrix, A(G) and D(G) are the adjacency and distance matrices, respectively.

Published under licence by IOP Publishing Ltd

The vertex-adjacency matrix A=A(G) of a labeled graph G of a molecule with n vertices is defined as

$A_{ij} = 1$ (if vertices are adjacent) or

$=0$

And distance matrix is defined as

$$D(G) = \sum_{i<j} d(i,j)$$

Here, d_{ij} is the element of distance matrix (D).

3. Results and Discussion

Table 1 contains molecular formulae. Molecular graphs along with their Schultz indices while chemical shifts with respect to cobalt and effective charge[4] of the complexes used for correlation are reported in Table2

Table 1. Molecular formulae. Molecular graphs and Schultz indices of substituted Hydroxypyridine ligands used in the present work.

No.Name	Molecular formulae	Molecular graphs	Schultz indices
1. 2.3-dihydroxypyridine (DHP)	$C_5H_5NO_2$		820
2. 2-amino 3-hydroxypyridine(AHP)	$C_5H_6N_2O$	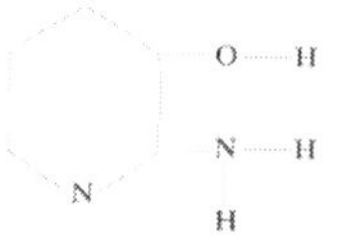	911
3. 3-hydroxypyridine 2-carboxylic acid (HPC)	$C_6H_5NO_3$	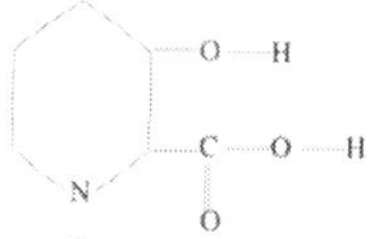	1451
4. 3-hydroxy 2-piconilamide (HPC)	$C_3H_6N_2O_2$	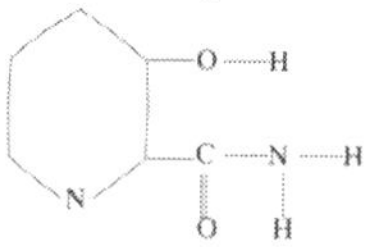	1578

Table 2. Chemical shift (Δ E) and effective charge for the complexes used for correlation

No.	Complex	Chemical shift ($\Delta E \pm 0.3$eV)	Effective nuclear charge Z_{eff}
1	Co-DHP	13.68	1.52
2	Co-AHP	12.56	1.40
3	Co-HPC	10.50	1.24
4	Co-HPA	9.82	1.20

The X-ray K-absorption edge structure of the cobalt (II) complexes show several features of interest. The shift of the X-ray absorption edge ΔEi (i=K.L.M.....) of an element in a compound with respect to that of pure element is given by

ΔEi=Ei (compound)- Ei (element)

The position of X-ray absorption edge depends on the valence of the absorbing ion, the effective charge on the central atom [5,6] and also upon the geometry of the complex[7]. According to Agarwal and Verma[8] the chemical shift is towards the higher energy side of the metal edge. It increases progressively with increase of the valence of the cations, unless the shift is either suppressed by the covalent character of the bond or enhanced by the formation of a metal-metal bonding. In the present work, the K-absorption edge of all the four complexes is found to be on the higher energy side. The order of chemical shift as indicated by their values has been found to be as follows.

Co-DHP > Co-AHP > Co-HPC > Co-HPA

The larger values of chemical shifts in all the complexes show that the complexes are more ionic in character. As the shift is taken to be proportional to the ionic character[6,7,9] , the role of various bonds attached to the central cobalt ion is prominent in determining the overall chemical shift in a metal complex. Further, not only ligation of a particular type, e.g Co- O or Co-N is responsible in imparting ionic character to the complex but their position in the planar ring/axial position also contributes to the ionic character As it is not possible to assess individually the contribution of these factors to the chemical shift values. It is only the relative ionic character that can be assessed from the observed values of chemical shift.

It has already been reported by a number of workers that chemical shift depends on the effective charge and different approaches are available for the estimation of effective charge[4,10,14] We have estimated the effective charge by using Gianturco and Coulson's method[4] which can be applied to the complex system. The details of the method have already been reported in our earlier work[13].

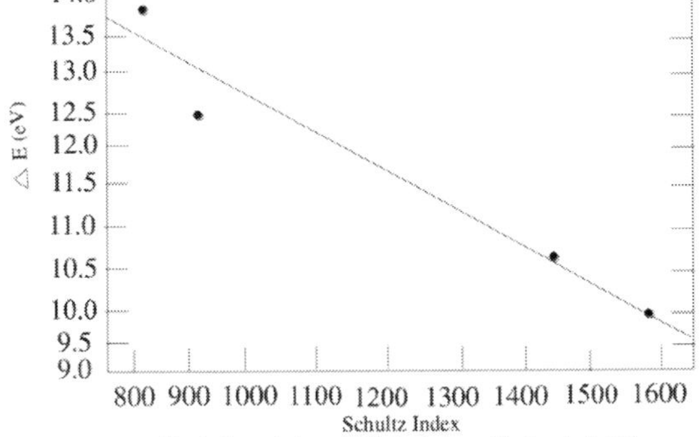

Fig 1. Co rrelation of Schultz Index with chemical shift for Co (II) hydroxypyridine complexes

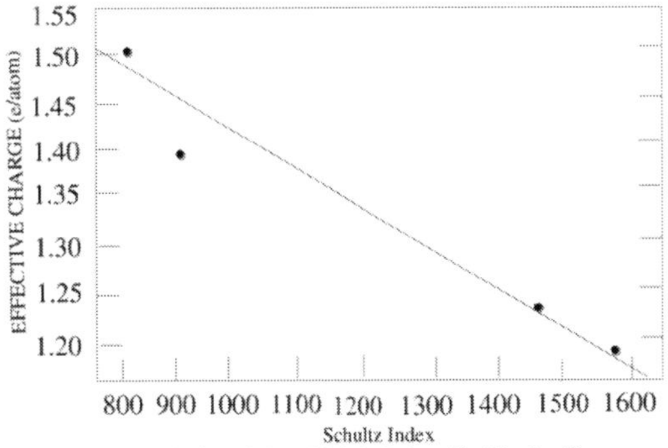

Fig 2. Co rrelation of Schultz Index with Effective Charge for Co (II) hydroxypyridine complexes

Fig 1 &2 shows correlation between Schultz indices of substituted hydroxypyridine ligands and corresponding X-ray absorption parameters. The correlation assumes the general form

$$\Delta E = - m (TI) + C$$

Where TI stands for a particular topological index.

It is evident from the plot (Fig 1 & 2) that ΔE and Z_{eff} correlate according to the family of the complex to which they belong. The Fig. 1 represents the correlation for Schultz indices and chemical shift while Fib.2 demonstrates the correlation for Schultz indices and effective charge. Both the plots yield excellent straight lines.

The correlation factors are 0.989 and 0.977 respectively. It is gratifying that the correlations with the Schultz index are better than that with the Wiener index[1]. The results obtained in this study show that the correlation of ΔE and Z_{eff} with Schultz indices is very effective. We can say that the chemical shift thus estimated indicates that topological indices of the organic molecules acting as ligands can be used for estimating chemical shift theoretically. However, this is possible only for a particular metal complex of closely related ligands (coordinating species)

4 Conclusion

Topological indices contain valuable structural information as evidenced by the success of their widespread applications. Topological indices are not, however, a panacea; users of totpological indices must be aware of their limitations and the pitfalls which can accompany with their use. The correlation between Schultz indices and X-ray parameters of molecules can lead to the development of a new area of present and future interest. It would be of great help to chemists as well as physicists.

References

1 Kekre P A. Joshi K P. Khatri S & Mishra A. *1996, Bull Soc Belg.* ***105*** *169.*
2. Mishra A, Khatri S, Joshi K. P. & Kekre PA. *2000, Asian Journal of Physics,* ***9*** *345.*
3. Schultz H.P. *1989 J. Chem. Inf. Comput Sci.* ***29*** *227.*
4. Gianturco F A & Coulson CA, 1968 *Mol Phys.* ***14*** *223.*
5. Becker M V 1964 *Naturwiseenschaften.* ***51*** *633.*
6. Barinskii R.L. 1967 *J Struct Chem USSR* ***8*** *805.*
7. Meisel A & Keilacker H.Z. 1971 *Phy Chem (Leipzing)* ***32*** *247.*
8. Agrawal B K & Verma L.P. 1970 *J Phys.* ***C3*** *535.*
9. Nadzhakov E G & Barinskii R.L. 1960 *Sov Phys Dokl* ***4*** *1319.*

10. Murugenson J., Sarode P K, Krishna Gopal & Rao C N R. 1980 *J Chem Soc. 832.*
11. Leonhardt G & Mersel A. 1970 *J Chem Phys.* ***52*** *6189.*
12. Pauling L. 1960 *Nature of Chemical Bond (cornell Univ Press, Ithaca. N Y)*
13. Mishra A., Dubey R., Pandey D., Kekre P. A. & Shrivastava B D. 1998 *Indian J Pure & Appl Phys.* ***76*** *537.*

Comparative X-Ray crystallographic studies of systemic fungicides hexaconazole and tricyclazole

Jyotsna Chauhan , Ashish Kumar Sharma, Gargi bhattacharya
School of Physics, DAVV, Indore (M.P), India

E-mail: jyotsnachauhan2006@gmail.com

Abstract. Chemical structure of a chemical is useful for the synthesis of new compounds with more specific actions and fewer adverse reactions, to increase/decrease the duration of action of the original drug or to get a more potent compound, to restrict the action to a specific system of the body and to reduce the adverse reactions, toxicity and other disadvantages associated. Recently it has been observed that some of the fungicides are loosing their effects. So analogous compounds can be designed as substitute, if their structures are known. A rational approach to test these fungicides is to know the three dimensional structure of these compounds and macromolecular receptor sites as well as their molecular complex .The structures of these compounds can be obtained by X-ray diffraction method in crystalline form and they will invariably be similar to their structure in solutions. The composition of crystal (RS)-2-(2,4-dichlorophenyl)-1-(1H-1, 2,4-trizole-1-yl) hexane-2-ol or hexaconazole and Tricyclazole are confirmed by comparing the infra-red spectra of two components. The unit cell parameters of Hexaconozole are a=10. 9068(7) Å, b=10.9895(7) Å c=13.6124(8) Å, α=90o, β=106.554(2)o, γ=90.000(5)o.The Crystal system is Monoclinic, and space group P21/c . The unit cell parameters of Tricyclazole are a=14.896(5) Å, b=7.410(5)Å, c=7.556(5) Å, α=90(5)o, β=90.000(5)o, γ=90.000(5)o.The Crystal system is Orthorhombic and space group is Pca21.

1. Introduction

A systemic fungicide is defined as systemic fungi toxic compound that controls a fungus pathogen remote from the point of application and that can be detected or identified. These compounds are absorbed by the plant and translocated within it, thus providing protection as well as eradicating already established infection.

Triazoles compounds contain a ring composed of two carbon atoms and three nitrogen atoms. We can understand the basic chemical groups responsible for drug action.These compounds possess a trityl carbon that can act as a carbonium ion, an organic ion carrying a positive at a carbon location. In Hexaconazole ORTEP figure 1we can see Triazole compound N1, N2, N3 and C9 and C10 .In Tricyclazole ORTEP figure 2 we can see Triazole compound N1, N2, N3 and C7 and C8.

2. Experimental

First grow the crystals of existing fungicides available and synthesize their derivatives in lab. Colorless well formed crystals of Tricyclazole and Hexaconazole of size 0.30 x 0.20 x 0.20 mm are obtained by slow evaporation from a solution of methanol at 297 o K temp. The crystals obtained are rectangular in shape. The crystal density is calculated by floatation method in the mixture of benzene and carbon-tetrachioride. The crystal is placed in RD bottle filled with carbon tetràchloride. Benzene is added to the solution until the crystal floated in the middle of mixture. Thus the crystal and solution are of same density and density of solution is measured with pyknometer. The determination of structural perturbation in fungicide derivatives and comparison of the result of their molecular association with other receptor sites by X-Ray crystallography techniques will be done. In parallel with these structural studies, spectroscopic studies carried out on them. The goal is then to tie together the structural and spectroscopic studies to have more comprehensive account of the precise shape of these molecules, the non-covalent interaction which are likely to be involved in and the changes introduced in molecular geometry and electronic structure of these compounds as a result of their molecular association with other compounds.

3. Data collection and structure solution

The three dimensional intensity data were collected on a computerized automatic 4-circle CAD-4 Enraf-Nonious Diffractometer using graphite filtered MoKα (Å) radiation's at SAIF Madras.. Temperature of both the crystals during data collection was 293°K. All the data were corrected for Lorentz and Polarization effect. The structure was solved using SHELXS (program for crystal structure solution)

3.1 *Refinement:*

The positional co-ordinates, which were obtained from SHELXS 97 and isotropic temperature factors, were subjected to refinement by SHELXL refinement program. After so many cycles of refinement the R factors dropped .Further refinement of the structure was carried out with individuals an isotropic temperature factors of the exponential form.-

$$2P_1\text{^ }2[h\text{^}2a\text{*^}2U_{11}+\text{-----------}+2hKa\text{*}bxU_{12}$$

The hydrogen atoms were fixed at this stage by geometrical considerations and were not refined. Refinement of the structure was terminated after two more cycles when the derivations in the parameters become much smaller than the corresponding estimated standard deviation.

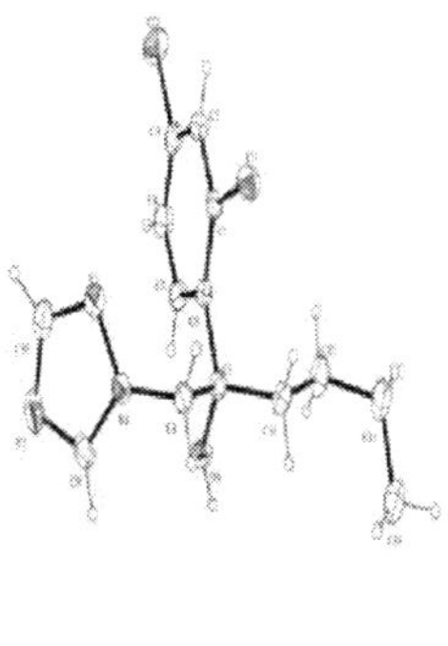

Figure 1. Ortep diagram of Hexaconazole.

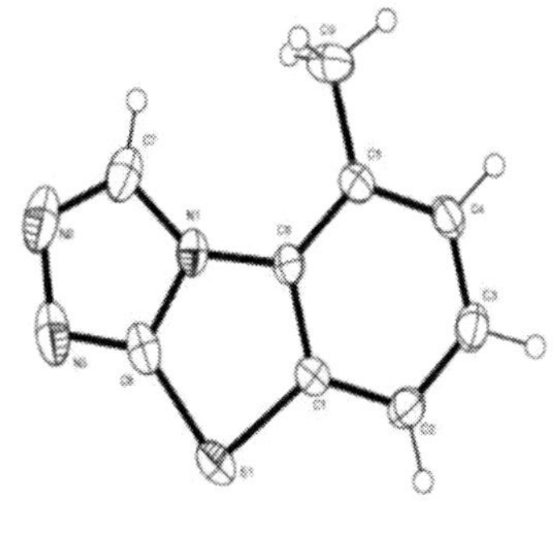

Figure 2. Ortep diagram of Tricyclazole

4. Results and discussion

In both the crystal the average bond distances of C-H is 0. 96(2)Å. In Triazole ring of Tricyclazole figure 2 the bond distances of C(7)-N(1) is 1.368Å , C(8)-N(1) is 1.360 Å., N (2)-N (3) is 1.396Å, C(7)-N(2)is 1.294(3), C(8)-N(3) is 1.305(2).In Triazole ring of Hexaconazole in figure 1 the bond distances of C(10)-N(3) is 1.305Å , C(10)-N(2) is 1.344 Å., N (1)-N (3) is 1.349Å, C(9)-N(2)is 1.313(3), C(9)-N(1) is 1.320(2) . In both the crystals triazole ring is distorted in shape. The average bond distances for C-N and N-N bonds are 1.354Å and 1.396Å. The bond lengths and angles in the benzene ring show regular features in the molecule. C-C distances are short and shortening may be due to delocalization of electrons from the benzene rings. The whole molecules appeared to be twisted and folded and reason may be due to stacking constraints. The bond distance around C (7) is as usual shorter than single bond value. This may also appears to bear a partial double bond character The bond distances in the five member ring are comparable to corresponding distances in heterocyclic ring 1. 339 (Å). The average value of bond lengths and angles in the rings derived from most reliable set of data by Spencer are 1.377Å and 119° respectively.Table 1 shows the crystal data of Tricyclazole and Hexaconazole

Table 1

Parameters	Crystal Data of Hexaconazole	Crystal Data of Tricyclazole
Empirical formula	C14 H17 Cl2 N3 O	C9 H7 N3 S
Formula weight	314.21	189.24
Temperature	293(2) K	293(2) K
Wavelength	0.71073 A	0.71073 A
Crystal System	Monoclinic	Orthorhombic
Space group	P21/c	Pca21
a	10.9068(7) A	14.896(5) A
b	10.9895(7) A	7.410(5) A
c	14.896(5) A	7.556(5) A
α	90 °	90 °
β	106.554(2) °	90 °
γ	90 °	90 °
Volume/Z	1563.96(17) A^3 /4	834.0(8) A^3/4
Calculated density	1.334 Mg/m^3	1.507 Mg/m^3
Absorption cofficient	0.414 mm^-1	0.335 mm^-1
F(000)	656	392
Crystal size	0.30 x 0.20 x 0.20 mm	0.30 x 0.20 x 0.20 mm
Θ range (for data collection)	2.42 to 25.00 deg.	3.07to 34.03 deg.
Limiting indices	-12<=h<=12,-11<=k<=13,-16<=l<=10	-23<=h<=23, -9<=k<=11,- 10<=l<=11
Reflections collected/unique	14496 / 2746 [R(int) = 0.0205]	8479 / 3134[R(int) = 0.0241]
Completeness to Θ	99.9 %(Θ=25)	99.9 %(Θ=34)
Absorption correction	Semi-empirical from equivalents	Semi-empirical from equivalents
Max./ Min. Transmission	0.9218 and 0.8859	0.9360 and 0.9062
Refinement method	Full-matrix least-squares on F^2	Full-matrix least-squares on F^2
Goodness of fit of F^2	1.031	1.061
Final R indices [I>2sigma(I)]	R1 = 0.0354, wR2 = 0.0862	R1 = 0.0433, wR2 = 0.1105
R indices (all data)	R1 = 0.0458, wR2 = 0.0935	R1 = 0.0533, wR2 = 0.1187
Largest diff. Peak and hole	0.344 and -0.619 e.A^-3	0.619 and -0.290e.A^-3

5. Conclusion

In this case we find that Hexaconazole is more effective than Tricyclazole due to its some active groups as we mentioned in result and discussion. Thus we study the structure of variety of such compounds and correlate their structure with biological activity so that more safer and effective fungicides at reasonable price can be developed.

References

[1] Berg D, Kramer W, Regel E, Buechel K H, Holmwood G, Plembel M and Scheinp flug H 1984, *Mode of action of fungicides studies on ergosterol biosynthesis inhibitors,*Proc Br. Crop, proconf.-pest and Diseases, **3**: 887

[2] Shephard M C, Noon R A,Worthington R A, McClellan W D and Lever B G 1986 *Proc. Br.Crop Prot. Conf. –Pest and Diseases* **1**:19-26

[3] Jolmson, C.K. 1965, *ORTEP, Report ORNL-3794. Oak Ridge National laboratory, Temessee, U.S.A.*

[4] Sheldrich G M 1997, SHELXS-97, *Program for the solution of crystal structure.*

[5] Nowell I W and Walker P E,1982 *Acta Cryst.* **B38** 1857

[6] Spencer M 1959 *Acta Cryst.* **12** 50

Estimation of Bulk modulus and microhardness of tetrahedral semiconductors

Sanjay Kumar Gorai
Department of Physics, Tata College Chaibasa, Westsingbhum 833201, Jharkhand, India

E-mail: shreyagorai@gmail.com

Abstract. A general empirical formula was found for calculating of bulk modulus (B) and microhardness (H) from electronegativity and principal quantum number of II-VI, III-V semiconductors. Constant C1, appearing the in the expression of bulk modulus and constants C2 and C3, appearing in the expression of microhardness and the exponent M have following values respectively The numerical values of C1,C2 , C3 and M are respectively 206.6, 8.234, 1.291, -1.10 for II-VI 72.4, 31.87, 7.592, -0.95 for III-V semiconductors. Both electro-negativity and principal quantum number can effectively reflect on the chemical bonding behaviour of constituent atoms in these semiconductors. The calculated values of bulk modulus and microhardness are in good agreement with the reported values in the literature. Present study helps in designing novel semiconductor materials, and to further explore the mechanical properties of these semiconductors.

1. Introduction

Interests in the study of various properties of binary tetrahedral semiconductors have been increasing in recent years because of their increasing scientific and technological applications [1-5]. Bulk modulus and microhardness of these materials has, however, not been studied too much. Kumar et al [6-7] using plasmon energy, proposed empirical relation for binary tetrahedral semiconductors between bulk modulus, microhardness with plasmon energy. Reddy et al [8-9] proposed an empirical relation between electronegativity with bulk modulus of semiconductors. An empirical relation, using ionic charge model, was however, proposed by Verma and Sharma [10-12], while determining bulk modulus and microhardness of semiconductors, with the consideration that both plasmon energy and ionic charge model depend upon the number of valence electrons. In this study, Electronegativity has been used to determine the bulk modulus and microhardness of these materials considering that it delineates the capability of an atom in a molecule to attract bonding electrons [13-14]. The principal quantum number is also used as a input parameter to estimate the bulk modulus and microhardness of these materials.

2. Methodology

The bulk modulus of any material depends on the volume of its constituent atoms. The bulk modulus of any material depends on the valence electron gas pressure completely determine the resistance of atoms to compression. This compression of an atom is closely related to two parameters. One parameter is the electronegativity value which is the attracting power of the electronegativity value which is the attracting power of effective nuclear charge to valence electrons in outer orbitals. The larger the EN value the more tightly the nuclei hold the valence and smaller the bulk modulus. Since it is a measure of the stiffness or of the energy required to produce a given deformation. The higher the bulk modulus, the stiffer is the crystal. This deformation is the distortion of the outer electronic shell. The distortion produced in a crystal

Published under licence by IOP Publishing Ltd

Table 1 Bulk modulus of tetrahedral semiconductors.

Compo-unds	$\left(\frac{\eta_{av}}{\sqrt{\chi_A \chi_B}}\right)$	Bulkmodulus (B) in GPa								% deviation
		Cal. Eq. (1)	Exp. Ref. 17]	Ref. [6]	Ref. [16]	Ref. [17]	Ref. [18]	Ref. [19]	Ref. [20]	
1	2	3	4	5	6	7	8	9	10	11
$A^{II} B^{IV}$										
BeO	1	206.6		265.12	291					
BeS	1.57	125.78		111.81	132					
BeSe	1.98	97.45		97.29	109					
BeTe	2.48	76.07		71.54	80					
ZnO	1.67	117.52		139.78	149					
ZnS	2.45	77.09	77	77.8		90	44.7	61	75.21	0.1283
ZnSe	2.93	63.32	62	68.07		75	39.1	58	65.7	2.1373
ZnTe	3.55	51.27	51	58.24		59	32.3	54	61.46	0.5331
CdS	2.98	62.1	62	59.35		69	36.5	66	61.6	0.2538
CdSe	3.51	51.91	53	51.57		60	32.8	60	58.9	2.047
CdTe	4.19	42.72	42	44.01		47	27.5	56	41.6	1.7292
HgS	3.35	54.64		43.07						
HgSe	3.9	46.23		59.07						
HgTe	4.61	38.46		51.39						
MgTe	3.26	56.31		42.15						
Average % deviation						14.98	38.21	14.34	6.91	1.04
$A^{III} B^{VI}$										
BN	0.84	203.45		218.52						
BP	1.32	132.43		164.33						
BAs	1.64	107.75		137.61		87				
AlN	1.26	138.41				79				
AlP	1.89	94.16	86	88.47	87	57				9.4965
AlAs	2.29	78.46	77	77.71	79					1.9067
AlSb	2.81	64.60	58	56.32	57					1.388
GaN	1.51	116.54		169.14		84.5				
GaP	2.21	81.16	89	86.63	87	74.5				8.804
GaAs	2.62	69.04	75	73.2	77	56.9				7.936
GaSb	3.16	57.78	57	53.12	58					1.3817
InN	1.89	94.16		117.75						
InP	2.7	67.10	71	66.8	67	65.6				5.489
InAs	3.15	57.96	60	59.73	61	59.5				3.397
InSb	3.76	48.99	47	47.3	47	46.7				4.2344
Average % deviation			*2.15*	*2.84*		*2.35*				4.89

is related with the principal quantum number of constituent atoms in a material. The larger the principal quantum number, the longer the distance between the nuclei and valence electrons, and the larger electronic polarisability and hence higher bulk modulus. The electronegativity and principal quantum number both are very useful parameter in understanding the nature of chemical bonding and predict mechanical properties like bulk modulus and microhardness of tetrahedral semiconductors. For II-VI and III-V semiconductors the bulk modulus is assumed to be correlated to is the average principal quantum

Table 2 Microhardness of tetrahedral semiconductors

Compounds	$\left(\frac{\eta_{av}}{\sqrt{\chi_A \chi_B}}\right)$	Microhardness (H) in GPa Cal. Eq. (3)	Expt. Ref. [21]	Ref. [6]	Ref. [23]	% deviation
1	2	3	4	5	6	
$A^{II} B^{IV}$						
BeO	1	6.94	9.1-12.7	8.21		23.7363
BeS	1.57	3.72		3.01		
BeSe	1.98	2.59		2.52		
BeTe	2.48	1.74		1.64		
ZnO	1.67	3.39	3.9,4.8	3.96	4.6	13.0769
ZnS	2.45	1.78	1.7,2.8,3.5	1.86	3.3	4.706
ZnSe	2.93	1.23	1.3-1.8	1.52	1.7	5.3846
ZnTe	3.55	0.75	0.8-1.1	1.19	1.0	6.25
CdS	2.98	1.19	1.2,1.22	1.23	1.1	0.8333
CdSe	3.51	0.78	0.7-1.2	0.96	0.8	11.29
CdTe	4.19	0.41	0.4-0.64	0.71	0.5	2.5
HgS	3.35	0.89		1.22		
HgSe	3.9	0.55		0.96		
HgTe	4.61	0.24		0.65		
MgTe	3.26	0.95		0.68		
Average % deviation				25.41	30.75	8.48
$A^{III} B^{VI}$						
BN	0.84	30.02	34.3-73.0	29.57	45.8	12.4781
BP	1.32	16.89	31-40	20.81	32.3	45.5161
BAs	1.64	12.33	19.0	16.5	24.2	35.1053
AlN	1.26	18				
AlP	1.89	9.816	5.5	5.5	10.9	78.473
AlAs	2.29	6.914	*4.8,5.0*	5	6.3	44.042
AlSb	2.81	4.351	*4.0*	4		08.775
GaN	1.51	13.95				
GaP	2.21	7.412	9.45	9.45		21.5661
GaAs	2.62	5.172	7.50,7.21	7.5		28.1667
GaSb	3.16	3.091	4.48	4.48		31.0045
InN	1.89	9.816				
InP	2.7	4.813	4.10,4.30	4.3		17.39
InAs	3.15	3.123	3.30,3.84	3.84		5.3636
InSb	3.76	1.464	2.20	2.2		34.545
Average % deviation				25.898	25.88	21.42

number $\eta_{av} = \left(n_A + n_B /_2 \right)$ of valence electrons of atoms A and B and χ_A and χ_B are electronegativities of atoms A and B respectively. For II-VI and III-V semiconductors the bulk modulus is empirically expressed as

$$B=C_1 \left(\frac{\eta_{av}}{\sqrt{\chi_A \chi_B}}\right)^M \text{ .... (1)} \quad B=LH+O \text{ .... (2)} \quad H=C_2\left(\frac{\eta_{av}}{\sqrt{\chi_A \chi_B}}\right)^M -C_3 \text{ .... (3)}$$

where C1 and M have values 206.6,-1.09 for II-VI and 72.4,-0.95 for III-V semiconductors respectively. Plendl et al.[15] have shown that B=KH, where K is the constant. Instead of using single proportionality factor which has not been sufficient to yield acceptable values of B and H ,in this paper two proportionality factor have been used which gives better fit to measured values . In equation 2, L and O are constants. The numerical values of L and O are respectively 25.09,32.31 for II=VI,5.409,41.07for III-V semiconductors. Using above Equation(1) and Equation (2) the microhardness have been estimated by the following relation: In equation 3 where C2 , C3 and M are the constants. The numerical values of C2 , C3 and M are respectively 8.234,1.291,1.10 for II-VI 31.87,7.592,0.95 for III-V semiconductors respectively. Using Equation (3) the microhardness of materials have been estimated and listed in Table 2 along with the results of experimental values and earlier researchers.

3. Results and Discussions

Bulk modulus and microhardness is estimated using Equation (1) and Equation (3) for binary tetrahedral semiconductors. The estimated values are given in Table 1 and Table 2. along with the results of earlier researchers [16-20] for the comparison. It is observed that the calculated values are in better agreement with the experimental data. The results also agree closely with the values reported for these compounds by earlier workers [15-20]. The percentage deviation is given in Table 1 and Table 2. The average percentage deviation of B and H is 1.09, 8.48 and 4.89, 21.42 for II-VI and III-V binary tetrahedral semiconductors. The minimum average percentage deviation in present approach indicates that empirical relation is better than earlier approach [6, 15-23]. The higher value of average percentage deviation of bulk modulus of III-V semiconductors may be to scattered data of elastic constants and bond length.

4. Conclusion:

The proposed empirical expression helps to estimate the bulkmodulus and microhardness of binary and ternary semiconductors in terms of electronegativity value and the average principal quantum number of valence electrons.The calculated values of these parameters are presented in Table1 and Table 2 respectively. The inclusion of electronegativity and principal quantum number has some bearing on the concept chemical bonding. From this study it is possible to correlate bulk modulus and microhardness with the nature of bonding of the materials.These two parameters electronegativity and average principal quantum number delineates the structure –property relationship in solid state sciences

References

[1] Shay J L, Wernic J H 1975 Ternary chalcopyrite semiconductors growth, in: Electronic properties and application, pergamon press, Newyork
[2] Kazmerski L L, 1983 *NuovoCimento* **20** 2013
[3] Deab S K and Zunger A, 1987 Ternary Multinary Compounds.Mater.Res.SocConf.Proc.**37**
[4] Yan C and Xue D 2008 *Funct.Mater.Lett.*1 **37**
[5] Zhu Y 2008 *Funct.Mater.Lett.*1239
[6] Kumar V, Shrivastava A K and Jha Vijeta 2010 *J.Physics and Chemistry of Solids* **71** 1513
[7] Kumar V, Prasad G M and Chandra D 1992 *Phys. Status. Solidi* **b170** 77
[8] Reddy R R, Ahammed Y N, Rama Gopal K, Rao T V R 1998 *Infrared Phys. & Tech.* **39**55
[9] Reddy R R, Nazeer A. Y, Rama, Gopal K, Rao T V R, Abdul Azeem P and Reddy M P 1983 *Optical Materials* **14** 361
[10] Verma A S 2009 *Phys. Stat. Sol (b)* **246** 192
[11] Verma A S and Sharma D 2007 *Phys. Scr.***76** 22
[12] Verma A S 2009 *Solid State Commun***149** 1236
[13] K Li and Xue D 2007 *Phys. Status. Solidi* B **244**1982
[14] Xue D, Zuo S and Ratajczak H 2004 *Physica B* **352** 99
[15] Plendl J N, Mitra S S and Gielisse P 1965 *J. Phys. Status. Solidi B* **12** 367
[16] Neumann H 1988 *Cryst.Res. Technol.***23** 97
[17] Cohen M L 1985 *Phys. Rev. B* **35** 7988
[18] Lam P K and Cohen M L and Martinez G 1987 *Phys. Rev. B* **35** 9190
[19] Douri Y Al-, Abid H and Aourag H, 2001 *Physica B* **305** 186
[20] Douri Y Al-, Abid H and Aourag H, 2002 *Physica B* **322** 179
[21] Lide D R 1999-2000 *Hand book Chemistry and Physics,* 80 th ed., CRC Publication
[22] Weber M 2003 (Ed.) *Hand book of Optical Materials*, CRC Publication
[23] Garbato L and Rucci A 1977 *Philos . Mag.***35** 1685

Synthesis and photoluminescence of CdSe/PVA nanocomposites

Kamal Kushwaha[1], Nitendra Gautam[2], P Singh[3], Meera Ramrakhaini[3]
[1] Department of Applied Physics, Jabalpur Engineering College, Jabalpur,M.P., India.
[2] Department of Applied Physics, Govt.S.S.A. College Sihora, M.P., India
[3] Department of Post Graduate Studies and Research in Physics and Electronics, Rani Durgavati Vishwavidhyalaya, Jabalpur, M.P., India

E-mail:kamal_kushwah2005@yahoo.com

Abstract. There is a great deal of interest in finding new synthetic routes to prepare nano-crystallites. This paper reports in situ preparation of CdSe nanoparticles using poly (vinyl alcohol) [PVA] as matrix by using chemical route. XRD analysis revealed that CdSe-PVA nano- composite with a small size and narrow size distribution were obtained. The XRD studies show the formation of hexagonal CdSe/PVA particles with an average size of 11nm. Photoluminescence properties of CdSe/PVA with varying polymer concentration are reported. The PL peak was observed at 510 nm in photoluminescence spectrum.

1. Introduction

Polymer- inorganic nanocomposites have attracted much attention recently due to their unique size dependent chemical and physical properties [1].Recently there have been many efforts to produce nanosized materials because the electrical and optical properties can be varied via chemical control over size, stoichiometry and interparticle separation[2-5].The combination of Polymer with semiconducting Nanoparticles allows the fabrication of thin film light-emitting devices several approaches have been reported,using two layer structure of PVA and CdSe Particle. The CdSe nanocomposites exhibit size dependent tunable photoluminescence, therefore they are Potentially useful in a wide range of applications [6]. Studies have been under taken to prepare CdSe nanoparticles in PVA matrix and investigate their luminescence [7].

2. Experimental

CdSe/PVA Nanocomposites were prepared by chemical method. First PVA solution was prepared in distilled water and then 1 ml $CdCl_2$ solution was added to it. After setting the pH at 10, 1 ml of freshly Prepared Na_2SeSO_3 solution was added and stirred for 90 minutes to obtain CdSe/PVA nanocomposite. The solution was spread on glass plates and on solvent evaporation nanocomposites films were obtained. A number of samples were prepared with different PVA concentration and subjected to X-ray diffraction and photoluminescence investigations. All samples were characterized at Inter University consortium (IUC) Indore. XRD pattern have been obtained by Rigaku Rotating Anode (H-3R) diffractometer with irradiation from $K\alpha$, line of copper (λ=1.548A) and PL Studies have been Performed at MANIT, BHOPAL by F-7000 FL Spectrophotometer. The particle size was calculated using Debye scherrer formula.

3. Results and discussion:

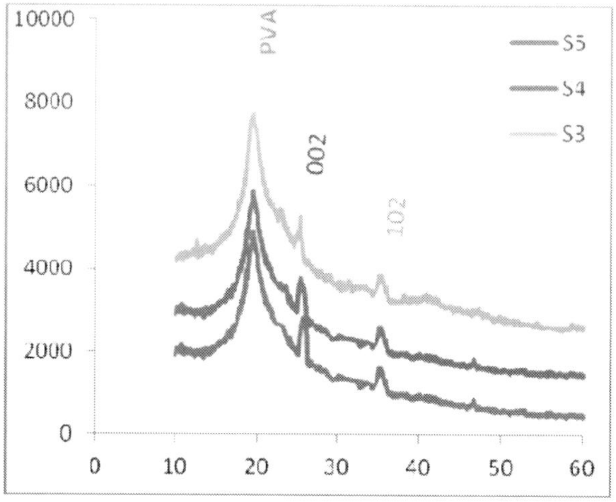

Figure 1. XRD pattern of CdSe/PVA nanocomposites.

XRD pattern of CdSe/PVA reveals hexagonal structure. Obtained Peaks are indicating reflection from (002) and (102) planes. The Particles size of samples computed from Debye Scherrer formula. PVA is acting as capping agent and by increasing its concentration smaller CdSe particles of average size of 11nm are formed in its matrix.

Photoluminescence (PL) spectra of CdSe/PVA samples excited by 380 nm UV light are shown in fig.2. A single peak is obtained at 510 nm and intensity is found to increase with increasing PVA concentration. Higher intensity is obtained for smaller CdSe particles in PVA matrix. The emission may be attributed to band to band transition of CdSe. The oscillator strength is increased by reducing the size which enhances the PL intensity. Due to proper passsivation of surface states nonradiative transitions are not increased. Thus increasing PVA concentration in CdSe/PVA composites, particle size is reduced and PL intensity is increased due to enhanced oscillator strength.

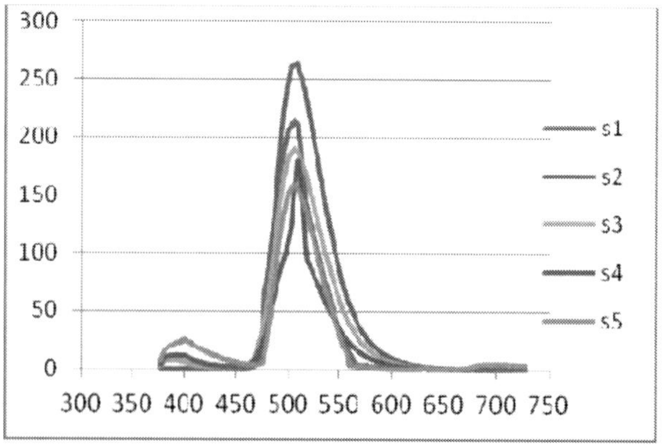

Figure 2. Photoluminescence Spectra of CdSe/PVA nanocomposites

4. Conclusion

CdSe/PVA Nanocomposites were successfully synthesized using simple and inexpensive techniques and it is shown in present work that average particle size reduces on increasing PVA Concentration and as a result PL intensity increases due to lowering in particle size.

5. Acknowledgement

The authors gratefully acknowledge the valuable assistant provided by Inter University consortium (IUC) Indore (INDIA) for characterizing samples by XRD and also very much thankful to Department of Nanotechnology, MAINAT, Bhopal (INDIA) for PL Studies.

6. References

[1] Indade T T, Neves M C and Barros A M V 2000 *Scr Mater* **43** 567
[2] Murray C B, Kaganand C R and Bavendi M G 1995 *Science* **270** 1335
[3] Fendler J H and Maldrum F C 1995 *Adv Mater* **7** 607
[4] Dabbouri B O, Viejo J R, Mikules F B, Heins J R, Mattonoro H, Ober R, Jonsen K F and Bavendi 1997 *J Phys Chem B* **101** 9463
[5] Platschek V, Schrader B, Herz K, Hilbert U,Ossan W Schottner G, Rahaush O, Bischof T, Lermann G,Malemy A, Keifer W,Bacher G, Forchel A, Su D, Glusig M, Muller G and Spanhal L 1997 *J Phys Chem B* **101** 8898
[6] Chan W C W and Nie S M, 1998 *Science* **281** 2016
[7] Murugadoss G, Rajamanan B and Ramasamy V 2010 *Dig. J.of Nano. and Bio* **5** 339

On the optimization of soft magnetic properties of high Bs $Fe_{83.7}B_{14.8}Cu_{1.5}$ nanocrystalline alloy

S. N. Kane[a, 1], Kalpana Singh[a], Nandkishor Ghodke[b], A. Gupta[b] and L. K. Varga[c]
[a]School of Physics, D. A. University, Khandwa Road Campus, Indore – 452001, India
[b]UGC-DAE CSR, University Campus, Khandwa Road, Indore – 452 001, India
[c]Wigner RCP, Hungarian Academy of Sciences, P. O. Box 49, 1525 Budapest, Hungary

E-mail: kane_sn@yahoo.com (S. N. Kane)

Abstract. Influence of thermal annealing as a function of temperature and time, in $Fe_{83.7}B_{14.8}Cu_{1.5}$ alloy was monitored using magnetic measurements, differential scanning calorimetery (DSC), X- ray diffraction (XRD) to get information on correlation between structure, formed nano-crystalline phases and soft magnetic properties. Thermal annealing dependence of coercivity (H_c) and saturation magnetization (B_{2000}) shows that these values are affected by the structural relaxation and the presence of the precipitated nanocrystals in the amorphous phase. For the studied samples lowest coercivity of 14.6 A/m and rather high saturation induction value of 1.99 Tesla was obtained after annealing at 390 °C/1h. Activation energy of crystallization is 1.82 ± 0.48 eV. Grain diameter and volume fraction of the nanograins varies between 16.5 – 29 nm and between 2 – 56 % respectively. Lattice parameter suggests the presence of α-Fe and $Fe_{23}B_6$ phase.

1. Introduction

Amorphous alloys exhibit excellent soft magnetic properties, which can be further improved by choosing a specific composition and also by post-preparation treatments e. g. - thermal annealing, known to relax the stresses (introduced during the fabrication process) and if the thermal annealing is done above crystallization temperature it can induce crystallization within the amorphous matrix, modifying the soft magnetic properties. These amorphous alloys are very important for energy saving applications e. g. – as a core material for transformers, (exhibiting losses ~ 30 % less as compared to their crystalline counterparts). The main requirements are low coercivity and high saturation induction. For the same, Fe-Cu-B-Si based amorphous alloys were reported by Ohta et al [1, 2], exhibiting excellent soft magnetic properties and also high saturation induction. Certainly, the soft magnetic behaviour of Fe-Cu-B-Si based alloys can be further improved by thermal treatments, which change the initial amorphous structure in a composite of nano-precipitates and residual amorphous phases. The research for high B_s nanocrystalline materials, aims at replacing the existing Fe-Si crystalline electric steel with low cost material with competitive properties.

Present work reports the influence of thermal annealing as a function of temperature and time, performed in order to optimize the magnetic properties in $Fe_{83.7}B_{14.8}Cu_{1.5}$ alloy. Thermal annealing induced changes in magnetic, structural properties were monitored using, differential scanning calorimetery (DSC), magnetic measurements and, x-ray diffraction (XRD) to get information on

[1] To whom any correspondence should be addressed.

correlation between structure, formed nano crystalline phases and soft magnetic properties. It is worth mentioning that the composition of $Fe_{83.7}B_{14.8}Cu_{1.5}$ alloy is at the edge of the glass forming ability, and is the first in a series of alloy [1, 2], where the casting (melt spinning) behaviour will be optimized by adding Si, which suppress the B_s a little but improves the ribbon properties, (width and thickness).

2. Experimental details

Ribbons (about 15 μm thick and 0.5 mm wide) of nominal composition $Fe_{83.7}B_{14.8}Cu_{1.5}$ were prepared using a planar flow casting technique. DSC measurements were done at a heating rate of 2, 5, 10 and 20 °C/min. Following DSC results, thermal annealing was done at 350, 370, 390, 410 °C for 60 minutes. Samples were also annealed at 390 °C for 15, 30, 45 and 60 minutes, to get information on time evolution of properties and, ultimately to get best soft magnetic properties in the studied specimens. Bruker D8 Advance X-ray diffractometer was used to carry out XRD measurements on as-cast and annealed samples. Cu-K_α radiation ($\lambda = 0.154$ nm) was used for the same. XRD patterns were analyzed by fitting a crystalline and an amorphous component using pseudo-Voigt profiles, where from the volume fraction (V_x) of nano-precipitates were determined. The average grain size D is obtained by Scherrer's formula using integral width of the (110) line and the lattice parameter a is calculated using Nelson-Taylor-Sinclair correction in order to take into account the peak shift due to sample offset. For amorphous phase, the first nearest neighbour distance X_m is calculated using the expression: $X_m = 1.227\ \lambda/\ (2\text{Sin}\theta)$ [3 - 4]. Hysteresis loops of as cast and annealed samples were measured at 50 Hz ($H_{max} = \pm$ 2000 A/m). Coercive field (H_c) and saturation induction (B_{2000}), measured at an applied magnetic field of 2000 A/m were determined by analyzing raw data using a Matlab® program, described elsewhere [5]. For each specimen two hysteresis loops - one each at low field (was used to measure coercivity 'H_c') and the high field (was used to measure Induction B_{2000} values), were measured.

3. Results and discussions

DSC studies reveal single step crystallization. DSC measurements done at various heating rates

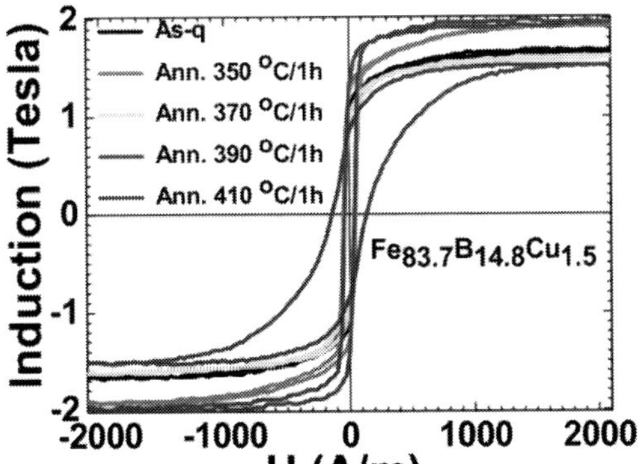

Figure 1. Thermal annealing induced changes in hysteresis loops of the studied specimens.

Table 1. Annealing temperature dependence of H_c and B_{2000}.

Sample details	H_c (A/m)	B_{2000} (T)
As-cast	18.6	1.9
Ann. 350 °C/1h	20.6	1.9
Ann. 370 °C/1h	16.1	1.6
Ann. 390 °C/1h	14.6	1.9
Ann. 410 °C/1h	138.4	1.5

Table 2. Annealing time dependence of H_c and B_{2000}.

Sample details	H_c (A/m)	B_{2000} (T)
Ann. 390 °C/15 min.	18.9	1.6
Ann. 390 °C/30 min.	14.7	1.9
Ann. 390 °C/45 min.	14.8	1.9
Ann. 390 °C/60 min.	14.6	1.9

Table 3. Temperature time transformation behaviour of grain size < ***D*** >, lattice parameter ***a***, volume fraction of nanograins V_x for nanocrystalline specimens, as obtained after analyzing XRD measurements. H_c values are also given, as obtained from magnetic measurements.

Sample details	<*D*> (nm) (±2)	*a* (nm) (±0.0001)	V_x (%) (±2)	H_c (A/m) (±0.5)
Temperature transformation				
As-cast	19.0	0.2859	15.0	18.6
Ann. at 350 °C /1h	22.0 19.0	0.2859 0.28505	14.5 11.0	20.6
Ann. at 370 °C/1h	21.5 20.0	0.2860 0.2951	23.0 13.0	16.1
Ann. at 390 °C /1h	29.0 23.1	0.2861 0.2949	23.0 9.0	14.6
Ann. at 410 °C /1h	23.9 19.0	0.2852 0.2939	44.0 12.0	138.4
Time Transformation				
Ann. at 390 °C /15 min.	16.5	0.2860	23.0	18.9
Ann. at 390 °C /30 min.	27.5	0.2859	25.6	14.7
Ann. at 390 °C /45 min.	27.0	0.2859	33.0	14.8
Ann. at 390 °C /1h	29.0 23.1	0.2861 0.2949	23.0 9.0	14.6

were used for calculating activation energy of crystallization. Calculated activation energy of crystallization (E_a) is 1.82 ± 0.48 eV.

Figure 1 depicts the annealing temperature dependence of hysteresis loops of the studied samples, showing variation of magnetic properties with annealing temperature. Table 1 and table 2 depicts the variation of H_c and B_{2000} as a function of annealing temperature and time respectively. Perusal of table 1 shows that for different annealing temperatures, H_c and B_{2000} range between 14.6 – 138.4 A/m and 1.5 – 1.9 Tesla respectively. Best H_c : 14.6 A/m and B_{2000}: 1.9 Tesla were obtained after annealing at 390 °C/1hour. To be able to further improve the magnetic properties, time evolution of magnetic properties after annealing at 390 °C for different time(s) was done and the results are depicted in table 2. Perusal of table 2 shows that H_c varies between 14.8 – 18.9 A/m and B_{2000} range between 1. – 1.9 Tesla. It is worth noting that after 30 min. annealing at 390 °C, best H_c and B_{2000} can be obtained.

XRD patterns, shows that after thermal annealing the alloy reveal the co-existence of amorphous and crystalline phases. It is worth noting that the as-cast specimen also shows features of crystallization. Table 3 depicts the parameters obtained by analyzing XRD data. For comparison, H_c values (obtained from magnetic measurements) are also shown, to be able to be able to see the variation of H_c with change in V_x and ***D***. Perusal of table 3 shows that ***D*** for the studied specimens varies between 19 – 29 nm, and depends on the annealing temperature. It should be noted that thermal annealing also affects the V_x (varies between 11 – 44 %), so that observed effect in the magnetic properties is due to both ***D*** and V_x . Within the experimental errors the first nearest-neighbour distance in the residual amorphous matrix (0.250 nm ± 0.001) almost remains constant for the studied specimens, showing that the studied samples have similar mass densities. The lattice parameter of the precipitated crystalline phase corresponds to α-Fe. Specimens annealed for temperature higher than at

350 °C, apart from α-Fe phase an additional phase with lattice parameter ~ 0.2950 nm is obtained, and JCPDS database shows that the obtained lattice parameter corresponds to $Fe_{23}B_6$ phase, which has high magnetocrystalline energy [6], thus affecting soft magnetic properties adversely, as can be seen in table 3. Volume fraction of $Fe_{23}B_6$ phase, range between 9 – 13 %. Presence of α-Fe in the studied specimens can be accounted for the obtained higher B_{2000} values, whereas the presence of $Fe_{23}B_6$ phase is accounted for the obtained higher H_c values, which was not observed in earlier reports [1, 2], showing lower H_c values. Average grain diameter D for the specimens annealed at 390 °C for different times, ranges between 16.5 nm - 29 nm. Volume fraction of the nanograins V_x ranges between 2 – 33 %. Obtained values of D and V_x suggest that, there is a counter-balance between these two parameters. This clearly shows that, the specimens exhibit grain growth as well a nucleation of new grains, accounted for the obtained D and V_x values. Lattice parameter values suggest the formation of α-Fe. Samples annealed for 1 hour does shows the existence of another phase, whose lattice parameter suggests this phase to be $Fe_{23}B_6$, as was also observed in earlier studies [6 - 7] showing soft magnetic behaviour and even high $B_s \approx 1.8$ Tesla (e. g. – as observed in [7]), ascribed to micro-structural changes. After annealing of specimens at 390 °C for different times, H_c ranges between 14.67 – 22.3 A/m and B_{2000} values range between 1.6 – 1.99 Tesla, with best H_c and B_{2000} value of 14.7 A/m, 1.99 Tesla is obtained, after annealing at 390 °C for 30 minutes.

4. Summary

DSC studies reveal single step crystallization of the specimen under investigation. Calculated activation energy of crystallization (from Kissinger's method) is 1.82 ± 0.48 eV. Thermal annealing dependence of coercivity H_c and saturation magnetization B_{2000} shows that, as expected, both B_{2000} and H_c are affected by the structural relaxation and the presence of the precipitated nanocrystals in the amorphous phase. For the studied samples as a function of temperature, H_c values range between 14.7 – 138.4 A/m, whereas the best H_c value of 14.64 A/m is obtained after annealing at 390 °C/1h. B_{2000} values range between 1.5 – 1.99 Tesla; best B_{2000} value of 1.99 Tesla is obtained after annealing at 390 °C/1h. Annealing time also shows appreciable changes in H_c and B_{2000} values. Observed variation in the coercivity is consistent with the structural relaxation, crystallization of the specimens. After annealing of specimens at 390 °C for different times, H_c ranges between 14.6 – 22.3 A/m and B_{2000} values range between 1.6 – 1.99 Tesla, with best H_c and B_{2000} value of 14.7 A/m, 1.99 Tesla is obtained. XRD analysis for the specimens as a function of temperature and time reveals noticeable changes in Scherrer's grain diameter D and volume fraction of the nanograins V_x . Obtained D for the studied specimens varies between 19 – 29 nm. V_x of the nanograins range between 2 – 33 %. The lattice parameter of the precipitated crystalline phase corresponds to the α-Fe, present in all the studied specimens (even in as-cast specimen). XRD analysis also shows the presence of $Fe_{23}B_6$ phase up to 13 %. XRD analysis also shows the presence of $Fe_{23}B_6$ phase up to 13 % in the over annealed samples with high H_c.

References

[1] Ohta M and Yoshizawa Y 2007 *J. Appl. Phys* **46** L477
[2] Ohta M and Yoshizawa Y 2008 *J. Appl. Phys* **103** 07E722
[3] Guinier A 1963 *X-ray Diffraction in Crystals, Imperfect Crystals and Amorphous Bodies.* (Freeman, San Fransisco) p 72
[4] Kane S N , Coisson M, Tiberto P, Vinai F and Mazaleyrat F 2011 *Current Appl. Phys.* **11** 981
[5] Kane S N , Gercsi Z, Ghodke N, Khinchi S S and Gupta A 2005 A digital hysteresis loop tracer for studying soft magnetic materials (Proc. of the DAE Solid State Physics Symposium vol. 50) ed. V K Aswal, K G Bhushan and J V Yakhmi, Prime (Time Education, Mumbai, India) (ISBN: 81-8372-019-6) 287
[6] Panda A K, Ravikumar B, Sahu S, Mitra A 2003 *J. Magn. Magn. Mater.* **260** 70
[7] Kong F, Wang A, Fan X, Men H, Shen B, Xie G, Makino A and Inoue A 2011 *J. Appl. Phys.* **109** 07A303

X-ray absorption near edge structure (XANES) study of some hydroxamic mixed ligand copper complexes

A Mishra[†], N Parsai[†], N.Soni and B D Shrivastava[*]
[†]School of Physics, Devi Ahilya University, Indore, 452 001
[*]School of Studies in Physics, Vikram University, Ujjain, 456 010

E-mail: parsai.neetu@gmail.com

Abstract. With the advent of modern bright synchrotron radiation sources, X-ray absorption spectra has emerged as a powerful technique for local structure determination, which can be applied to any type of material. The X-ray absorption measurements of four hydroxamic mixed ligand copper complexes have been performed at the recently developed BL-8 Dispersive EXAFS beamline at 2.5 GeV Indus-2 synchrotron at RRCAT, Indore, India. The X-ray absorption near edge structure (XANES) data obtained has been processed using data analysis program *Athena*. The energies of the K absorption edge, chemical shifts, edge-widths, shifts of the principal absorption maximum in the complexes have been determined. The values of the chemical shift suggest that copper is in oxidation state +2 in all of the complexes. The chemical shift data has been utilized to estimate effective nuclear charge on copper atom. The order of the chemical shifts has been correlated to the relative ionic character of the bonding in these complexes.

1. Introduction
Hydroxamic acids have been used as therapeutic agents in chelation theraphy and as metalloenzyme inhibitors [1]. Other medicinal applications of hydroxamates which utilize their affinity for high charge density metal ions include the possible use of their complexes as imaging agent [2]. With regards to the strong ability of hydroxamic acids to form chelates, clarification of their interaction with metal ions is of particular importance in terms of biological effects. In the present work, X-ray K-absorption spectra of copper in the four mixed ligand copper complexes having hydroxamic acid as primary ligand have been studied. The details of the four complexes are given in table 1.

2. Experimental
In the present investigation, the X-ray absorption spectra have been recorded using synchrotron radiation. The X-ray spectroscopic setup is available at Raja Ramanna Center for Advaced Technology (RRCAT), Indore, India and is called dispersive EXAFS beamline BL-8. This beamline has been recently commissioned at the 2.5 GeV Indus-2 synchrotron radiation source [3]. In this beamline a bent crystal (Si 111) polychromator is used to select a band of energy from the white synchrotron beam which is horizontally dispersed and focused on the sample. The crystal is bent in the shape of an ellipse in such a way that source and sample positions are at two focii of the ellipse. The transmitted beam intensity from the sample is recorded on a position sensitive CCD detector, thus enabling recording of the whole EXAFS spectrum around an absorption edge in a single shot (typical acquisition time of one spectrum is ~300 msec). The intensities I0 and It, are obtained as the CCD outputs without and with the sample respectively. Using the relation, $I_t = I_0e^{-\mu x}$, where μ is the absorption coefficient and x is the thickness of the absorber, the absorption $\mu(E)$ corresponding to the photon energy (E) are obtained. The experimental data has been analyzed using the available computer software package Athena.

Published under licence by IOP Publishing Ltd

Table 1. Copper (II) hydroxamic mixed ligand complexes, abbreviation and molecular formula.

S. No.	Complex	Abbreviation	Molecular formula
1.	Cu(II) 5 chloro-benzohydroxamic acid	Cu 5CBHA	$CuC_7H_6O_2NCl$
2.	Cu(II) malanohydroxamic acid	Cu MHA	$CuC_2H_4O_4$
3.	Cu(II) succenohydroxamic acid	Cu SuHA	$CuC_4H_8O_4N_2$
4.	Cu(II) acetylohydroxamic acid	Cu ActHA	$CuC_8H_{10}O_3N$

Table 2. XANES data for the copper K-absorption edge in copper (II) hydroxamic mixed ligand complexes.

Complex	E_{k1} (eV)	E_{k2} (eV)	E_A (eV)	Chemical shift (eV)	ENC	Shift of the principal ab.max. (eV)	Edge-width (E_A-E_k) (eV)
Cu 5CBHA	8986.7	8992.7	9001.0	6.2	0.67	6	14.3
Cu MHA	8987.1	8992.4	9000.9	6.6	0.70	5.9	13.8
Cu SuHA	8987.8	8992.9	9000.8	7.3	0.75	5.8	13.0
Cu ActHA	8988.4	8993.4	8999.5	7.9	0.80	4.5	11.1

3. Results

The normalized XANES spectra in the region $-30 < E < 50$ eV, indicating positions of the absorption edges K1 and K2 and principal absorption maximum A are shown in figure 1. The first derivatives of the spectra in this region, indicating positions of the absorption edges K1 and K2 and principal absorption maximum A are shown in figure 2. First peak in the derivative spectra gives the position of first inflection point, i.e., the K1-absorption edge (Ek1) and second peak gives the second inflection point, i.e., the K2-absorption edge (Ek2). The position where the derivative is zero, gives the position of principal absorption maxima (EA). The results of the energies of the K-absorption edges (Ek1 and Ek2) and the energies of principal absorption maximum A (EA) of copper in its four complexes are presented in table 2. The chemical shifts (in eV) of the K-absorption edge of copper in the complexes are also given in table 2. For all the complexes the distance (in eV) of the principal absorption maximum A with respect to the respective K-absorption edge have also been computed and are collected in the same table. It can be readily seen from table that copper K-edge is found to be shifted towards the high energy side in all the four complexes, as compared to the copper metal K-absorption edge.

4. Discussions

4.1 Chemical shift

The shifts of the K-absorption edge of copper in the complexes with respect to that of copper metal have been determined according to the eqn.

$$\Delta E_K = E_K\,(\text{complex}) - E_K\,(\text{metal})$$

The results are given in table 2. For computing the chemical shift the value of EK (Cu metal) has been taken as 8980.5 eV.

The compounds having copper in oxidation state +1 show chemical shifts less than 5 eV while those having copper in oxidation state +2 show chemical shifts more than 5 eV [4]. In table 2, all the four complexes have the values of chemical shifts between 6.2 to 7.9 eV. Hence, on the basis of values of the chemical shifts, all the complexes are found to have copper in oxidation state +2.

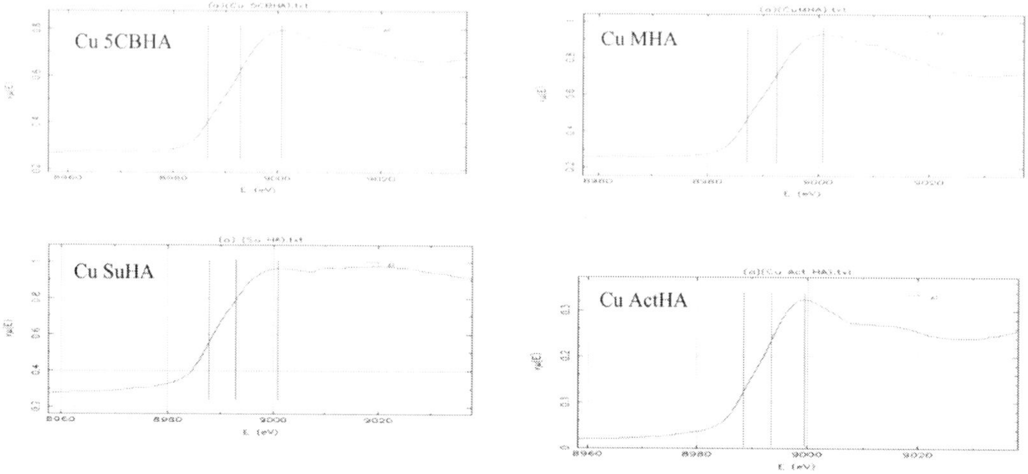

Figure 1. XANES spectrum for copper (II) hydroxamic mixed ligand complexes indicating position of absorption edges K1 and K2 and principal absorption maxima A.

The chemical shift is largest for CuActHA complex and the order for other complexes is as follows:

Cu ActHA > Cu SuHA > Cu MHA > Cu 5CBHA.

As is well known, an ionic bonding enhances the chemical shift, whereas a covalent bonding suppresses it. Hence, the above order may also be taken as representative of the relative ionic character of the bonding in these complexes.

4.2 Effective nuclear charge (ENC)

If the binding energies of K electron of copper in different oxidation states are determined, one can find from the difference in binding energies of the neutral atom and the ionized atom, the so called theoretical shifts in the X-ray absorption edge. In the present work, the values of binding energies have

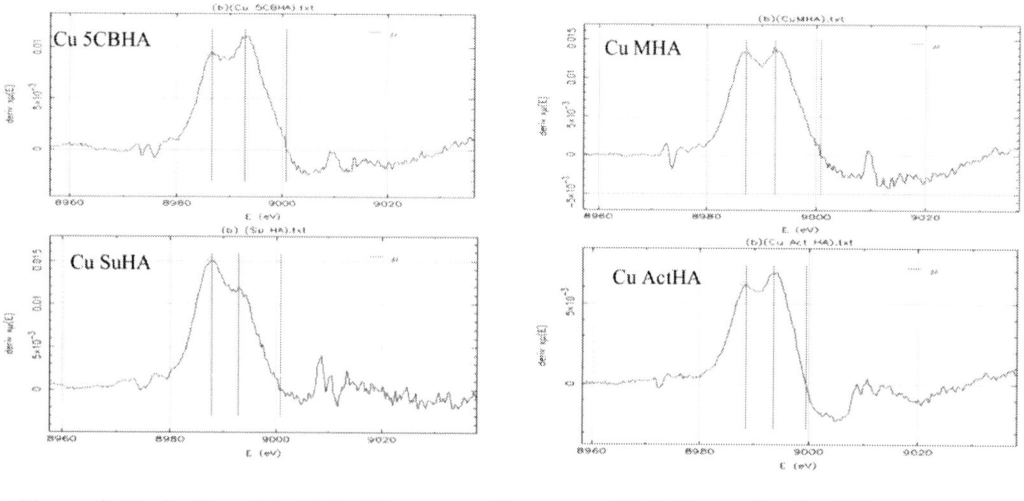

Figure 2. Derivative of the XANES spectrum for copper (II) hydroxamic mixed ligand complexes indicating position of absorption edges K1 and K2 and principal absorption maxima A.

been determined from the tables of Clementi and Roetti [5] using Koopman's theorem[6]. The theoretical shifts are plotted against the oxidation number. From the graph (not shown here), the effective nuclear charge on copper in its complexes has been determined and the results have been presented in table 2.

4.3 Principal absorption maximum

In table 2, the data for the principal absorption maximum EA for the complexes have also been included. It has been observed that with respect to copper metal, the value of EA is shifted towards the higher energy side. For the complexes mentioned in table 2, the energy range of chemical shift in these complexes is between 6.2 – 7.9 eV while the range for shift of principal absorption maximum is between 4.5 - 6.0 eV. Hence, on the basis of the shift of the principal absorption maximum also, it can be inferred that copper is in +2 oxidation state in these complexes. The shift of the principal absorption maximum for Cu (II) complexes are in the order

Cu ActHA < Cu SuHA < Cu MHA < Cu 5CBHA

The order of the shift of the principal absorption maximum is in the reverse order of the chemical shift of the complexes.

4.4 Edge-width

In table 2, the values of the edge-width (EA-EK1) have been reported. The experimental data of edge-width of Cu (II) complexes show that the edge-width decreases as follows:

Cu ActHA < Cu SuHA < Cu MHA < Cu 5CBHA

The order of the edge-width is in the reverse order of chemical shift of the same complexes. This represents that the edge-width is inversely proportional to ionic character for this series. The reverse trend for these complexes is justified on the basis of the criterion that, in general, edge-width of the K-absorption edge increases with the increase of covalent character of the bonds provided other factors like molecular symmetry etc., remain the same [7].

5. Conclusions

The copper K-edge has been found to split into two components, i.e., K1 and K2 in all of these complexes. The energies of K1 edge (EK1), K2 edge (EK2) and principal absorption maxima (EA) have been reported. From these, the shift of the K1-edge (chemical shift), shift of the principal absorption maximum and edge-width has been obtained. The values of the chemical shifts suggest that copper is in oxidation state +2 in all of the complexes. The chemical shift has been used to determine the effective nuclear charge on the absorbing atom. The order of the chemical shifts is representative of the relative ionic character of the bonding in these complexes.

Acknowledgement

Thanks are due to Drs. S. N. Jha, D. Bhattacharya, A. Poswal, Applied Spectroscopy Division, BARC, Mumbai and S. K. Deb, RRCAT, Indore for their help in recording the spectra.

References

[1] Rockwell M, Melden R A, Copeland K, Hardman C P, Decicco W W and Degrado 1996 *J. Am. Chem. Soc.* **18** 10337
[2] Miller M J and Lin Y M 1999 *J. Org. Chem.* **64** 7451
[3] Bhattacharya D, Poswal A K, Jha S N, Sangeeta and Sabharwal S C 2009 *Bull. Mater. Sci.* **32(1)** 103
[4] Kau L S, Spira-Solomon D J, Penner-Hahn J E, Hodgson K O, and Solomon E I 1987 *J. Am. Chem. Soc.* **109** 6433
[5] Clementi E and Roetti C 1974 *Atomic data and Nuclear data tables* **14** 177
[6] Koopmans T 1934 *Physica* **1** 104
[7] Kumar A, Nigam A N and Shrivastava B D 1981 *X-ray Spectrom.* **10** 25

Temperature dependence of the elastic constant of Borassus Flabellifier 'BF' material by acoustic response

Sushil Phadke[1], B DShrivastava[2], N Dagaonkar[3] and Ashutosh Mishra[4]
[1]Department of Physics Govt. Girls College, Dhar, (M.P.), India,
[2,3]Government P.G. College, Dhar, (M.P.), India &
[4]School of Physics, Devi Ahilya University, Indore, (M.P.), India.

[1] Corresponding Author email: sushilphadke5@gmail.com

Abstract. The homogeneous continuous materials are widely used for many structural applications. Migrations of atoms or molecules are the mechanism of mechanical and kinetic processes in materials for their synthesis processing as well as for their structural evolutions. The elastic constant of solids provides valuable information on their mechanical and dynamical properties. In particular, they provide information on the stability and stiffness of materials. In the present study author investigated relation between elastic constant and temperature in Borassus Flabellifier 'BF' wood part. Determination of elastic properties of material is based on the longitudinal wave's velocities via ultrasonic methods. The resonant frequencies of the specimens were measured by Ultrasonic Interferometer (for solids) dual frequency using longitudinal cubic piezoelectric crystal of quartz of frequency 123.62 KHz. The temperature variations from room temperature were done by PID control unit, Mittal Enterprises, New Delhi, India. Characterization of the samples was done by scanning electron microscope (SEM) Model JEOL JSM5400 at 5.0kvx750, 10 μm.

Key words: homogeneous, elastic, ultrasonic, piezoelectric, longitudinal.

1. Introduction

Determination of the elastic properties of homogeneous continuous materials by means of measurements of ultrasonic velocity has a long history for different materials [1.2]. Acoustic methods [3] based on the phenomena accompanying propagation of longitudinal stress waves demonstrate their particular usefulness for such testing. Traditional methods required a long time to perform tests and provide only an approximate evaluation of a large batch of material on the basis of testing of small sample population. Development of non-destructive testing [4] made it possible to use the technique for the purpose of diagnosis and assortmentation of raw material. The longitudinal orientation of tracheids or fires is partially disturbed by medullary rays. The ultrasonic energy injected into a fibrous materials couple several modes on each fibre. The physical properties, shape and size of the fibres affect the transmitted ultrasonic field. Each structural element acts independently as an elementary resonator. The spatial distribution of Ultrasonic velocities and frequencies that matched the frequency of the natural fibres could explain the acoustical behaviour of wood as illustrated by its overall parameters [5]. The interfacial discontinuities, cracks, porosity, or density variations may be detected by non-destructive techniques such as ultrasound. This method can help in the understanding of material behaviour under different environment conditions. It is important to relate the non-destructive measurements to the mechanical properties of wood and wood-based composites'. Ultrasonic techniques are however qualified for non destructive measurement of all the elastic constants of such materials. The elastic modulii Cij identification from the measurement of longitudinal ultrasonic velocity were often deducted from the low-frequency, using piezoelectric ultrasonic interferometer. In the present study the material was taken from District Dhar of Madhya Pradesh. Author used the part of Borassus Flabellifier female and male tree wood because from where the toddy or neera collected in pots around 30 litters per day.

2. Material and method

The preparation of material and measurement of density and ultrasonic velocity was reported by Sushil Phadke et al [6] in their earlier communication at room temperature. In the present study the author measured ultrasonic velocity with temperature variation from room temperature to 400K.The variation was made by proportional integral derivative control unit. Single shell PID controlled hot air oven with digital temperature indicator is complete in all respect ready to use on 230-volt input. It consists of electrically operated heating source, air source, & PID controller, Mittal Enterprises, New Delhi, India. Characterization of the samples was done by scanning electron microscope (SEM) Model JEOL JSM5400 at 5.0kvx750, 10 µm by Diya Lab, Mumbai. The technical specifications and main features of PID control unit used in this study as follows:

TECHNICAL SPECIFICATION	SALIENT FEATURE
Input supply 230volt & +/-10% stabilizer controlled supply.	Temperature controller to control the temperature. Max. 300 deg C Approx.
Temperature controller, West – 6100	Heater is available producing heat in oven.
Solid-state relay.	Solid-state relay is available to control the temperature.
Air pump.	
Heater of 600 watts rings type.	
Cell dimension are 150/100mm (app.)	
Temperature sensors of PT 100.	

3. Theory and calculation

The resonant frequency of the specimen fs is evaluated using the relation:

$$\mathbf{f_s = f_c + [m_q/m_s](f_c - f_q)} \quad (1)$$

Where fq, fc, mq ,ms are resonant frequency of quartz, composite resonant frequency, mass of the quartz rod and mass of the specimen.

Using the value of fs and the length of specimen L, velocity of ultrasonic waves on the specimen can be calculated by the relation:

$$\mathbf{V_l = 2\,L\,f_s} \quad (2)$$

We can find the Young's modulus for the specimen from the relation.

$$\mathbf{E = V_l^2 \rho} \quad (3)$$

Where ρ is the density of the specimen.

4. Result and Discussion

The measured and calculated values of frequency, ultrasonic velocity and elastic constant with temperature for female and male Borassus Flabellifier wood parts were given in the table 1.The relationship and between

Figure1 Variation of frequency with temperature for female BF wood part.

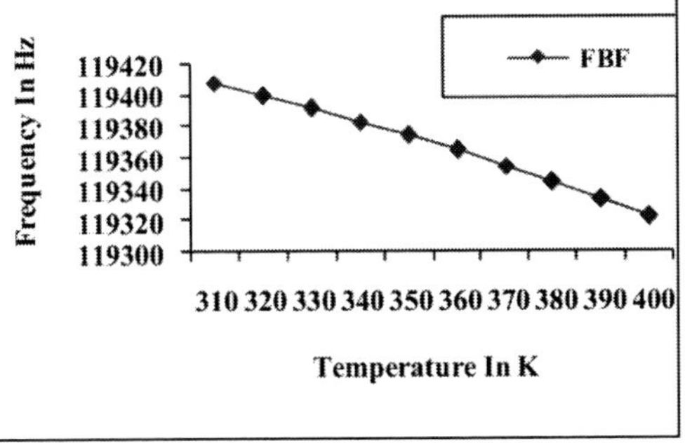

Figure 2 Variation of elastic constant with frequency for female wood part.

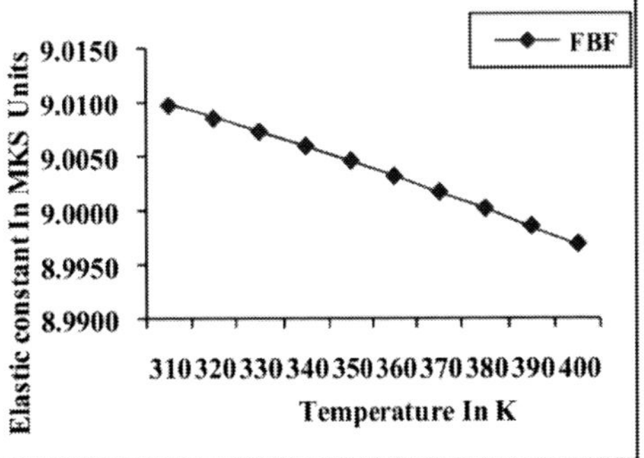

Figure 3 Variation of frequency with temperature for male BF

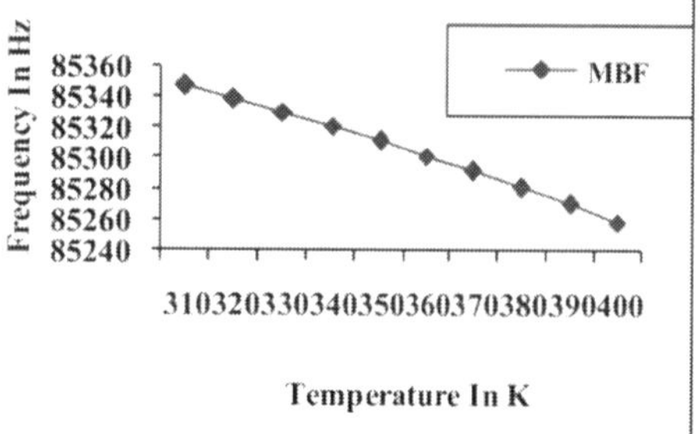

Figure 4 Variation of elastic constant with temperature for male wood part.

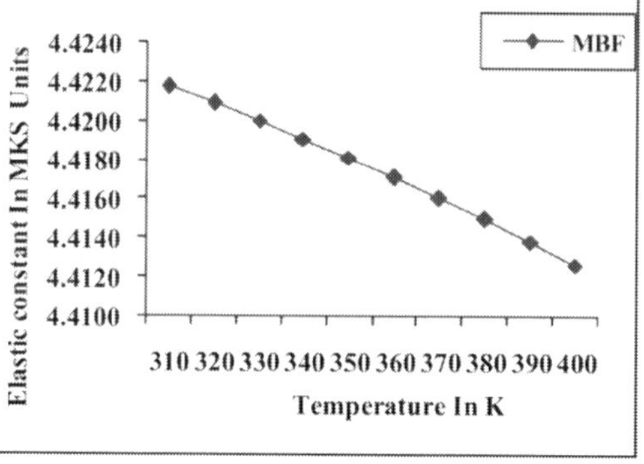

Table 2-The measured and calculated values of mass, density, ultrasonic velocity and elastic constant.

Mass of the quartz rod m_q= 0.924 X 10^{-3}Kg.					Thickness of specimen=0.004m		
Natural frequency of quartz rod f_q =123620Hz.					Breath of the specimen=0.004m		
Volume of the specimen=2.272x10^{-07}Kg/m^{-3}					Length of the specimen L =0.0142m		
S. No.	**Mass of specimen m_sX10^{-3}**	**Density ρ**	**Temperature**	**Frequency of composite system f_c**	**Frequency of specimen f_s**	**Ultrasonic velocity v_l**	**Elastic constant E x 10^9**
In MKS UNIT							
In the longitudinal direction of fibre Female FBF							
1.	0.178	783.450704	310	122939.598	119407.625	3391.1766	9.0097
			320	122938.31	119399.651	3390.9501	9.0085
			330	122936.968	119391.34	3390.714	9.0073
			340	122935.568	119382.676	3390.468	9.0060
			350	122934.11	119373.645	3390.2115	9.0046
			360	122932.589	119364.231	3389.9442	9.0032
			370	122931.004	119354.419	3389.6655	9.0017
			380	122929.352	119344.191	3389.375	9.0002
			390	122927.63	119333.53	3389.0722	8.9986
			400	122925.835	119322.417	3388.7566	8.9969
In the longitudinal direction of fibre Female MBF							
2.	0.171	752.640845	310	111642.598	85346.6966	2423.8462	4.4218
			320	111641.31	85338.4497	2423.612	4.4209
			330	111639.968	85329.8534	2423.3678	4.4200
			340	111638.568	85320.8929	2423.1134	4.4191
			350	111637.11	85311.5527	2422.8481	4.4181
			360	111635.589	85301.8169	2422.5716	4.4171
			370	111634.004	85291.6686	2422.2834	4.4161
			380	111632.352	85281.0904	2421.983	4.4150
			390	111630.63	85270.064	2421.6698	4.4139
			400	111628.835	85258.5705	2421.3434	4.4127

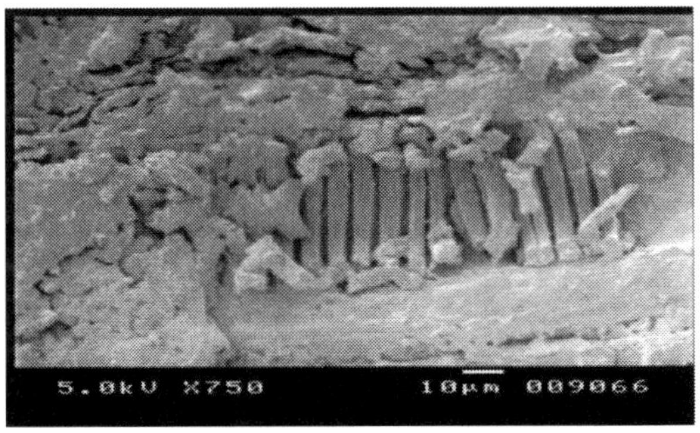

Figure 5 SEM of female BF wood at 5.0kvx 750 &10μm.

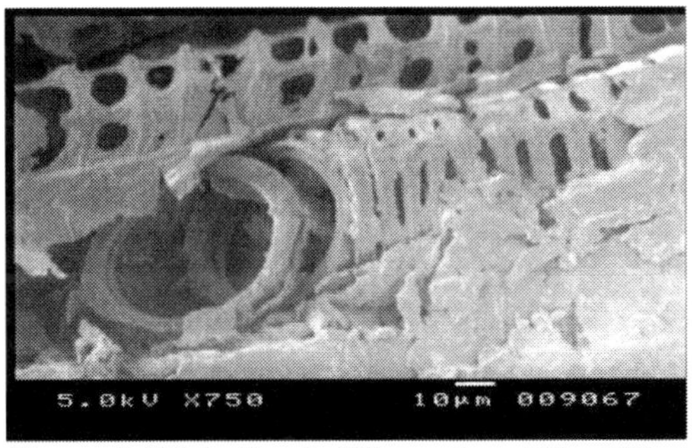

Figure 6 SEM of male BF wood at 5.0kvx750 & 10μm.

temperature and frequency are shown in figure 1 figure 3 for female and male Borassus Flabellifier wood parts. The figures (1&3) show nearly negative exponential relationship between temperature and frequency. It seems that as frequency increases the structure of the material also change. Figure 2 and figure 4 shows the nearly negative exponential relation between frequency and elastic constant of female and male Borassus Flabellifier wood parts. The figure 5 and figure 6 shows the SEM picture of both the samples at 5.0kvx 750 &10μm.

5. Conclusion

The temperature affect the frequency and elastic constant exponentially were decreasing. As temperature increases the frequency and elastic constant decreases in both the female and male Borassus Flabellifier wood parts because the ultrasonic velocity were increased with temperature in solid material. The SEM picture shows that both the wood parts were highly rough.

6. Acknowledgement

One of the authors (Sushil Phadke) is highly thankful to University Grants Commission, New Delhi for awarding a Minor research Project. Our sincere thanks to Dr. S.N. Mandloi, Principal Govt. Girls College Dhar (MP) India for providing facility to bring samples from Dist. Dhar (M.P.) India. We are very thankful to Mr. Ritusing Kalesh of Dist. Dhar (M.P.) India for providing sample. I also thankful to Devendra Jain of Diya Lab, Mumbai for sending SEM reports in time.

7. Reference

[1] Hearmon, R F S, An introduction to applied anisotropic elasticity, Oxford University Press, Oxford 1961.
[2] Musgrave M J P Crystal acoustics, *Holden Day*, San Francisco 1970.
[3] Bucur V *Acoustics of Wood, CRC Press Inc., New York* 1995.
[4] Hillier C, Handbook of Nondestructive Evaluation, McGraw-Hills Professional, 2001.
[5] Kothari D S , Determination of elastic constants of solids by pulsed method, Vol.**XVI**-No. **4**, p. 227-235. 1949.
[6] Phadke S, Shrivastava B D, Mishra A , Dagaonkar N, Comparison of elastic constant for *Borassus Flabellifier* in the longitudinal and perpendicular direction of fibres with frequency, to be published in the Journal of Trans Tech Publication, *Diffusion in Solids and Liquids*. 2012.

Study of Optical Band Gap of CuO Using Fermi's Golden Rule

K. R. Nemade, *S. A. Waghuley
Department of Physics, Sant Gadge Baba Amravati University,
Amravati 444 602, India.
*Corresponding author e- mail: krnemade@gmail.com,
sawaghuley@yahoo.co.in

Abstract. Quantum size effect where the electronic and optical properties of solids are altered due to changes in the band structures, enhanced the surface/volume ratio in nano dimensions forces more than 33% of the atoms to be on the surface (for 10nm dot 35), which drastically altering the physical properties such as having lower melting temperature and lower sintering temperature, and higher diffusion force at elevated temperatures. Consequently, its Fermi's golden rule analysis becomes crucial. Cupric oxide (CuO) is an important transition metal oxide with the basis of several high temperature superconductors and giant magnetoresistance materials. In present investigation, optical Band Gap from UV data using Fermi's golden rule for single step chemically synthesized CuO was computed.

1. Introduction

Cupric oxide (CuO) is an important transition metal oxide with a narrow bandgap (Eg 1.9 eV) and forms the basis of several high temperature superconductors and giant magnetoresistance materials. Furthermore, because of the complex magnetic phases, CuO has been used for the preparation of a wide range of organic–inorganic nanostructure composites that possess unique characteristics such as high thermal and electrical conductivities as well as high mechanical strength and high-temperature durability. Therefore, on the basis of the fundamental and practical importance of CuO nanomaterials, well-defined CuO nanostructures with various morphologies have been fabricated, and many synthesis techniques have been developed to prepare different assemblies of CuO building blocks [1]. CuO nanowires (NWs) gas sensors with p-type semiconducting behavior have scarcely been investigated, although a few studies have reported the gas sensing characteristics of CuO nanopowders, nanoribbons, and nanorods. Many physico-chemical routes to prepare Cu2O or CuO 1D nanostructures have been explored. Some examples include a polyolmethod, seed-mediated solution growth, a hydrothermal reaction, anodization, electrospinning, and thermal oxidation. The CuO NWs prepared by thermal oxidation exhibit higher crystallinity and longer aspect ratios compared to those prepared via solution-based routes. Thus, in this study, the CuO NWs were grown by the thermal oxidation of Cu foil. Subsequently, the CO and NO_2 sensing characteristics of the CuO NWs were studied [2].
In this present article, synthesis of CuO nanopartials synthesized through chemical single step method by varying concentration of hexamethylenetetramine (HMT) in constant 1M cupric nitrate ($Cu(NO_3)_2.3H_2O$) at room temperature. Material were characterized by XRD and UV-visible.

2. Experimental

For present work AR Grade chemicals used. CuO nanoparticals was prepared by varying concentration of hexamethylenetetramine (HMT) in constant 1M Cupric nitrate ($Cu(NO_3)_2.3H_2O$) at room temperature. For our present work we varies mole concentration of hexamethylenetetramine (HMT) from 0.25-1M [3]. The samples of CuO were characterized by using XRD techniques. The X-ray diffraction pattern of powder was recorded on Phillips-1730 X-ray diffractometer using CuKα radiation (λ=1.54Å). The diffractogram was in terms 2θ in range 200-750. Samples of CuO were also characterized by UV-Visible (Model: LAMBDA 25, M/S PERKIN ELMER, USA) for bang gap analysis.

3. Results and Dissection

3.1. XRD Analysis

Fig. 1 shows XRD Pattern of CuO (0.25M) nanoparticals. It is observed that XRD pattern contains 2-4 prominent peaks between position 2θ= 300-700.The average crystallite size, determined from XRD pattern using Scherrer formula. The average crystallite size from a sharp peak at 380 for CuO is 105nm estimated by using the Scherrer's formula eq. (1) [4]

$$D = K^*\lambda/\beta \ ^*\cos\theta \qquad (1)$$

where D is the crystallite size, K is the shape factor, which can be assigned a value of 0.89 if the shape is unknown, θ is the diffraction angle at maximum peak intensity, and β is the full width at half maximum of diffraction angle in radians.

3.2. UV-VIS analysis of CuO nanoparticals

Transmition spectrum of synthesized CuO material shows in fig.2 clear shift to the lower wavelength side over bulk crystallite at 650 nm [1] as shown in following UV visible spectrum of sample.

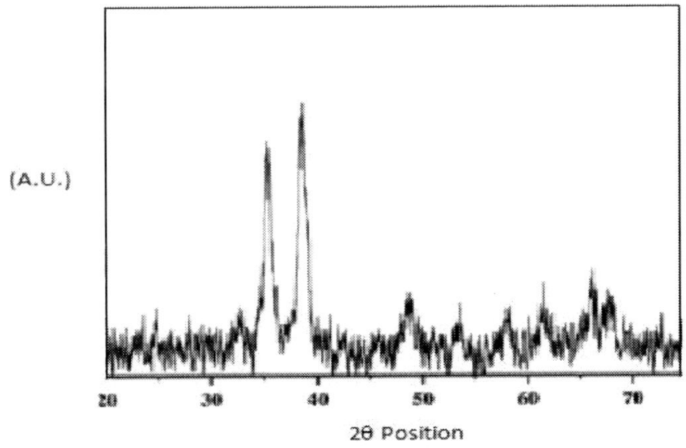

Figure 1. XRD Pattern of CuO (0.25M).

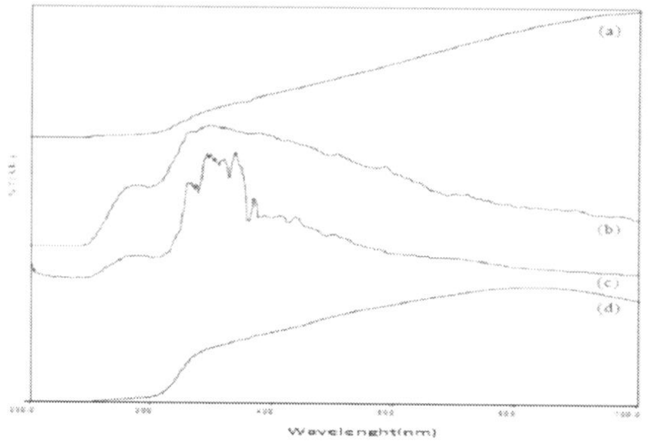

Figure 2. UV-visible spectra of CuO synthesized with (a) 1 M (b) 0.75M (c) 0.50M and (d) 0.25 M concentration of HMT

3.3. Optical band gap analysis with Fermi's golden rule

The shift of transmission spectrum shown in fig.2 towards lower wavelength indicates that during synthesis as the concentration of HMT increases; there is decrease in the size of CuO particles. Optical band gap of synthesized CuO nanoparticals was determined by using eq. (2). Band gap of material also increase with the concentration of HMT as shown in fig.3.

$$E(eV)*\lambda(\mu m)=1.24 \qquad (2)$$

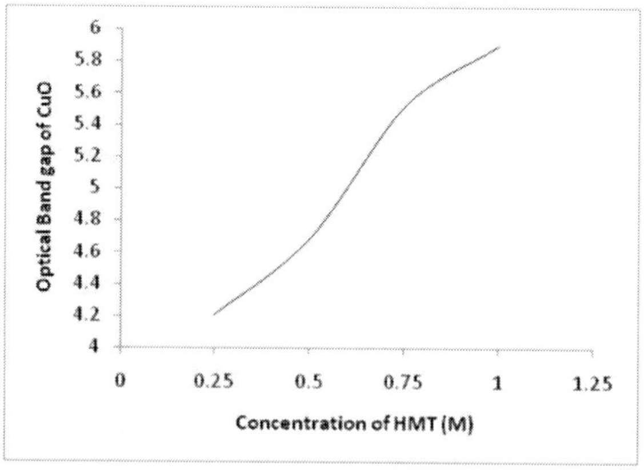

Figure 3. Variation of optical band gap with molar concentration of HMT (0.25-1M)

4. Conclusions

Fermi's golden rule is also a simple method for band gap analysis with UV data. From results and discussion following conclusions can be drawn, Transmition spectrum of synthesized CuO material shows clear shift to the lower wavelength side over bulk crystallite at 650 nm, consequently particles size of CuO goes on decreasing. Band gap of material increase with the concentration of hexamethylenetetramine (HMT).

5. Acknowledgements

Authors are thankful to Head, Department of Physics Sant Gadge Baba Amravati University, Amravati for providing necessary facilities. One of the authors S.A. Waghuley is thankful to University Grant Commission (New Delhi), for financial assistance in term of minor research project F.No. 37-565/2009 (SR).

References

[1] Gu A, Wang G , Zhang X and Fang B 2010 *Bull. Mater. Sci.* **33** 17
[2] Kim Y, Hwang I, Kim S, Lee C and Lee J 2008 *Sensors and Actuators B* **135** 298
[3] Pike J, Chan S, Zhang F, Wang X and Hanson J. 2006 *Applied Catalysis A* **303** 273
[4] Waghuley S A, Yenorkar S M, Yawale S S and Yawale S P 2008 *Sensors & Actuators B* **128** 366

dc Electrical conductivity of V_2O_5-P_2O_5 binary glassy systems

R. V. Barde[a] **and S.A. Waghuley**[b]
[a]Dept. of Engg. Physics, H.V.P.M. College of Engg. & Technology, Amravati 444 601, India.
[b]Department of Physics, Sant Gadge Baba Amravati University, Amravati 444 602, India.

Corresponding author(s) e- mail: rajesh_barde@indiatimes.com, sawaghuley@indiatimes.com

Abstract. Electrical conductivity of V_2O_5-P_2O_5 binary glassy systems prepared by melt quenching technique was studied in temperature range 303-470 K. The mole of P_2O_5 was varies from 0.2-1 M with constant 1 M V_2O_5 in the preparation of glass samples. The dc electrical conductivity of samples was measured and found to be higher for sample 1:0.2 M V_2O_5-P_2O_5. Using the Arrhenius equation of conductivity, the activation energy of conduction is evaluated. The conduction in these glasses takes place by the transfer of electron from V4+ to V5+ ions.

1. Introduction

Glassy electrolytes are of particular importance because of their inherent advantages such as isotropic conductivity, ease of preparation, better thermal stability, the large available composition ranges which make them potential candidates for technological applications (new solid-state batteries, fuel cells, chemical sensors, "smart windows"), and also the search for an general theory of ion-transport in glassy materials [1]. Transition metal oxides such as V_2O_5, P_2O_5 etc can form glasses when mixed with glass forming oxides. Electrical conduction of glasses containing transition metal oxides (TMO) is known to be semi-conducting. The dc conduction in TMO glasses is the reason of transfer of electron from lower valance to higher valance state of transition metal ions. Small polarons are charge carriers trapped by self-induced lattice distortions that extend over the near surrounding. Generally small polarons hopping model can explain the dc conduction and transport properties of vanadate glasses. The strong interaction of electrons with ions produces the localization and the formation of polarons in the vanadium glasses [2-5]. The small polarons theory has been extensively used to describe the electrical properties of a wide range of TMO glasses, the best agreement between experiment and theory has been found for vanadate glasses such as V_2O_5-P_2O_5. It appears probable that the electrical properties of all the 3d transition-metal oxide phosphate glasses will not succumb to a single theory. The experimental determination of activation energies from the electrical conductivity (σ), as a function of temperature (T), is sometimes different; the Arrhenius plot (ln σ versus 1000/T) may not be linear. This may be due to process involving several similar activation energies, conduction by polarons or variable range hopping of carriers [6].
In the present work, the electrical conductivity of V_2O_5-P_2O_5 binary glassy systems prepared by melt quenching technique was studied in temperature range 303-470 K by varying the mol% P2O5. The glass sample was characterized by XRD.

2. Experimental

The glass sample of V_2O_5-P_2O_5 in which mole of P_2O_5 was varies from 0.2-1 M with constant 1 M V_2O_5 was prepared by the melt quenching technique. The required quantities of V_2O_5 and P_2O_5 were mixed together by grinding the mixture repeatedly to obtain a fine powder. The mixture was dried at 373 K for 30 min. Then mixture was melted in a silica crucible in an electrically heated furnace under ordinary atmospheric conditions at a temperature of about 1100 K for 3 h to homogenize the melt. The melted mixture was poured on mercury leveled 2×1 cm^2 silica substrate to form bulk glass. For electrical contact, the two opposite faces of the sample were coated with silver paste. The sample was kept at 400 K for drying the electrical contacts. For dc characteristics measurement, the sample was mounted on a sample holder, the temperature of which can be varied over the range 303-473 K. The X-ray diffraction pattern of sample was recorded on Rigaku X-ray diffractometer using Cu Kα radiation (λ = 1.54 Å). The diffractogram was in terms of 2θ in the range 10°–90°.

3. Results and Discussion

3.1. X-Ray diffraction (XRD) analysis

Fig. 1 shows the X-ray patterns of V_2O_5-P_2O_5 glass ceramic. The XRD spectra of sample was recorded between 2θ= 10-90°. The X-ray pattern of sample shows noisy spectra and the peak was observed at the position 25° which is attributed to the amorphous halos nature of glass ceramics.

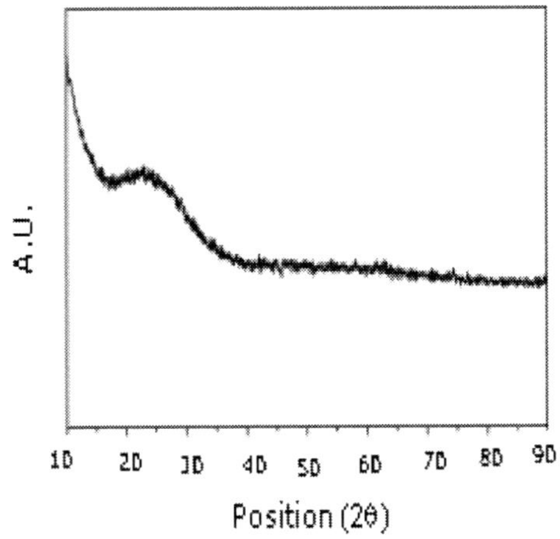

Figure 1. XRD spectra of 0.2 mol% P_2O_5 glass ceramic.

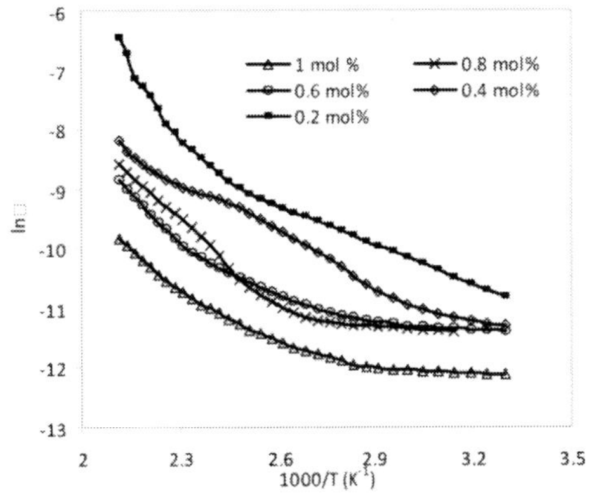

Figure 2. Temperature dependence of dc conductivity (σ) for different mole % of P_2O_5

3.2. DC conductivity

The DC electrical conductivity of the sample was measured in temperature range of 303–345K and the conductivity curve is shown in Fig. 2. The conductivity data may be fitted to straight lines, where conductivity can be analyzed by an Arrhenius equation of the form

$$\sigma = \sigma_0 \exp(-\Delta E/kT) \tag{1}$$

where, σ_0 is the pre-exponential factor and ΔE is the activation energy for conduction. The activation energy and the pre-exponential factor were determined for each conduction region as shown.

In many cases, the electrical conduction in glasses has been proved to be electronic in nature. The conduction process is believed to occur by electron hopping between the ions existing in different valence states in the glass. Fig. 2 clearly gives the linear relationship between ln (σ) and 1000/T. The slope of the curves, which gives the activation energy for conduction, however, has two different values, and increases towards higher temperatures. According to Austin and Mott [7], the electrical conductivity and temperature for the glass is related by Arrhenius equation. In Fig. 2, the dc conductivity varied linearly against 1000/T at the two temperature regions, indicating that the conduction of the present system is mainly electronic [8]. The higher value of conductivity at room temperature (303 K) was found to be 2.53×10-5 S cm-1 for 0.2 mole % P_2O_5 glass. This is the lower value of activation energy as compare to others. This shows the agreement with theoretical aspects. The conductivity of all the samples increases with the temperature. The values of activation energies are calculated at two temperature regions. The region I and region II are assign for the low and high temperature. The activation energy for 0.2 mole % P_2O_5 for region I was found to be 0.38 eV and that for region II, 0.19 eV.

These two activation energies are associated with the intramolecular and the intermolecular conductivity process. Particularly, the lower values of ΔE are associated with the intermolecular conduction process, while the higher values are related to the intramolecular conduction process. In this sample, there are two stages in the movement of carrier motion within the sample, which are the intramolecular and intermolecular transfer of the current carrier. In the intramolecular transfer of electrons, electrons can hop from one atomic site to another if orbitals exist at these sites with the same energy levels. In the case of intermolecular orbital overlap, electrons or holes can travel from one kind of macromolecule to another [9]. The electrical conduction is control by strong electron phonon interaction and result in the formation of small polarons between V^{4+} to V^{5+} ions.

4. Conclusions

In binary glassy system V_2O_5-P_2O_5, the value of conductivity at room temperature was found to be higher for 0.2 mole % P_2O_5. The activation energy value was observed to be lower as compare to others. The XRD study revels that the amorphous nature of 0.2 mole % P_2O_5 glass. The region I and region II are assign for the low and high temperature. The activation energy for 0.2 mole % P_2O_5 for region I was found to be 0.38 eV and that for region II , 0.19 eV.

Acknowledgement

Authors are thankful to Head, Department of Physics Sant Gadge Baba Amravati University, Amravati for providing necessary facilities

References

[1] Nascimento M L F, Nascimento E, Pontuschka W M, Matsuoka M and Watanabe S 2006 *Ceramica* **52** 22
[2] Sreenivasu D and Chandramouli V 2000 *Bull. Mater. Sci.* **23** 281
[3] Shukla D K and Mollah S 2007 *Indian journal of pure and applied physics* **45** 52
[4] Al-Hajry A, Soliman A A and El-Desoky M M 2005 *Thermochimica Acta* **427** 181
[5] Simockov J, Miklos P and Saly V 2000 *Acta physica slovaca* **50** 685
[6] Al-Shukri A M, Khattak G D and Salim M A 2000 *J. Of Materials Science* **35** 123
[7] Al-Hajry A, Soliman A A and El-Desoky1 M M 2005 *Thermochimica Acta* **427** 181
[8] El-Desoky M M and Kashif I 2002 *Phys. Stat. Sol. A* **194** 89
[9] Yakuphanoglua F , Aydogdua Y, Schatzschneiderb U and Rentschlerb E 2003 *Physica B* **334** 443

EXAFS and XRD studies of copper and cobalt complexes of amino acid

A Mishra[1], A Yadav[2], S Ninama[1] and A Trivedi[1]
[1]School of physics, Devi Ahilya University, Khandwa Road, Indore 452001
[2]Govt P.G. College, shajapur

Email: amishra1960@yahoo.co.in, mrsanitayadav@rediffmail.com

Abstract .The present paper deals with the synthesis of transition metal schiff base complexes of Para amino benzoic acid (PABA) by chemical root method. The synthesized metal complexes were characterized by XRD analysis, Extended X-ray absorption fine structure (EXAFS) is a technique that has been used for determining the metal ligand bond length using conventional X-ray source and also by EXAFS analysis using IFEFFIT programming. Bond lengths determined from these data analysis methods are compared with the bond lengths obtained from LSS, Levy's and Lytle's methods. All the complexes show the crystalline simple cubic structure. Bond lengths of the complexes are in good agreement obtained by theoretical and experimental methods.

1. Introduction

The transition metal Schiff base complexes of amino acid have been playing an important role in the development of coordination chemistry. Schiff base metal complexes have been widely studied, because of their antifungal and biological applications. X-ray diffraction studies provide information about the shape of the unit cell and the size of the particles [1]. The structure extending from 40 to 1000eV from the edge is called extended X-ray absorption fine structure (EXAFS) [2, 3] and gives information about the local structure of the complexes. The aim of the present investigation is to measure the X-ray K-absorption spectra and X-ray diffraction studies of copper and cobalt complex of Para amino benzoic acid (PABA), and to compare first shell inter atomic distance using the programmed called IFEFFIT.

2. Preparation of Schiff base ligand

In order to prepare these complexes, we have taken an an ethanolic solution of the prepared benzoin (0.01 mol) and amino acids (0.01mol) in a 500cm3 round bottom flask, 3.0gm of anhydrous sodium acetate was added. The mixture was refluxed for an hour on steam bath. To precipitate the product, the content was poured into an ice-cold water which was separated and recrystallized from rectified spirit, suction filtered, washed with water, and dried in a vaccum dessicator .

2.1 Synthesis of Complexes

Ethanolic solutions of $CuCl_2$ (0.025mol) and Schiff base (0.05mol) were mixed and the resulting mixture followed by drop wise addition of ammonia until the chelate separated, which were then suction filtered, washed with ethanol and ether before dried in vaccum dessicator. The crystals were recrystallized from rectified spirit and dried. Then the complex was in powered form and was ready for our spectroscopic studies.

3. Experimental Technique

In the present study, X-ray diffraction studies were carried out using Bruker D8 Advace `X-ray diffractometer. The X-ray were produced using a sealed tube and the wavelength of X-ray was 0.154nm (Cu K-alpha). The X-ray were detected using a fast counting detector based on silicon strip technology (Bruker LynxEye detector).

X-ray absorption fine structure studies were carried out using a conventional Siefert sealed X-ray tube with Tungsten target operating 20 kV and 40 mA. After this process, the scanning of the X-ray films was completed on Carl-Ziess microdensitometer coupled with computer to convert the data into IFEFFIT analysis.

4. Results and discussion

4.1. XRD analysis

The crystal structure and lattice parameter were analyzed by Bragg's law , $2d\sin\theta = n\lambda$ and particle size is calculated by Debye Scherer's formula, $t = 0.9\lambda/B\cos\theta$.. The XRD pattern of the complexes is reported in figure 1. The particle size and lattice parameter are shown in table 1 . The X-ray analysis shows the complexes are crystalline in nature [4].

Table 1 Particle size and lattice parameter by XRD.

Complex	Particle size(nm)	Lattice parameter(Å)
Cu(PABA)	9.44	5.67
Co(PABA)	20.04	6.55

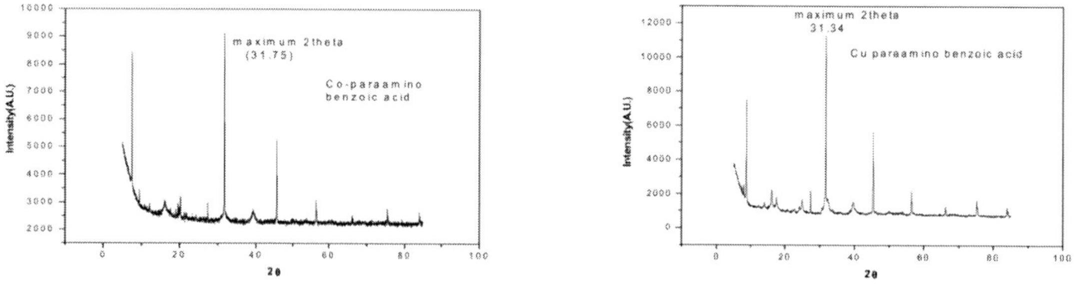

Figure 1 : XRD pattern of copper and cobalt complexes

4.2. EXAFS analysis

The bond lengths of copper and cobalt complexes were calculated using IFEFFIT method and compared with LSS, Levy's, lytle's methods [5,6,7].

According to LSS method, we have determined the bond length using the slope of n vs k plot. The phase parameter $\alpha_1$and β_1, the metal ligand bond length R_1 have been estimated with expression $(1/2+n)\pi=2k(R_1-\alpha_1)+2\beta_1-\pi$. The EXAFS curves are shown in figure 4.

Using the Levy's method, the bond length has been obtained from the expression $r=(151/\Delta E)^{1/2}$ where, ΔE is the energy difference between first maxima and first minim [8].

Lytle method is modification of LSS method and the bond length is given by the expression $R=(37.60/M)^{1/2}$ where M is slope between E vs Q plots.

In FEFFIT programming the EXAFS data were analyzed by the standard procedure. The pre-edge background absorption was subtracted to yield the atomic absorption spectrum of the atom of interest background above the edge was removed from the spectrum using spline fit. The data converted to energy space and then to k space (figure 3). A Fourier transform is applied to these data to convert into the r-space is shown in figure 2. The bond lengths of copper and cobalt complexes calculated by IFEFFIT programming is compared with LSS, Lytle and Levy's methods as depicted in table 2.

Table 2: Average value of metal ligand bond length in Åfor Cu and Co complexes

Name of complex	R_{LSS}	R_{Lytle}	$R_{Levy's}$	$-\beta$	IFEFFIT
Cu(PABA)	2.46	1.78	1.97	2.16	1.2
Co(PABA)	2.10	1.61	1.85	2.06	2.06

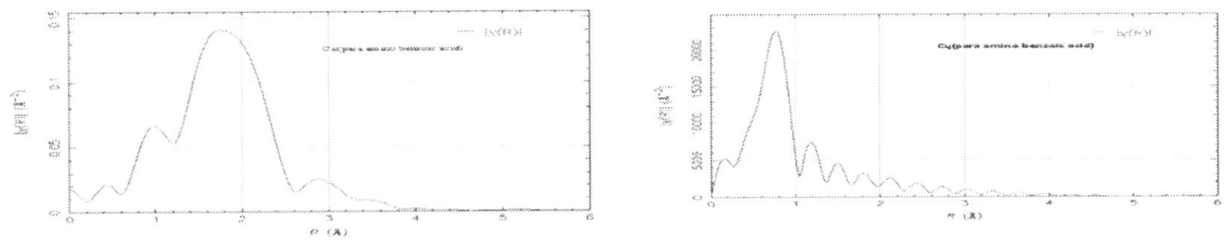

Figure2: Magnitude of Fourier transform of $k2\chi(k)$ for copper and cobalt complexes obtained from FEFFIT program

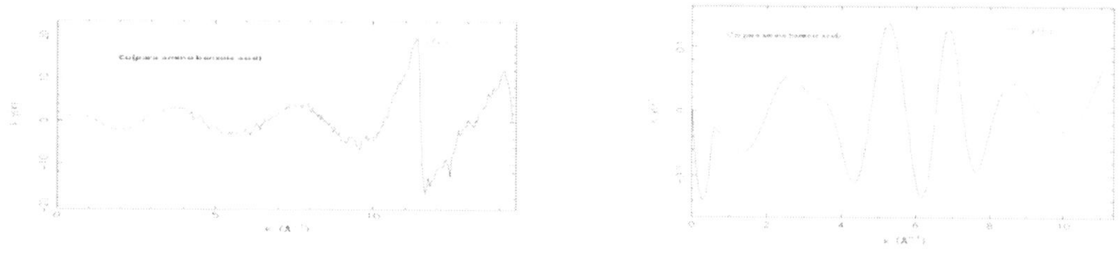

Figure3: EXAFS function for copper and cobalt complexes obtained from FEFFIT program.

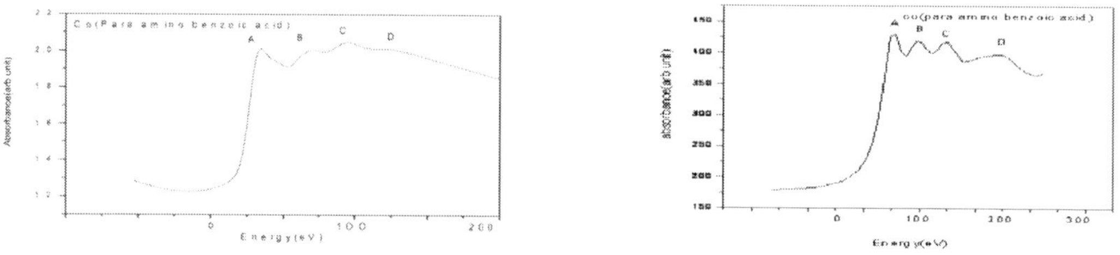

Figure 4 X-ray absorption discontinuity for copper and cobalt complex

5. Conclusion

As is evident from the analysis of X-ray diffraction pattern, one can conclude that the samples exhibit crystalline nature and shape of unit cell is simple cubic. Results indicate that the bond lengths determined by Fourier transforming the FEFFIT programming are comparable with the bond length obtained by LSS, Levy's and Lytle methods. The theoretical and experimental values agree well with each other. This means that the parameterized theoretical calculation of the EXAFS spectra of copper and cobalt complexes described here is in good agreement with physical reality.

6. References

[1]. Cullity B D, Element of X-ray Diffraction, *Second Edition*, 1978.
[2]. Kekre P A, Joshi K P and Mishra A, 1996 *Indian Journal of Pure and applied Physics* **34**306
[3]. Konningsberger D C, *R. Prins (Eds.)*, "X-ray Absorption Principle, Technique of EXAFS, SEXAFS, XANES" *Wiley, New York*, 1998.
[4]. Newville M 2001 *J Synchrotron Rad*.**8** 322
[5]. Stern E A, Sayers D E and Lytle F W 1975 *Physical Review* B **11** 4836
[6]. Stern E A, 1974 Physical Review B **10** 3027
[7]. Stern E A, Sayers D E and Lytle F W 1975 *Physical Review* B **11**, 4825
[8]. Levy R M 1965 *J. Chem. Phys*. **43,** 1846

Journal of Physics: Conference Series **365** (2012) 012021 doi:10.1088/1742-6596/365/1/012021

Determination of bond lengths from EXAFS spectra using different methods.

Ashutosh Mishra[1], **Samrath Ninama**[1]
[1]School of Physics, Vigyan Bhavan, Devi Ahilya University, Khandwa Road Campus, Indore [MP] - 452001, India

E-mail: mishraa1960@yahoo.in, samrathn@gmail.com.

Abstract. The EXAFS spectra have been recorded at the K-absorption edge of copper acetate (Cu (CH3COO)2 on a bent crystal 0.4m Cauchois type transmission spectrograph employing a mica crystal oriented to reflect from (100) planes for dispersion. EXAFS maxima and minima positions have been obtained. The bond length have been determined using four different methods, i.e., Levy's, Lytle , Lytle, Sayers and Stern's (LSS) and Fourier transformation methods. Both LSS and Fourier transformation methods give phase uncorrected bond lengths and hence results obtained from them have been compared with each other.

1. Introduction

The extended X-ray Absorption Fine Structure (EXAFS) in the X-ray absorption spectrum is the normalized oscillatory part of the absorption coefficient above the absorption edge to approximately 1000 eV or higher energy. These EXAFS oscillations are of interest as they contain structural and chemical information specific to the scattering atomic shells , including the number of coordinating atoms and their distance from the absorber. The Fourier transform of an EXAFS spectrum provides information on the distribution of atomic shells as a function of distance from the target absorber. The distances found in Fourier transform are, however, shorter by 0.2-0.5 Å than the actual distances due to energy dependence of the phase factors in sine function of the EXAFS equation. These distances are called phase uncorrected bond lengths. The bond lengths can also be determined by three other methods, namely, Levy's, Lytle's, and Lytle, Sayers and Stern's (LSS) methods. These methods are semi empirical and graphical methods. The LSS method also gives phase uncorrected bond length of the first shell. Hence, it is worth comparing the bond lengths obtained from LSS method with those obtained from Fourier transformation method. To make such a comparison we have recorded the EXAFS spectra at the K-edge of copper in copper acetate. The results of the study have been reported in this paper.

2. Experimental

X-ray absorption spectra were recorded using a Sheifert X-ray generator having tungsten target sealed X-ray tube operating at 20 kV and 40 mA and a bent crystal 0.4m Cauchois type transmission spectrograph employing a mica crystal oriented to reflect from (100) planes for dispersion The spectra was recorded on the Kodak X-ray films. After obtaining a number of good spectrograms, the micro-photometer records were obtained on a Carl Zeiss GIII micro-photometer. The microphotometry has been done by using the program DEPICTS. The micro-photometer was set for the speed 10 sec/cm and magnification 10X. The digital spectral data was thus obtained on a computer. The computer software Athena version 0.8.056 was used for data analysis.

3. Results and Discussion

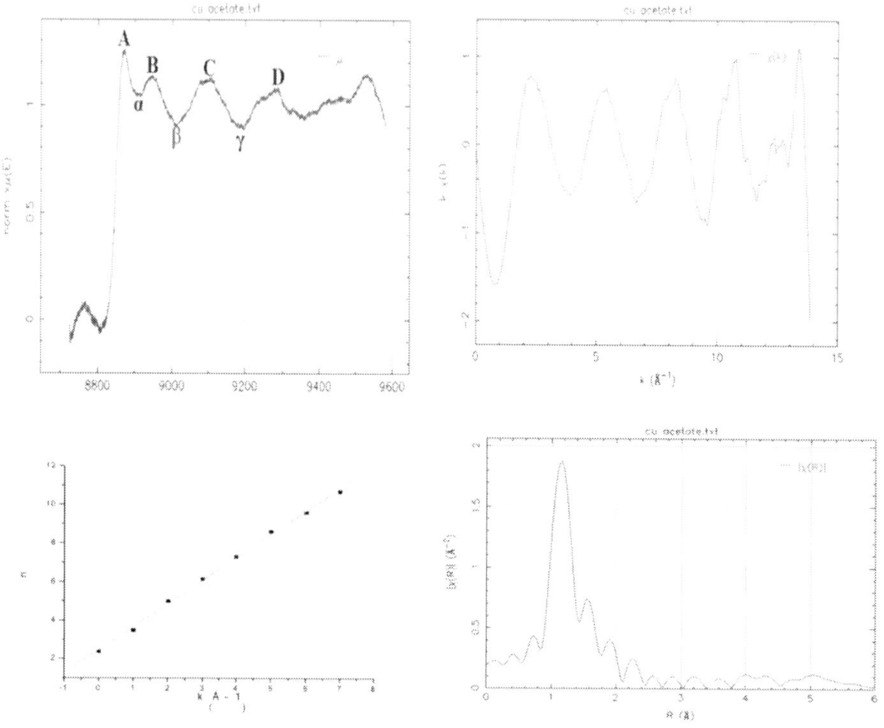

Figure.1 (a): The normalized XAFS spectrum (b) χ (k) versus k curve (c) n versus k curve (d) Magnitude of Fourier transform of χ (k) versus k curve and for copper acetate.

The X-ray absorption spectra, i.e., normalized μ(E) vs. E curve of copper acetate (Cu $(CH_3COO)_2$ is shown in fig.1(a). Fig.1 (b) shows χ (k) vs. k curve obtained from fig. 1(a). In Table 1, it gives the data for EXAFS maxima and minima, which are marked as A, B, C,..... and α, β, γ,........ respectively.

Table 1 Energy E (eV) and wave vector k (Å-1) for EXAFS maxima and minima at the K-absorption edge of copper acetate and their corresponding values of n and energy level Q.

Structure	**A**	**α**	**B**	**β**	**C**	**γ**	**D**
n	0	1	2	3	4	5	6
Q	2.04		6.04		12		20
E	21.90	46.58	95.05	143.81	202.62	281.21	350.4
k	2.4	3.5	5	7.3	8.6	9.6	10.7

Table 2 Values of first shell bond lengths (in Å) calculated from Levy's, Lytle's, L.S.S. and Fourier transform methods for copper acetate.

S. No.	Cu salt	Phase corrected		Phase uncorrected	
		Levy's method R_1	Lytle's method R_s	L.S.S. method R_1-α_1	F.T. method R
1.	Cu acetate	1.76	1.43	1.86	1.16

The bond lengths have been determined using the following four methods, i.e., Levy's, Lytle, Lytle, Sayers and Stern's (LSS) and Fourier transformation methods and the results obtained are given in table2.

3.1 Levy's method
In Levy's method [1], the bond lengths are given by R1 = [151/ΔE]1/2 Å, where ΔE is the difference in eV of the energies of the EXAFS maximum B and minimum β and R1 is the radius of the first coordination sphere. The positions of B and β are given in table 1. The bond length thus determined is given in table 2.

3.2 Lytle's method
The energy values (E) of the EXAFS maxima, given in table 1, are plotted against the Q values given by Lytle [2] for p symmetry, i.e., Q = 2.04, 6.0, 12.0, and 20.0. The (E, Q) plots have been found to be linear .The slope M of the E versus Q plot have been used to evaluate the radius Rs of equivalent polyhedron through the relation $Rs = [37.60 / M]^{1/2}$. The values of Rs calculated with the help of this method are reported in table 2.

3.3 Lytle, Sayers and Stern's (LSS) method
The values of the wave vector k (Å-1) for EXAFS maxima (n = 0, 2, 4....) and minima (n =1, 3, 5....), are presented in table 1. In the LSS method [3], for determination of the nearest neighbor distances, n versus k graph is plotted. The plot has been found to be linear and is shown in fig. 1(c). The slope of n versus k plot, gives the value of $2(R_1 - \alpha_1) / \pi$ where R_1 is the bond length. The value of $(R_1 - \alpha_1)$ thus obtained is given in table 2. $(R_1 - \alpha_{un})$ is the phase uncorrected bond length.

3.4 Fourier transformation method
The Fourier transform of the χ(k) versus k spectra is given in fig.1(d). The Fourier transform peaks at the radial distances of the neighboring atoms from the absorbing atom. The distances found in Fourier transform are, however, shorter by 0.2-0.5 Å than the actual distances due to energy dependence of the phase factors in sine function of the theoretical expression for EXAFS, which is known as the EXAFS equation [4]. The peaks in the Fourier transform are shifted towards the origin by an amount αj and hence the peaks are at distances Rj-αj. For the first peak j=1 and hence the position of the first peak determines the distance R1-α1. This distance is given in table 2.

As pointed out above, both the LSS method and the Fourier transformation method give the value R1-α1, i.e., both the methods give the value of bond lengths which have not been corrected for the phase shifts. This distance is called the phase uncorrected bond length. It can be seen from table 2 that the phase uncorrected bond lengths obtained from these two methods agree with each other within the limits of experimental error which is ± 0.05 Å.

References

[1] Levy R M, 1965 *J. Chem. Phys.* **43** 1846
[2] Lytle F W, 1966 *Advances in X-ray Analysis* **9** 398
[3] Lytle F W, Sayers D E and Stern E A 1975 *Phys. Rev. B* **11** 4825
[4] Stern E A, Sayers D E and Lytle F W 1975 *Phys. Rev. B*, **11**, 4836

Effect of Cu doping on LPG sensing properties of soft chemically grown nano-structured ZnO thin film

Arindam Ghosh[1], **Anil Ghule**[2], **Ramphal Sharma**[1,*]
[1]Thin-film and Nanotechnology laboratory,
[2]Department of Nanotechnology,
Department of Physics, Dr. Babasaheb Ambedkar Marathwada University, Aurangabad, India

E-mail:rps.phy@gmail.com

Abstract. Zinc oxide (ZnO) thin film was grown by soft chemical route. Structural studies of ZnO thin film reveal hexagonal structure. The c/a ratio increase as the concentration of Cu (dopant) enhanced. Elemental analysis reveals the presence of Cu ions into ZnO matrix, which substitute at Zn site into the nano-structured ZnO crystal as it was presented in ball and stick model. Successful tailoring of band gap by doping was found in optical study and presented in band diagram model. The gas response was found to be 1.15 for 200ppm of gas concentration. The response and recovery times were found to be ~ 10 and 8 sec respectively.

1. Introduction

Amongst various II-VI compounds, ZnO is widely used metal oxide semiconductor for fabricating various types of gas sensors due to their versatile properties like wide energy band gap (3.3 eV), strong binding energy of 60 meV, higher transparency and low electron affinity. Due to these properties, ZnO is one of the promising material for many optoelectronic applications, such as ultraviolet lasers, light emitting diodes, transparent conductors, thin film transistor [1]etc. Doping is a widely used method for improving the opto-electrical properties of various semi conducting materials. Various materials like Al, Ga etc. on ZnO have been well studied for various types of applications. In general, the role of Cu as a luminescence activator and as a compensator of n-type materials is of considerable significance for II–VI compound semiconductor. Cu doped ZnO thin films has been used as phosphor and active material varistor. There are several reports available on the effect of Cu doping in ZnO and the variation in photoluminescent and magnetic properties. The effect of LPG sensing properties of spray deposited ZnO thin film at the operating temperature of 400 °C [2]. Mitra et al[3] have reduced the operating temperature upto 250°C after Pd doping. Further more, to reduce the operating temperature and to increase the response Cu was doped into the ZnO crystal. Generally, Cu doped ZnO is a widely used for CO sensor [4]. Cu doped ZnO thin-film is rarely used as liquid petroleum gas (LPG) sensor. With these motivations, in the present study we have doped Cu into ZnO thin film by SILAR technique for (LPG) sensor which enhances its importance for device grade application point of view. Also, we are introducing the energy band diagram to explain the role of Cu in the tailoring of band gap and a ball and stick model to represent how Cu doping controls the LPG sensing.

[*] To whom any correspondence should be addressed.

2. Experimental details

The detailed description of the SILAR technique was mentioned in our earlier report [5]. In this paper we extend our results on physical modifications in ZnO thin films after Cu doping. For doping of Cu into the ZnO thin films, $CuCl_2$ was added into the solution at the rate of 2.5 % and 5 % respectively. Then the solution was stirred at constant speed at room temperature for about two hours for getting good quality thin films. The deposition was carried out by successive immersion and rinsing in the precursors and the double distilled water respectively for about 20 sec and 40 sec respectively. Cu doped ZnO thin films (~450 nm.) were annealed in air for about 2 hours.

The identifications of the crystal structure was carried out by CuKα (1.5405 Å) radiation using Burker D8 Discover in the scanning range of 200 to 600. The optical characterization of the films was done in the range 350–800 nm with the help of UV–Vis spectrophotometer lambda 25. The elemental analysis was done by EDAX an instrument which is in attachment with the SEM instrument. The x-ray photo electron spectroscopy study was performed in Indus 1, RRC CAT. The gas response was studied using an indigenous homemade gas sensor unit. The change in current before (air atmosphere) and after the insertation the test gas was measured using a digital picoammeter (model: DPM-111).

3. Result and discussion

Figure 1 shows the X-Ray diffraction (XRD) (Bruker D8 Discover) pattern of the pristine and Cu doped. The diffraction peaks observed at 2θ = 32.04o, 34.43o, 36.15o, 47.84o and 56.91o which attributes the (100), (002), (101), (102) and (110) planes for both (pristine and Cu doped) the cases respectively. Significantly, it is noted that the peak intensity increased with the increase in doping concentration. The peak (*) corresponds to the formation of copper oxide (JCPDS card no: 80-1917). As the doping concentration reaches 5% the CuO phase was found clearly. This attributes the thermal solubility limit of Cu in ZnO is below at lower concentration of Cu doping [6]. The comparative study of XRD pattern and the JCPDS card (75-0576) reveals the hexagonal structure (the pristine samples).

The calculated values of the lattice constant, bond length and the crystallite size calculated with the help of Sherrer formula, summarized in the table 1. The lattice constansts a and c were calculated to be 3.25 Å and 5.20 Å, respectively refined by the least square method (c/a = 1.59) which is in well agreement with the reported value (c/a = 1.60). From table 1 it was found that as the doping concentration increased, the crystallite size increased while the anionic displacement decreased

Figure 2. shows the results of the optical absorption spectra for both the cases. From this figure the energy band gap (Eg) was found to be 3.22 eV for the pristine sample which is well matched with the values reported earlier [5]. But as the doping concentration increased, the Eg goes on decreasing. The reduction in band gap, with the increase in doping concentration, suggests that the Cu ions substitute

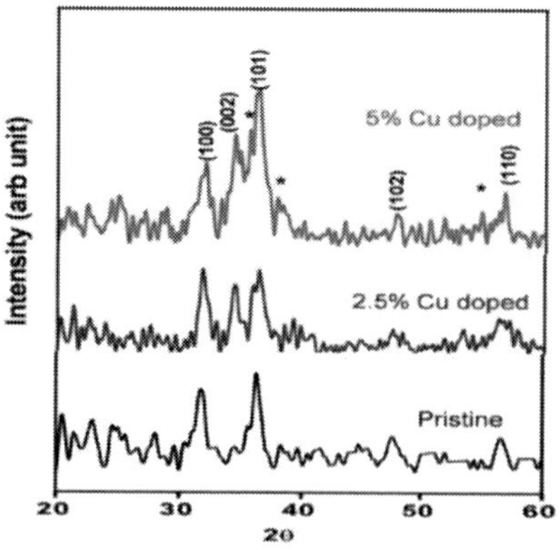

Figure 1. XRD pattern of pristine and Cu doped ZnO thin film.

Table 1. Structural parameters of Cu doped ZnO thin films

Sample type	Lattice cons. a	c	Crystallite size
Pristine	3.2587	5.2016	10.35
2.5% Cu doped	3.2312	5.1825	12.48
5% Cu doped	3.2603	5.2636	14.65

Zn in the ZnO [6].Inset of fig 2 shows the band diagram model for Cu doped ZnO film. The 3d state of Cu^{2+} is located at 0.1 to 0.19 eV below the conduction band [7]. The acceptor level (Cu^{+}, h) is situated at about 0.45 eV above the valance band [7a] is formed after the excitation of Cu^{2+} due to charge transfer. Zinc interstitial (Znoi) lie at 0.05 eV below the conduction band[8]. Considering these above mentioned facts, the change in band gap confirms the proper substitution of Cu in this case.

Furthermore, the compositional verification was studied by using EDAX and XPS analysis. Figure 3 a confirms the proper deposition of Cu in ZnO thin film. Other elements such as Si, Na etc. which are not (don't mentioned here in the EDAX spectra) expected, may appear due to the glass substrate [9]. As the doping concentration enhanced, the peak position has shifted from the earlier reported peak positions. There was a shift in peak position was observed. These shifts in binding energy can be explained in terms of electronegativity (χ)[10]. Figure 3.b shows the presence of a strong peak at BE 932.1, which signifies the peak of Cu 2p3/2 [11].

In case of Cu doped ZnO the valancy of Cu could be assumed to be +1 or +2 [12].The radius of Zn^{2+},Cu^{2+} and Cu^{+} were found to be 0.74 nm, 0.72 nm and 0.96 nm respectively[13]. Considering the above mentioned facts we can conclude Cu^{2+} might replace the Zn as shown in the ball and stick model (figure 3.c) of Cu doped ZnO crystal. This increase in crystallite size (confirms from the structural analysis) results in a lower density of grain boundaries which may behave as trap for free carriers and barrier for carrier transport which may attribute to the enhancement in conductivity [14]. So, in this case, the Cu may work like a bridge which may cause the improvement in gas response than the pristine sample.

Response is defined as the ratio of the current in gas atmosphere versus the current in air atmosphere. Figure 3 d shows the plot of response versus operating temperature of Cu doped ZnO thin film at 200 ppm of LPG gas concentration. This plot shows the increasing trend of gas response as a function of operating temperature. At 200°C it shows the best response then after it decreases. At first the Cu doped ZnO samples were kept into an airtight gas chamber and starts heating. As the temperature increases, the surface of the test sample absorbed the oxygen. Further more, as the temperature increases ($\geq$ operating temperature) the oxygen molecule extract the electrons from the conduction band and the O^2 converted into O^{2-}. After attaining the operating temperature O^{2-} converted into O^-. The thermal energy of the gas molecule at lower temperature is insufficient to react with the O^{2-} spices rendering the possibility of chemisorptions to be negligible [15].

The response and recovery time of a sensor can be defined as the time required for a film resistance to reach its 90% of its saturation value from the starting value of the gas exposure. Figure 6.b shows the response curve of the Cu doped ZnO thin film in the presence of 200 ppm of gas concentration. There was a sudden increase was found (upto 1.14). The response and recovery times for the Cu doped ZnO thin films were found to be 10 sec and 8 sec respectively.

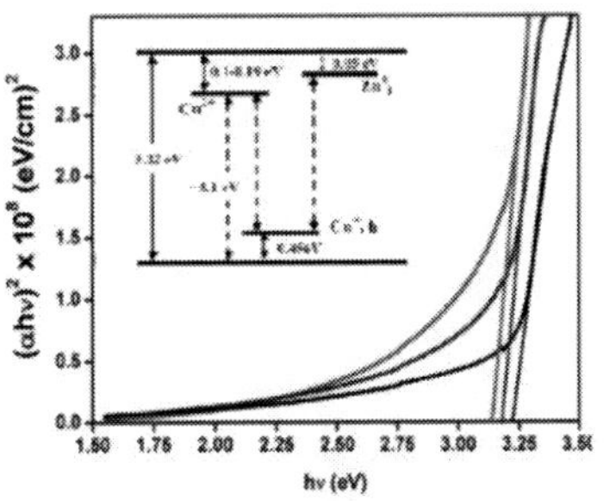

Figure 2. Optical studies of pristine and Cu doped thin film

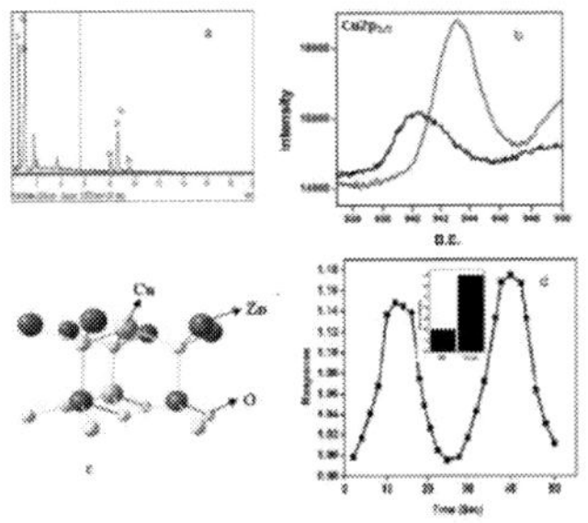

Figure 3.(a) EDAX and XPS spectra (c) Ball-stick model (d) gas sensing properties of Cu doped ZnO thin film

The inset of figure shows the comparative study of the gas response of ZnO and Cu doped thin film. This figure clearly depicts the enhancement in gas response after doping than the previously reported ZnO thin film[15].

4. Conclusion

Cu doped ZnO films with hexagonal phase were successfully deposited by SILAR technique at room temperature onto the glass substrates. The presence of CuO phase confirms the proper doping. The band diagram shows how the band gap was decreased from 3.22 eV to 3.13 eV with the increase in doping concentration which is well agreement with the previously reported values. The EDAX spectra and XPS study confirms the presence of Cu in the samples. The ball and stick model shows possible Cu substitution of into ZnO crystal. With these observations we can finally conclude that we have deposited the good quality Cu doped ZnO thin films at room temperature by cost effective chemical technique named SILAR.

5. Acknowledgement

The authors are thankful to We are also thankful to Dr. D. M. Phase, Dr. V.R.Reddy and Dr. V. Ganesan, UGC–DAE Consortium for Scientific Research, Indore for helpful discussions. In addition to this, we would like to thank to the Head of the Department, Department of Physics, Dr. B. A. M. University for providing the laboratory facility.

References

[1] Carcia P F, Melean R S, Reilly M H, Nunes G J, 2003 *Appl. Phys. Lett.* **82** 1117
[2] Shinde V R, Gujar T P, Lokhande C D 2006 *Sens. Actuators B* **120** 551
[3] Mitra P, Maiti H S 2004 *Sens. Actuators B* **97** 49
[4] Gong H, Hu J Q, Wang J H, Ong C H, Zhu F R 2006 *Sens. Actuators, B* **115** 247
[5] Ghosh A, Deshpande N G, Gudage Y G, Joshi R A, Sagade A A, Phase D M, Sharma R 2009 *J. Alloys and Compd.* **469** 56
[6] Lee H J, Kim B S, Cho C R, Jeong S Y *Phys. Stat. Sol (b)* **7** (241)1533
[7] (a) Dingle R 1969 *Phys. Rev. Lett.* **23** 579, (b) Mollwo E, Muller G, Wagner P, 1973 *Solid State Commun.* **13** 1283
[8] Lima S A M, Sigoli F A, Jafelicci M J, Davolos M R 2001 *Int. J. Inorg. Mater.* **3** 749
[9] Bilgin V, Akyuz I, Kose S, Alay F 2006 *Semicond. Sci. Technol* **21** 579
[10] Ahmad M, Zhao J, Iqbal J, Miao W, Xie L, Mo R, Zhu J 2009 *J. Phys. D: Appl. Phys.* **42** 165406
[11] Chen H Y, Lau S P, Chen L, Lin J, Huan C H A, Tan K L, Pan J S 1999 *Appl Surf. Sci.* **152** 193
[12] Wahl U, Rita E, Correia J G, Alves E 2004 *Phys. Rev. B* **69** 012102
[13] Peng X, Xu J, Zang H, Wang B, Wang Z 2008 *J. Lumin.* **128** 297
[14] Zhang H, Liu H, Zhou A, Yuan C 2009 *J. Semicond.* **30** 113002
[15] Ghosh A, Sharma R, Ghule A, Taur V S, Joshi R A, Desalea D J, Gudage Y G, Jadhav K M, Han S H 2010 *Sens. Actuators B* **146** 69.

Growth and characterization of ferrite film prepared by pulsed laser deposition

S. N. Kane[a, 1], M. Satalkar[a], A. Ghosh[a], D. M. Phase[b], R. J. Chaudhary[b], A. Pasko[c], M. LoBue[c], and F. Mazaleyrat[c]
[a]School of Physics, D. A. University, Khandwa road Campus, Indore – 452001, India
[b]UGC-DAE CSR, University Campus, Khandwa Road, Indore – 452 001, India
[c]SATIE, UMR CNRS 8029, ENS de Cachan, 61, Avenue du President Wilson, 94235 Cachan, France

E-mail: kane_sn@yahoo.com (S. N. Kane)

Abstract. $Ba^{2+}Fe_{12}O_{19}$ ($BaO.6Fe_2O_3$) films have been grown on (0001) sapphire substrate using pulsed laser deposition (PLD) technique. Prepared films were characterized by X-ray diffraction (XRD), X-ray photoelectron spectroscopy (XPS), X-ray reflectivity (XRR), atomic force microscopy (AFM) and magnetic measurements. XRD confirms the formation of $Ba^{2+}Fe_{12}O_{19}$ phase, whereas presence of Fe_2O_3 was also identified. X-ray photoelectron spectroscopy (XPS) shows the signature of all the elements present in the barium ferrite. XRR data show the thickness of the film is ~ 24.2 nm with top layer roughness of ~ 3.7 nm. AFM measurement suggests the formation of clusters with average roughness of ~ 3.9 nm. Magnetic measurements of the target reveal its anisotropic nature with coercivity values: 0.23 and 0.24 T, obtained along two orthogonal directions. For the prepared thin film, the obtained coercivity ~80 mT and saturation induction 0.13 T are lower than the usual values obtained for $Ba^{2+}Fe_{12}O_{19}$ compound (coercivity and saturation induction values are 0.19 Tesla and 0.48 T respectively), can be ascribed to the presence of Fe_2O_3 phase.

1. Introduction

Prominent feature of M-type hexaferrites is their chemical inertness, large uniaxial anisotropy inherent in the hexagonal crystal structure where the crystallographic c-axis is the magnetic easy axis, and high coercivity [1-2]. Hexaferrites have been the subject of intensive studies due to an appealing combination of good magnetic properties and low cost. This large family of oxides with hexagonal crystal structure contains ferrimagnetic compounds with easy axis of magnetization (*e.g.* M-type ferrites) and easy plane of magnetization (*e.g.* Y-type ferrites). Hence, hexaferrites have been widely adopted in two distinct fields: permanent magnets and microwave technology components [3]. In exchange spring media, two magnetically hard layers and a magnetically softer layer are strongly exchange coupled. The technological realization of such materials shows that they should be crystallographically coherent and consequently they are magnetically exchange coupled [4]. Objective of the present work is to obtain proper preparation conditions for growing $Ba^{2+}Fe_{12}O_{19}$ films using pulsed laser deposition (PLD), a pre-requisite in realizing exchange-spring magnets, for various applications. Although, we have grown $Ba^{2+}Fe_{12}O_{19}$ phase in our films, but our film also shows the

[1] To whom any correspondence should be addressed.

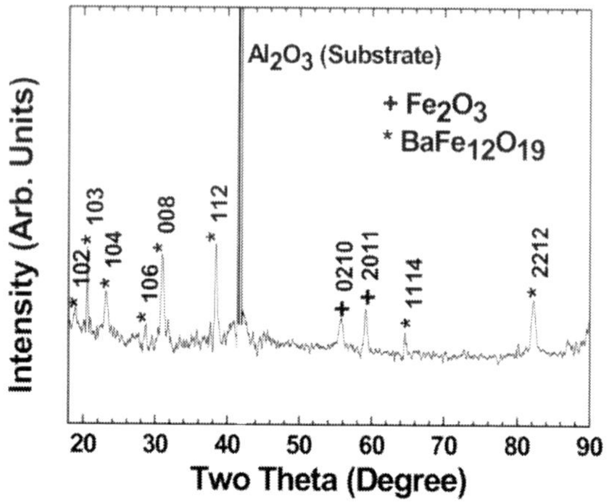

Figure 1. X-ray diffraction pattern of the prepared film.

presence of Fe_2O_3 phase. Further studies are under way, in order to grow single phase $Ba^{2+}Fe_{12}O_{19}$ films.

2. Experimental details

Thin film was deposited using a commercial barium hexaferrite target mounted on a rotating axel, and deposition was done for 30 min on (0001) sapphire (Al_2O_3) substrate. PLD set-up uses a KrF excimer laser (Lambda Physik model COMPEX-201) of wavelength 248 nm and pulse duration of 20 ns. During deposition the laser energy was kept 200 mJ/pulse, with pulse repetition rate of 10 Hz. The substrate temperature and oxygen pressure were maintained respectively at 800 °C and ~ 50 Pa (base pressure ~ 0.001 Pa). Focussed laser beam was made incident on the target surface at an angle of 45°. The substrate was mounted opposite to the target at a distance of 4 cm on a heater plate. After deposition, the sample was cooled slowly to room temperature. During cooling, the oxygen pressure was raised to 101325 Pa. It is of value to note that similar experimental conditions as well as same substrate (sapphire) was used to deposit barium hexaferrite thin films as reported in [5]. X-ray reflectivity and X-ray diffraction (ϑ-2ϑ configuration) measurements were done using a Bruker D8 Advance diffractometer, with Cu-K_α radiation (λ = 0.154 nm). The X-ray photoelectron spectroscopy (XPS) study was carried out using an Omicron EA-125 photoelectron spectrometer at a base pressure better than 6.7×10^{-8} Pa. Al K_α radiation was employed for recording the spectra. Au 4f7/2 at 84.7 eV served as an external reference. The sample surface was cleaned using 1 keV Ar ions before performing measurements to remove surface contamination. Atomic force microscope (AFM) image of sample was taken by Nanoscope-E using a pyramidal silicon nitride tip in contact mode. Magnetization measurements on target and deposited thin film were done using LakeShore Model 7410 VSM with a maximum applied field of ± 1.7 Tesla.

3. Results and discussions

Figure 1 depicts the X-ray diffraction pattern of $Ba^{2+}Fe_{12}O_{19}$ thin film. XRD of the substrate (Al_2O_3) was also done and the contribution of the substrate was subtracted from the diffraction pattern of the film. But, still a strong peak corresponding to substrate is seen in figure 1. Analysis of XRD pattern reveals the formation of $Ba^{2+}Fe_{12}O_{19}$ phase with lattice parameters: $\boldsymbol{a}$ = 0.5936 nm, $\boldsymbol{c}$ = 2.3297 nm, matches well as reported in [3]. XRD pattern also shows the presence of Fe_2O_3 phase (Lattice parameters: $\boldsymbol{a}$ = 0.5572 nm, $\boldsymbol{c}$ = 2.2506 nm), also agrees well with the values given in JCPDS database.

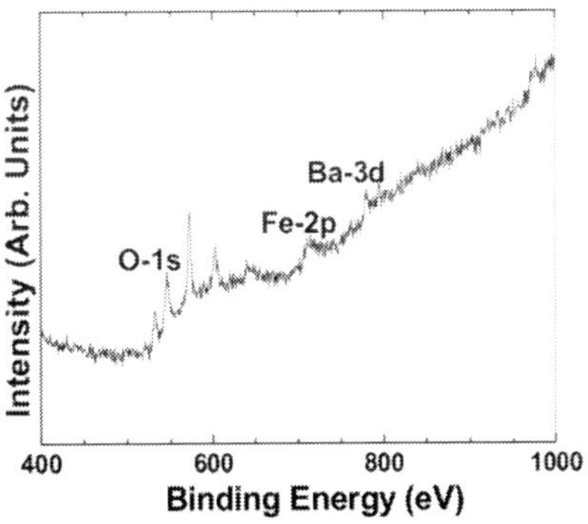

Figure 2. XPS spectrum of the prepared film.

Figure 2 depicts the XPS spectrum corresponding to the pulsed laser deposited thin film prepared from barium ferrite target on sapphire (Al_2O_3) substrate. Perusal of figure 2 shows the signature of all the elements present in barium ferrite.

Figure 3 shows the X-ray reflectivity pattern of the prepared film. Experimentally obtained XRR data were fitted using software based on the formalism developed by Parratt [6]. The thickness value obtained by fitting the X ray reflectivity pattern is ~ 24.2 ± 0.5 nm with top layer roughness ~3.7 nm.

Figure 4 shows the AFM image of the prepared film. AFM image shows cluster formation with average roughness ~ 3.9 nm, matches well with reflectivity measurements. Observed surface roughness (~ 3.9 nm) can be attributed to the higher substrate temperature during deposition.

Figure 5 shows the hysteresis loops of the target in two orthogonal directions, suggests the presence of in-plane anisotropy, with coercivity values 0.23 and 0.24 T respectively.

Figure 6 depicts the hysteresis loops of the prepared thin film. The obtained coercivity ~80 mT and saturation induction 0.13 T are lower than the usual values obtained for $Ba^{2+}Fe_{12}O_{19}$ compound

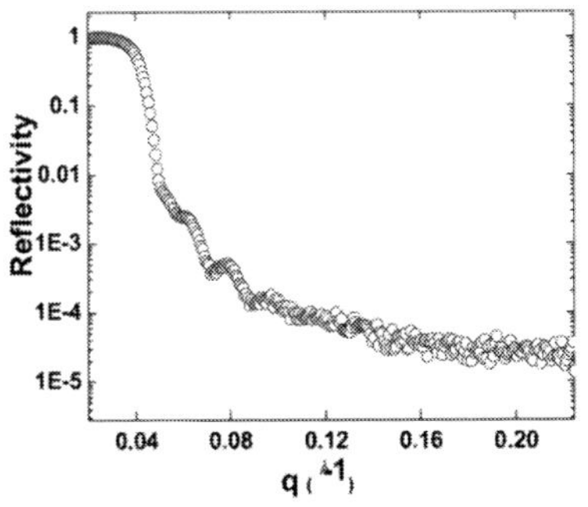

Figure 3. X-ray reflectivity of the prepared film.

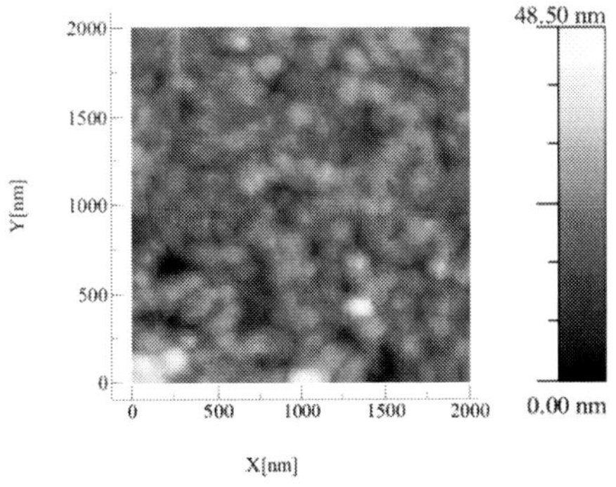

Figure 4. AFM image of the prepared film.

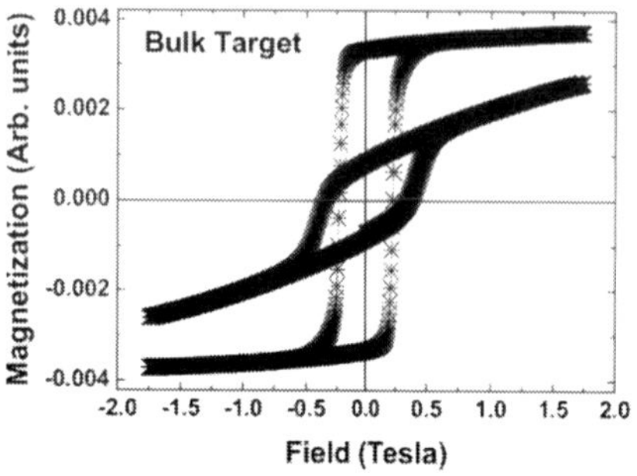

Figure 5. Hysteresis loops of the target in two orthogonal directions, showing anisotropic nature.

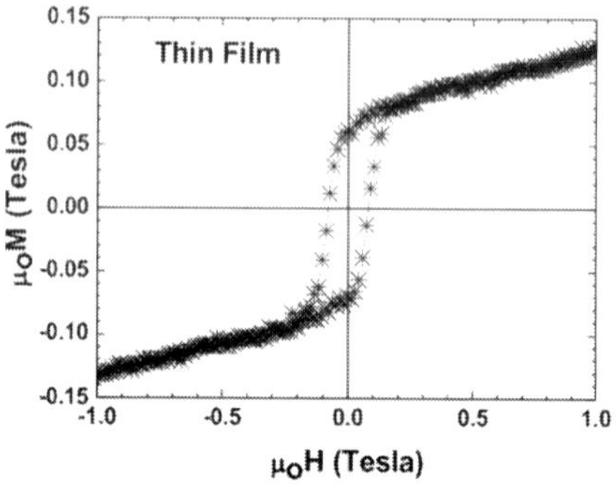

Figure 6. Hysteresis loop of the prepared film.

(coercivity and saturation induction values are 0.19 T and 0.48 T respectively), can be ascribed to the presence of Fe_2O_3 phase. Observed positive slope in the M-H curve shown in figure 6 could be due the contribution from the Al_2O_3 substrate.

4. Summary

Pulsed laser deposition technique was used to grow $Ba^{2+}Fe_{12}O_{19}$ film on (0001) sapphire substrate. Formation of $Ba^{2+}Fe_{12}O_{19}$ phase was confirmed by XRD. Presence of Fe_2O_3 phase in the film was also confirmed by XRD. XPS measurements show signatures of all the elements present in the barium ferrite. XRR data give the thickness of the film to be ~ 24.2 nm with roughness of the layer ~ 3.7 nm. AFM images suggest the formation of clusters with average roughness of ~ 3.9 nm. Hysteresis loops of the target reveal its anisotropic nature with coercivity values 0.23 and 0.24 T. Hysteresis loop of the prepared thin film reveals its coercivity to be ~ 80 mT and saturation induction 0.13 T. These values are lower than the usual values obtained for $Ba^{2+}Fe_{12}O_{19}$ compound (coercivity and saturation induction values are 0.19 T and 0.48 T respectively). This behaviour can be ascribed to the presence of Fe_2O_3 phase, confirmed by XRD measurements.

Acknowledgments

Support in the form of UGC-DAE CSR project No. CSR-I/CSR51/2011-12 is gratefully acknowledged.

References

[1] Heczko O, Gerber R and SĬimsĬa Z 2000 *Thin Solid Films* **358** 206; Su Hui-Chia, Jeng-Yi Dai, Liao Yen-Fa, Wu Yu-Han,. Huang J C A and Lee Chih-Hao 2010 *Thin Solid Films* **518** 7275
[2] Ozgur U, Alivov Y and Morkoc H 2009 *Mater Sci: Mater Electron* **20** 789, Gal H J, Schäfer W, Will G, Hillberg M, Wäppling R 1996 *Phys. Rev. B* **53** 9143
[3] Smit J and Wijn H P J 1959 *Ferrites* (Eindhoven: Philips Technical Library)
[4] Kneller E F, Hawig R 1991 *IEEE Trans. Magn.* **27** 3588
[5] Zhang X Y, Ong C K, Xu S Y and Fang H C 1999 *Appl. Surf. Sci.* **143** 323
[6] Parratt L G 2000 *Phys Rev* **95** 359

Pressure induced structural phase transition of XC (X = Si, Ge, Sn)

N. Kaurav[1*], S. Verma[1], K. K. Choudhary[2], U. Sharma[2] and S. Shah[3]
[1]Department of Physics, Govt. Holkar Science college, A. B. Road, Indore (MP)-452001 India.
[2]Department of Physics, Shri Vaishnav Institute of Technology and Science, Baroli, Sanwer Road, 453331 Indore, India.
[3]Department of Physics, P. M. B. Gujarati Science College, Indore-452001, India

*E-mail: netramkaurav@yahoo.co.ok

Abstract. An effective interionic interaction potential (EIOP) is developed to investigate the pressure induced phase transitions from ZnS-type (B3) to NaCl-type (B1) in XC [X = Si, Ge, and Sn] compounds. The long range Coulomb, van der Waals (vdW) interaction and the short-range repulsive interaction up to second-neighbor ions within the Hafemeister and Flygare approach with modified ionic charge are properly incorporated in the EIOP. The vdW coefficients are computed following the Slater-Kirkwood variational method, as both the ions are polarizable. The estimated value of the phase transition pressure (Pt) and the magnitude of the discontinuity in volume at the transition pressure are consistent as compared to the reported data. The vast volume discontinuity in pressure volume phase diagram identifies the structural phase transition from B3 to B1 structure.

1. Introduction

The pressure induced structural phase transition in binary SiC, GeC, and SnC compounds with the ZnS- type structure have received considerable attention in recent past [1 - 5]. The complexity of these exotic new materials poses a challenge to the physicist to investigate the fundamental properties of these compounds, such as, lattice constant and structural stability. Pressure is believed to be an attractive thermodynamical variable to reveal the mechanical properties of most of the solids and alloys. It has been shown that the pressure induced phase transition of SiC from *B*3 to *B*1 structure occurred at 100 GPa or higher [1]. Considerable theoretical studies of *XC* (*X* = Si, Ge, Sn) compounds have been published during the last few years [2 - 5]. In particular, Zhang *et al.* [2] have performed first-principles calculations to investigate the structural stability of *XC* compounds. The transition pressures at which these compounds undergo the structural phase transition from *B*3 to *B*1 phase are 62, 87, and 33 GPa for *XC* (*X* = Si, Ge, Sn), respectively.

The first principles density functional theory and microscopic tight binding models have been used successfully to address the electronic, magnetic and structural properties of binary compounds. On the hand, phenomenological lattice models [6, 7] have proved very successful in obtaining qualitative and quantitative understanding of binary compounds with proper parameterization. In the present paper, we have chosen two-body interaction potential because it includes vdW attractions, which are not well described by the standard methods currently used in first principle microscopic calculations.

2. Theory and method of computation

The understanding of thermodynamical properties for XC compounds needs the formulation of an effective interionic potential. The idea we have in mind follows: the change in force constants is small, the short range interactions are effective up to the second-neighbour ions, and the atoms are held together with harmonic elastic forces without any internal strains within the crystal. The effective interionic potential between pair of ions (i and j^{th}) is expressed as

$$U(r)=\sum_{ij}\frac{Z_m e^2}{r_{ij}}+\sum_{ij} b\beta_{ij}\exp\left(\frac{r_i+r_j-r_{ij}}{\rho}\right)+\sum_{ij} c_{ij}r_{ij}^{-6}+\sum_{ij} d_{ij}r_{ij}^{-8}. \quad (1)$$

Here, long-range Coulomb energy is represented by first term. Second term corresponds to Hafemeister and Flygare [8] kind of short-range overlap repulsive energy. van der Waals multipole energies are represented by third and fourth terms, respectively. Here, the Pauling coefficients β_{ij} are defined as: $\beta_{ij} = [1+(z_i/n_i)/(z_j/n_j)]$ with z_i (z_j) and n_i (n_j) as the valencey and number of outermost electrons in the anions (cations), respectively. The symbols: c_{ij} and d_{ij} are representing the dipole-dipole (d-d) and dipole-quadrupole (d-q) van der Waals coefficients. $Z_m e$ is the modified ionic charge and parametrically includes the effect of Coulomb screening effects. b (hardness) and ρ (range) are short-range parameters. Thus, the effective interionic potential contains only three free parameters (Z_m, b and ρ) which can be determined from the crystal properties [2].

An isolated phase is stable only when its free energy is minimum for the specified thermodynamic conditions. As the temperature or pressure or any other variable acting on the systems is altered, the free energy changes smoothly and continuously. A phase transition is said to occur when the changes in structural details of the phase are caused by such variations of free energy. The test materials transform from their initial $B3$ to $B1$ structure under pressure. The stability of a particular structure is decided by the minima of Gibbs's free energy, $G = U + PV - TS$, U is internal energy, which at 0 K corresponds to the cohesive energy. S is the vibrational entropy at absolute temperature T, pressure P and volume V. The Gibbs's free energies $G_{B3}(r) = U_{B3}(r) + 3.08Pr^3$ for $B3$ phase and $G_{B1}(r') = U_{B1}(r') + 2Pr'^3$ for $B1$ phase become equal at the phase-transition pressure P and at zero temperature i. e., ΔG ($= G_{B1} - G_{B3}$). Here, U_{B3} and U_{B1} represent cohesive energies for $B3$ and $B1$ phases, and are

$$U_{B3}(r)=-1.6381\frac{e^2Z_m^2}{r}+4V_{ij}(r)+6V_{ii}(r)+6V_{jj}(r), \quad (2)$$

$$U_{B1}(r')=-1.7475\frac{e^2Z_m^2}{r'}+6V_{ij}(r')+6V_{ii}(r')+6V_{jj}(r'). \quad (3)$$

r and r' are nearest-neighbor (nn) separations corresponding to $B3$ and $B1$ phases, respectively. Here, $V_{ij}(r)$ and $V_{ii}(r)$ [$V_{jj}(r)$] are the overlap potentials between the nearest (anion - cation) and the next nearest neighbors (anion -anion or cation-cation), respectively, and defined as:

$$V_{ij}(r)=b\beta_{ij}\exp\left(\frac{r_i+r_j-r_{ij}}{\rho}\right)-c_{ij}r_{ij}^{-6}-d_{ij}r_{ij}^{-8}; \qquad i,j=1,2. \quad (4)$$

Table 1 van der Waals coefficients of XC (X= Si, Ge, Sn) compounds.(c_{ij} in units of 10^{-60} erg cm^6 and d_{ij} in unit of 10^{-76} erg cm^8). C and D are the overall van der Waals coefficients.

Compound	c_{ii}	c_{ij}	c_{jj}	C	d_{ii}	d_{ij}	d_{jj}	D
SiC	28.76	0.64	0.023	13.77	14.28	0.25	0.003	2.82
GeC	73.76	0.93	0.023	32.15	50.14	0.51	0.003	8.41
SnC	168.73	1.27	0.023	69.83	151.14	0.94	0.003	22.97

3. Results and Discussion

Knowledge of the force constants is crucial to the understanding of the stability of the different structures at different volumes. Two different factors determine the response of any crystal structure to pressure. First, changes in nearest-neighbor distances, which affect the overlapping of adjacent ions and bandwidth of the conduction band. Second, changes in symmetry, which affect the hybridization and bond-repulsion. The formalism described above is applied to *XC* compounds belonging to the cubic crystal system. For such purposes we have then three free parameters, namely, modified ionic charge, range and hardness parameters (Z_m, ρ and b). To estimate the free parameters, we first deduce the vdW coefficients from the Slater-Kirkwood variational method [9] and are listed in table 1.

We consider that the *XC* compound is to be partially ionic. It is perhaps worth to remark that we have deduced the values of free parameters, modified ionic charge (Z_m), range (ρ) and hardness (b), from the knowledge of equilibrium distance and the bulk modulus [1, 2] following the equilibrium conditions [7]. The deduced model parameters for *XC* compounds are given in table 2. In an attempt to reveal the structural phase transition of the test materials, we minimise the Gibbs's free energies $G_{B3}(r)$ and $G_{B1}(r')$ for the equilibrium interatomic spacing (r) and (r'). Fig. 1 shows the Gibbs's free energy difference ΔG [$= G_{B1}(r') - G_{B3}(r)$] as functions of pressure (P) using the interionic potential discussed above for *XC* compounds. Let us summarize the results of the plot. The pressure corresponding to ΔG approaching to zero is the phase transition pressure (P_t) [indicated by arrows in figure]. At zero pressure, the *B*3 crystal phase is thermodynamically and mechanically stable, while the *B*1 is not. As pressure increases, beyond the phase transition pressure (P_t), the *B*1 system becomes mechanically and thermodynamically stable.

Eventually, at a pressure higher than the theoretical thermodynamic transition pressure, the *B*3 crystal becomes thermodynamically unstable while the *B*1 phase remains stable up to the greatest pressure studied. In yttrium compounds, a crystallographic transition from *B*3 to *B*1 occurs. The phase-transition pressure (P_t) thus obtained are listed in table 2 and compared with available other experimental [1] and theoretical [2 - 5] results. It is inferred from the table that the calculated phase transition pressure of SiC is similar to the previous theoretical results. However, it is differing considerably from the experimental data [1]. Further, we found that the P_t of GeC is higher than that of SiC. Such an observation could be due to existence of the d-core states in Ge as compare to Si, results in an extra repulsion, which leads to a higher transition pressure [3]. The consistency between band structure calculations data and the present lattice model calculation is attributed to the formulated effective interionic potential which considers the various interactions.

Let us now estimate the values of relative volumes associated with various compressions following Murnaghan equation of state [7] as

$$\frac{V}{V_0} = \left(1 + \frac{B'}{B_0} P\right)^{-1/B'}, \qquad (5)$$

Here V_0 being the cell volume at ambient conditions. The estimated value of pressure dependent radius for both structures, the curve of volume collapse with pressure to depict the phase diagram is

Table 2 Model parameters, calculated transition pressures and volume collapse at transition for *XC* (X = Si, Ge, Sn) compounds.

Compo-und	Model parameters			Calculated transition pressures (P_t) and volume collapse (V_t)	
	Z_m^2	$b(10^{-12}$ erg)	ρ (10^{-1} Å)	P_t (GPa)	V_t (%)
SiC	3.0	12.96	3.3	65 (62 [2], 65.1 [3], 66.5 [4], 65.3 [5], 100 [1])	27 (18.2 [3])
GeC	3.0	14.23	3.59	85 (87 [2], 89.4 [3])	25 (13. 8 [3])
SnC	2.2	7.46	3.37	34 (33 [2], 32.5 [3])	28 (14.7 [3])

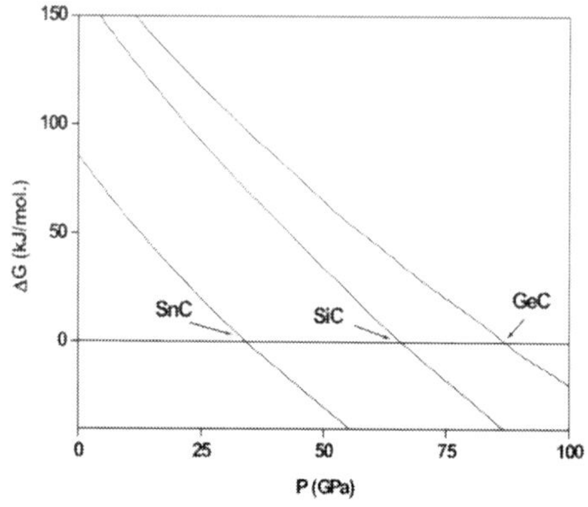

Figure 1. Variation of Gibbs free energy difference (ΔG) with pressure (P) for XC.

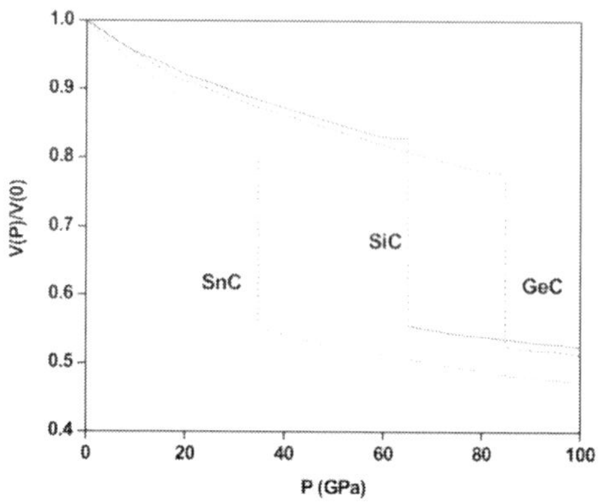

Figure 2. Variation $V(P)/V(0)$ as a function of pressure for XC.

illustrated in Fig. 2 for XC (X = Si Ge, Sn). It is noticed from the plot that our approach has predicted correctly the relative stability of competitive crystal structures, as the values of ΔG are positive. The magnitude of the discontinuity in volume at the transition pressure is obtained from the phase diagram and tabulated in table 2, which is in good agreement with those revealed from other theoretical [3] works for all the compounds.

4. Conclusion

An effective interionic interaction potential is formulated to analyzing the structural properties of XC (X = Si Ge, Sn) compounds. The obtained values of free parameters allow us to predict phase transition pressure and associated volume collapse. We have found vast volume discontinuity in pressure volume phase diagram, identifies the structural phase transition from ZnS ($B3$)to NaCl ($B1$) structure. From our calculated results, it can be emphasized that the present approach reproduced the structural properties at high pressure consistently in terms of the screening of the effective Coulomb potential through modified ionic charge (Z_m^2).

Acknowledgements:
Financial assistance from Department of Science and Technology, New Delhi and Madhya Pradesh Council of Science and Technology, Bhopal, India is gratefully acknowledged.

References

[1] Yoshida M, Onodera A, Ueno M, Takemura K and Shimomura O 1993 *Phys. Rev. B* **48** 10587
[2] Zhang X, Quan S, Ying C and Li Z 2011 *Solid State Commun.* **151** 1545
[3] Hao A M, Yang X C, Wang X M, Zhu Y, Liu X and Liu R P 2010 *J. Appl. Phys.* **108** 063531
[4] Miao M S, Prikhodko M and Lambrecht W R L 2002 *Phys. Rev. B* **66** 064107
[5] Karch K, Bechstedt F, Pavone P and Strauch D 1996 *Phys. Rev. B* **53** 13400
[6] Varshney D, Kaurav N, Sharma P, Shah S and Singh R K 2004 *Phase Trans.* **77** 1075
[7] Kaurav N, Kuo Y K, Joshi G, Choudhary K K and Varshney D 2008 *High. Pres. Res.* **28** 651
[8] Hafemeister D W and Flygare W H 1965 *J. Chem. Phys.* **43** 795
[9] Slater J C and Kirkwood J G 1931 *Phys. Rev.* **37** 682

Analysis of heat transport in the iron oxyarsenide $TbFeAsO_{0.85}$.

N. Kaurav[1*], K. K. Choudhary[2] and Y. K. Kuo[3]
[1]Department of Physics, Govt. Holkar Science college, A. B. Road, Indore (MP)-452001 India.
[2]Department of Physics, Shri Vaishnav Institute of Technology and Science, Baroli, Sanwer Road, 453331 Indore, India.
[3]Department of Physics, National Dong Hwa University, Hualien 97401, Taiwan.

*E-mail: netramkaurav@yahoo.co.ok

Abstract. Thermal conductivity $\kappa(T)$ of a high-quality polycrystalline $TbFeAsO_{0.85}$ sample is theoretically investigated. The lattice contribution to the thermal conductivity (κ_{ph}) is discussed within the Debye-type relaxation rate approximation in terms of the acoustic phonon frequency and relaxation time below superconducting transition temperature (T_c = 42.5 K). The theory is formulated when heat transfer is limited by the scattering of phonons from defects, grain boundaries, charge carriers, and phonons. The κ_{ph} dominates in $TbFeAsO_{0.85}$ and is an artifact of strong phonon-impurity and -phonon scattering mechanism. Our result indicates that the maximum contribution comes from phonon scatters and various thermal scattering mechanisms provide a reasonable explanation for maximum appeared in $\kappa(T)$.

1. Introduction

The recent discovery of high superconducting transition temperatures in FeAs based compounds has generated great interest in the scientific community [1], and because the strong interplay between the structural, magnetic and superconducting properties has led to close comparisons with the physics of the superconducting copper oxides [2, 3]. It has been also recognized that FeAs-based compounds can be promising thermoelectric materials in refrigeration applications around liquid nitrogen temperatures [4]. These aspects have generated several motivating studies in this direction by replacing O atoms by F atoms at a concentration of 3-13 atoms%, with enhancing of critical temperature from 26 K to highest of about 56 K [5-7]. These quaternary superconductors have a rather simple structure of alternating $[Fe_2As_2]^{2-}$ and $[Ln_2O_2]^{2-}$ layers with eight atoms in a tetragonal unit cell (P4/nmm) [8], where the [Fe-As]2- layer is thought to be responsible for superconductivity, while the $[Ln_2O_2]^{2-}$ layer provides charge carriers, resemblance to that of high-T_c cuprates.

Very recently, the superconductivity has been realized by producing oxygen vacancies instead of F atoms doping [9 - 10], which can create more carriers in the charge reservoir layer, as considered necessary to tune the physical properties. Thus, the stacking of these layers and ionic states of constituent ions have decisive role in the transport and superconducting properties. Evolution of structural, electronic and thermal properties would be an important step toward realizing the potential technological scenario apart from richness of physics of this material class. Previously, we have reported the synthesis of highly pure single phase of $TbFeAsO_{0.85}$, x-ray absorption spectroscopy, and anomalous thermal transport properties [11]. The electrical (with $T_c = 42$ K) and magnetic properties of the polycrystalline $TbFeAsO_{0.85}$ sample have been described elsewhere [10]. In this paper, we report the theoretical investigation of thermal conductivity κ(T) of a high-quality polycrystalline $TbFeAsO_{0.85}$ sample experimental data. The κ(T) measurement shows an anomaly in the vicinity T_c. The κ_{ph} is discussed within the Debye-type relaxation rate approximation in terms of the acoustic phonon frequency and relaxation time below transition temperature. Our result indicates that the maximum contribution comes from phonon scatters and various thermal scattering mechanisms provide a reasonable explanation for maximum appeared in κ(T).

2. The Model

We start with a model Hamiltonian that follows [12]

$$H = \sum_k \varepsilon_k a_k^+ a_k + \sum_q \omega_q b_q^+ b_q + \sum_{k_1,k_2} \varphi(k_1,k_2) a_{k_1}^+ a_{k_2} + D\sum_{k,q} q\left[\frac{\hbar}{2\rho\omega_q}\right]^{1/2} a_{k+q}^+ a_k (b_q + b_{-q}^+)$$
$$+\frac{R}{2N}\sum_{q_1,q_2} e^{i(q_1+q_2)r_i}\left[\frac{\hbar\omega_{q_1}\hbar\omega_{q_2}}{4}\right]^{1/2}(b_{q_1} - b_{-q_1}^+)(b_{q_2} - b_{-q_2}^+) + H_{ph-ph}. \quad (1)$$

Here, initial two terms is holes as carriers and phonon excitation. The third and fourth terms represent the carrier-impurity interactions and carrier-phonon interactions, respectively. The fifth term denote the phonon-impurity interactions and the last term stands for the phonon-phonon interaction.
The notations a (a^+) and b (b^+) are the creation (annihilation) operators for holes and phonons. Further, ε_k is hole free energy, phonon frequency of a wave vector q is ω_q, D_p is the deformation-potential constant. R is the relative ionic-mass difference $[(M''-M)/M'']$, respectively. N is number of cells. ρ is the mass density of ions and r_i stands for the position of defects. We begin with phonon contribution to thermal conduction. The thermal conductivity can be calculated from the Kubo formula [12]. It has contributions from both the phonons and the carriers. We first look for the lattice part and in the continuum approximation it follows

$$\kappa_{ph} = \frac{k_B\hbar^2}{2\pi^2 v_s}\int_0^{\omega_D} d\omega\omega^2\tau(\omega)(\beta\omega)^2\frac{e^{\beta\omega}}{(e^{\beta\omega}-1)^2}. \quad (2)$$

with k_B is the Boltzman constant, v_s is the sound velocity, ω_D is the Debye frequency and $\beta = \hbar/k_BT$.
The relaxation time is proportional to the imaginary part of the phonon self-energy. In the weak interaction case, it has been calculated to the lowest order of the various interactions. The relaxation times are expressed as

$$1/\tau(\omega) = 2\left|\mathrm{Im}\,P(\omega/v_s,\omega)\right| = 1/\tau_{ph-d} + 1/\tau_{ph-e} + 1/\tau_{ph-gb} + 1/\tau_{ph-ph}, \quad (3)$$

with the various relaxation times are defined as

$$\tau_{ph-d}^{-1}(\omega) = (A/k_B^3)\omega^4\hbar^3, \quad (4)$$

$$\tau_{ph-e}^{-1}(\omega) = B\omega n_F, \quad (5)$$

$$\tau_{ph-gb}^{-1}(\omega) = v_s/L, \quad (6)$$

$$\tau_{ph-ph}^{-1} = (\omega) = C(T\omega\hbar/k_B)^3. \quad (7)$$

where L is the sample size, n_F is the Fermi-Dirac distribution function and Δ is the gap function. The notation $\tau_{ph\text{-}d}$, $\tau_{ph\text{-}e}$, $\tau_{ph\text{-}gb}$, and $\tau_{ph\text{-}ph}$ are the phonon scattering relaxation time due to defects, carriers, grain boundaries, and phonon-phonon interactions respectively. We note that to this order Mathiessen's rule holds namely, that the inverse of the total relaxation time is the sum of the various contributions for the different scattering channels.

Further we define $A = \left[\frac{3n_i R^2}{4\theta_D^3}\right]$ and $B = \frac{9\pi}{4}\left[\frac{m}{3M}\right]^{1/2}\frac{D_P{}^2}{E_F^2}$, are parameters, which characterize the strengths of the phonon-defect and phonon-carrier, scattering process. Here, n_I is the density of impurities or defects, E_F is the Fermi energy of holes as carriers, and m is its mass.

3. Results and Discussion

The experimental results of the thermal conductivity on polycrystalline $TbFeAsO_{0.85}$ used in this work were reported previously [11] and is shown in Fig. 1 for comparison. At low temperatures, κ increases with temperature and a broad peak appear around 27 K and then a steep fall as temperature increases further. The occurrence of such a low temperature maximum is a typical feature of solids due to the reduction of thermal scattering, i.e., the phonon mean free path is approximately equal to the crystal site distance and ascribed to the generalized Umklapp process at low temperatures. Further, κ is found to be almost saturating as the temperature increases above 75 K. In order to elucidate these distinguishing features of the $TbFeAsO_{0.85}$ polycrystalline sample, we have analyzed our data in the framework of Kubo formula as described earlier. Here, we have neglected the contribution from electrons below T_c because of the condensed superconducting pairs at low temperatures. In the calculation of temperature dependent thermal conduction of $TbFeAsO_{0.85}$, we use the parameters which characterize the strengths of the phonon-defect, phonon-electron, and phonon- phonon scattering process as $A = 3.9 \times 10^{-7}\, K^{-3}$, $B = 0.12$ and $C = 0.5\, K^{-6}\,\text{sec}^{-1}$, respectively. The length of the sample is about 3 mm and $v_s = 4.76 \times 10^5$ cm sec^{-1}.

We first qualitatively discuss the properties of thermal conductivity due to phonons. Figure 2 shows our results for phonon thermal conductivity of $TbFeAsO_{0.85}$. As the temperature is lowered through T_c, the quasi particle excitation condensed into the ground state and they cannot scatter phonons. The phononic thermal conductivity, thus, increases exponentially as the temperature increases in the absence of the other scattering mechanism. Although the phonon thermal conductivity experiences an exponential increase below the transition temperature, the presence of the defect, and

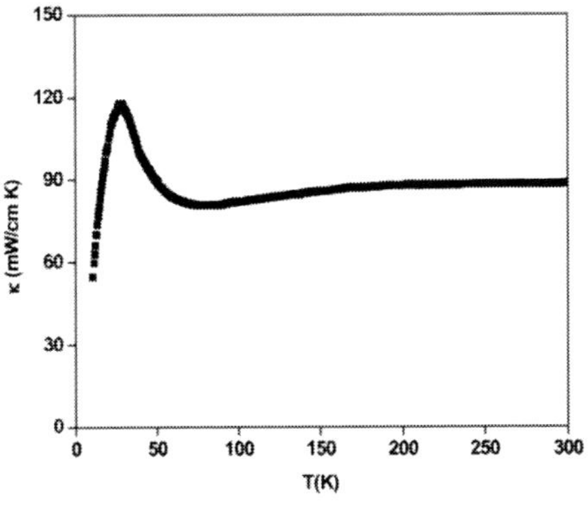

Figure 1. Variation of thermal conductivity as a function of temperature.

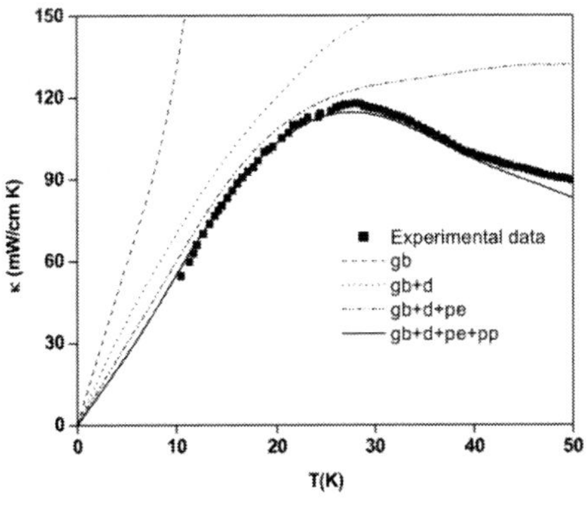

Figure 2. Variation of Phononic thermal conductivity with temperature: Effect of various phonon scattering mechanism.

the electron scatterings set a limit on its growth, as a consequence the phononic thermal conductivity diminishes as the temperature increases. At much higher temperature phonon-phonon scattering becomes more effective and decreased mean free path of phonon is responsible for decrease in thermal conductivity at higher temperature. The maximum position depends on the relative magnitudes of the phonon-electron, phonon-defect and phonon-phonon scattering processes.

It is meaningful to comment that the contribution of the electron-impurity towards thermal conductivity to the phonon thermal conductivity is about 1% in the temperature domain ($0 < T < T_c$). Phonons are, then, the sole carriers of heat in this temperature domain. Further the all the contributions are clubbed together and temperature dependent of total thermal conductivity is shown in Figure 2 along with experimental data. The present analysis on thermal conductivity shows good agreement with the experiments results.

4. Conclusion

Thermal conductivity $\kappa(T)$ of a high-quality polycrystalline $TbFeAsO_{0.85}$ sample is investigated. The $\kappa(T)$ measurement shows an anomaly in the vicinity of superconducting transition temperature (T_c = 42.5 K). The lattice contribution to the thermal conductivity (κ_{ph}) is discussed within the Debye-type relaxation rate approximation in terms of the acoustic phonon frequency and relaxation time. The theory is formulated when heat transfer is limited by the scattering of phonons from defects, grain boundaries, charge carriers, and phonons. Our result indicates that the maximum contribution comes from phonon scatters and various thermal scattering mechanisms provide a reasonable explanation for maximum appeared in $\kappa(T)$. The numerical analysis of heat transfer below T_c shows similar results as revealed from experiments.

Acknowledgements:
Financial assistance from Department of Science and Technology, New Delhi and Madhya Pradesh Council of Science and Technology, Bhopal, India is gratefully acknowledged. The authors would like to thank for financially supporting this research to the National Science Council of Taiwan under Contract No. NSC- 97-2628-M-259-001-MY3 (YKK).

References

[1] Kamihara Y, Watanabe T, Hirano M, and Hosono H 2008 *J. Am. Chem. Soc.* **130** 3296
[2] Chen X H, Wu T, Wu G, Liu R H, Chen H and Fang D F 2008 *Nature* **453** 761.
[3] Chen T Y, Tesanovic Z, Liu R H, Chen X H and Chien C L 2008 *Nature* **453** 1224.
[4] Pinsard-Gaudart L, Berardan D, Bobroff J and Dragoe N 2008 *Phys. Stat. Sol. (RRL)* **2** 185
[5] Chen G F, Li Z, Wu D, Li G, Hu W Z, Dong J, Zheng P, Luo J L and Wang N L 2008 *Phys. Rev. Lett.* **100** 247002
[6] Ren ZA,Yang J, Lu W, Yi W, Shen X L, Zheng-Cai Li, Guang-Can Che, Xiao-Li Dong, Li-Ling Sun, Fang Zhou and Zhong-Xian Zhao 2008 *Eurphys. Lett.* **82** 57002.
[7] Liu R H, Wu G, Wu T, Fang D F, Chen H, Li S Y, Liu K, Xie Y L, Wang X F, Yang R L, Ding L, He C, Feng D L and Chen X H 2008 *Phys. Rev. Lett.* **101** 087001
[8] Quebe P, Terbüchte L J and Jeitschko W 2000 *J. Alloys Compd.* **302** 70
[9] Ren Z A, Che G C, Dong X L, Yang J, Lu W, Yi W, Shen X L, Li ZC, Sun L L, Zhou F and Zhao Z X 2008 *Eurphys. Lett.* **83** 17002.
[10] Shi Y G, Yu S, Belik A A, Matsushita Y, Tanaka M, Katsuya Y, Kobayashi K, Yamaura K, Takayama-Muromachi E *J. Phys. Soc. Jpn.* **Suppl. C 77** 155.
[11] Kaurav N, Chung Y T, Kuo Y K, Liu R S, Chan T S, Chen J M, Lee J F, Sheu HS, Wang X L, Dou S X, Lee S I, Shi Y G, Belik A A, Yamaura K 2009 *Appl. Phys. Lett.* **94** 192507
[12] Callaway J 1991 *Quantum theory of the Solid State* (Academic Press, London).

X-ray diffraction and absorption spectroscopic studies of copper mixed ligand complexes with aminophenol as one of the ligands

A. Mishra, Garima Jain & H. Patil
School of Physics, Devi Ahilya University, Khandwa Road, Indore, 452001, India

Email: garimasop@gmail.com

Abstract. X-ray diffraction (XRD) and X-ray absorption spectroscopic (XAS) studies have been conducted on two copper complexes, i.e., copper macrocyclic complex of succinic acid and ortho aminophenol (complex-1) and copper macrocyclic complex of pthalic acid and ortho aminophenol (complex-2). The diffraction pattern of the complexes have been recorded using Rigaku RINT-2000 X-ray diffractometer equipped with rotating anode X-ray tube operated at 40 kV and 100 mA. The X-ray absorption spectra of the complexes have been recorded at the K-edge of copper on Cauchois type bent crystal spectrograph having radius 0.4 m employing a mica crystal, oriented to reflect from (100) planes, for dispersion. The X-ray absorption near edge structure (XANES) parameters, viz., chemical shift, energy position of the principal absorption maximum and edge-width have been determined and discussed. From the extended X-ray absorption fine structure (EXAFS) data, the bond lengths have been calculated using three methods, namely, Levy's method, Lytle, Sayers and Stern's (LSS) method and Fourier transformation method. The results obtained have been compared with each other and discussed.

1. Introduction

X-ray absorption spectroscopy (XAS) refers to the details of how X-rays are absorbed by an atom at energies near and above the core-level binding energies of that particular atom. The X-ray absorption spectra is typically divided into two regimes: X-ray absorption near edge structure (XANES) and extended X-ray absorption fine structure (EXAFS). While XANES refers to the structure on the high energy side of the main edge extending up to ~40 eV from the edge, EXAFS refers to the undulations of the absorption coefficient on the high energy side of the main edge extending from ~40eV to ~1000eV from the edge. XANES is strongly sensitive to formal oxidation state and coordination chemistry of the absorbing atom, while EXAFS is used to determine the distances, coordination number, and species of the neighbors of the absorbing atom. As a result, XAS provides a practical and simple way to determine the chemical state and local atomic structure for a selected atomic species. In the present paper, we have studied X-ray absorption spectra (XAS) at the K-edge of copper in two copper complexes, viz., copper macrocyclic complex of succinic acid and ortho aminophenol (complex-1) and copper macrocyclic complex of pthalic acid and ortho aminophenol (complex-2). XANES parameters, e.g., chemical shift, i.e., shift of K-edge of copper in complex with respect to K-edge of copper metal, energy position of the principal absorption maximum and edge-width have been determined and discussed. The positions of EXAFS maxima and minima have also been determined. Using the EXAFS data, the bond lengths have been calculated using three methods, namely, Levy's

Published under licence by IOP Publishing Ltd

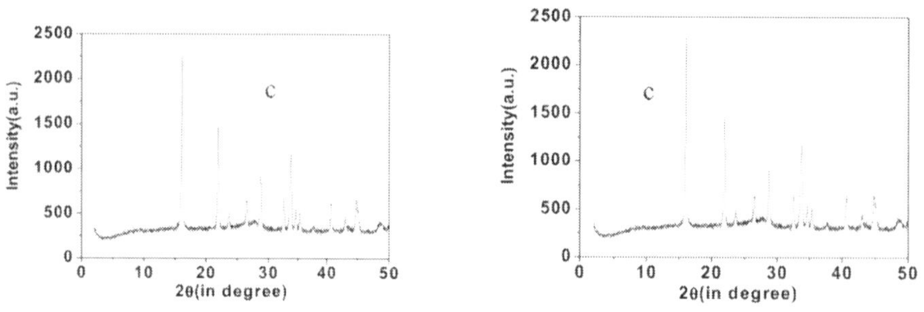

(a) Complex-1 (b) Complex-2

Figure1. X-ray diffractometer records of complexes

method, Lytle, Sayers and Stern's (LSS) method and Fourier transformation method. The X-ray diffractometer records of the complexes have been obtained from which the particle size and the lattice parameter have been determined.

2. Experimental

All the complexes have been prepared by chemical root method. The X-ray diffraction patterns of copper complexes have been recorded using Rigaku RINT-2000 X-ray diffractometer with a rotating anode operated at 40 kV and 100 mA. Copper target was used as the source of X-rays at wavelength λ=1.54Å.

The X-ray K-absorption spectra of the copper complexes have been recorded using a Rich Seifert X-ray generator with tungsten target operating at 20 kV and 40 mA. A 0.04 m Cauchois type bent crystal transmission spectrograph employing a mica crystal, oriented to reflect from (100) planes for dispersion, has been used to record the spectra.The spectra were recorded on Kodak X-ray photographic films. The exposure time varied from 2 to 4 hours. Digital records of the spectra were obtained on Carl-Zeiss microphotometer coupled to a computer. The computer software Athena version 0.8.056 has been used for analysis of the digital spectral data.

3. Results and discussion

3.1. X-ray diffraction studies:

The X-ray diffractometer records of the complexes are given in figure 1. Particle size was calculated using Scherer's formula given by: $t=0.9\lambda /B\cos\theta$

here t is the crystal thickness (in the same units as λ), B is full width half maxima (in radians) of diffraction line, θ is the Bragg angle and λ is the wavelength. The values of particle size and the lattice constant calculated from the diffractometer records are given in table 1.

3.2 X-ray absorption near edge structure:

The X-ray absorption spectra at the K-edge of copper in the complexes are shown in figure 2. The X-ray absorption near edge structure (XANES) parameters, viz., chemical shift, energy position of the principal absorption maximum and edge-width have been determined. The results are given in table 1 and discussed below.

3.2.1.Chemical shift:

The shift of X-ray absorption edge i (i=K, L, M…) of an element in a complex with respect to that of the pure metal is written as:

ΔE = Ei (Compound) – Ei (Element)

The chemical shifts of copper complexes in the present investigation are given in Table 1. The values of shifts are 8.1 and 7.4 eV for complexes 1 and 2 respectively. The compounds studied earlier [1]

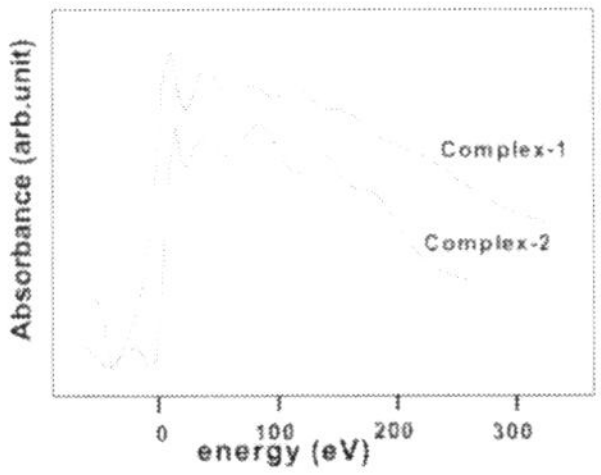

Figure2. X-ray absorption spectra of the complexes

THE GRAPHIC ON THIS PAGE WAS NOT AVAILABLE ON THE E-MEDIA VERSION OF THIS PAPER.

Figure3(a). Fourier transform from figure 2 for complex 1

THE GRAPHIC ON THIS PAGE WAS NOT AVAILABLE ON THE E-MEDIA VERSION OF THIS PAPER.

Figure3(b). Fourier transform from figure 2 for complex 2

CuO, $CuCl_2.2H_2O$ and $CuSO_4.5H_2O$ have been reported to have chemical shifts 4.3, 5.3 and 6.8 eV respectively, which are all +2 oxidation state compounds. Since the chemical shifts in the present work are less than 10 eV, on this basis, the conclusion is that, the two complexes studied are copper (II) complexes. Therefore, copper exists in oxidation state +2 in the complexes studied in the present investigation. The sequence of variation of chemical shifts of the complexes is as follows:
Complex-1>complex-2
Recalling that the chemical shift value for a compound is largely dependent on the effective nuclear charge on the central metal ion, the decrease in the chemical shift in complex-2 with respected to complex-1 suggested that pthalic acid has lesser field strength than succinic acid.

3.2.2. Shift of principal absorption maxima:
The values for the shifts of the principal absorption maxima with respect to the absorption edge for the complexes are given in Table 1. The shift in the principal absorption maximum in transition metal complexes is known to arise from the transition of an electron from the 1s level to a vacant orbital of appropriate symmetry. This parameter is sensitive to the valency of the metal ion. It is worth comparing the present values for the shift of principal absorption maxima with the values reported by earlier workers for other copper (II) complexes which lie in range of 17.7 - 25.4 eV. [3,4] Hence on the basis of this comparison, we may say that all the samples in the present investigation posses an oxidation state +2. The order of the shifts of the principal absorption maxima is the same as the order of the chemical shifts indicating that the metal ligand bond is stronger in complex 1 as compared to complex 2.

Table 1. Results of X-ray diffraction and absorption spectroscopic studies of complexes

Complex	**XRD results**		**XANES results**			**Bond lengths from EXAFS data**		
	Particle size (nm)	Lattice param eter (nm)	Chemical shift (ev)	Shift of principal absorption maximum (ev)	Edge-width (ev)	$R_{Levy's}$ (Å)	R_{Lss} (Å)	RFT (Å)
Complex-1	38.64	0.951	8.1	19.40	11.2	1.84	2.08	1.1
Complex-2	28.44	0.954	7.4	15	7.6	2.29	1.97	1.8

3.2.3. Edge-width:

The values of the edge width of complexes are given in table 1. Edge-width of the K-absorption edge increases with the increase of covalent character of the bonds [2]. The relative order of the values of edge- width are not supported by the values of the chemical shift or the shift in principal absorption maxima in the present work. The reason may be that the values of edge-width not only depend on the factors responsible for the chemical shift but also on the geometry of the complexes.

3.3 Bond lengths from extended X-ray absorption fine structure data

From the extended X-ray absorption fine structure (EXAFS) data, the bond lengths have been calculated using three methods, namely, Levy's method, Lytle, Sayers and Stern's (LSS) method and Fourier transformation method as follows.

3.2.3.1. Levy's method

In Levy's method [5], the bond lengths are calculated by using the relation: $R1 = (151/\Delta E)^{1/2}$ Å, where ΔE is the difference in eV of the energies of the EXAFS first maximum and first minimum and R1 is the radius of the first coordination sphere. The bond lengths thus determined are given in table1.

3.2.3.2. LSS method

The values of the wave vactor k (Å-1) for EXAFS maxima (n=0, 2, 4, ---------) and minima (n=1, 3, 5, ----), for all copper complexes have been determined. In the Lytle, Sayers and Stern's (LSS) method [6] for determination of the nearest neighbour distances, n versus k graph is plotted. The plots have been found to be linear for all the complexes. The slope of n versus k plot, gives the values of $2(R1 - \alpha1)/\pi$ where R1 is the bond length. The parameters α1 depends to a large extent on the central absorbing atom. It is found that for chemically similar system, the values of α1 remains more or less the same. The values of (R1 - α1) thus obtained are given in table 1. This is phase uncorrected bond length.

3.2.3.3. Fourier transformation method

The Fourier transform of the oscillatory part of the X-ray absorption spectra are given in figures 3(a) and 3(b) for the two complexes. The Fourier transform peaks at the radial distances of the neighbouring atoms from the absorbing atom. However, the distance found from Fourier transform is about 0.2 Å – 0.5 Å shorter than the actual distance due to energy dependence of the phase factors in the sine function of the EXAFS equation [6]. The peaks in the Fourier transform are shifted towards the origin by an amount αj and hence the peaks are at distance Rj- αj. For the first peak j=1 and hence the position of the first peak in the Fourier transform determines the distance $R_1 - \alpha_1$.

It is important to note here that the distance R1- α1 should be equal to the distance found from the LSS graphical method outlined above. Hence, both the LSS method and the Fourier transformation method give the value R1- α1, i.e., both the methods give the value of bond lengths which have not been corrected for the phase shifts. We have called this distance as the phase uncorrected bond length.

It is seen from this table that the values of R1- α1 as determined from LSS method and those determined from the Fourier transformation method are in good agreement with each other, i.e., both the LSS method and Fourier transformation method give nearly the same value of the phase uncorrected bond length, i.e., $R_1 - \alpha_1$.

4. References

[1] Hemchandran K and Chetal A R 1986 *Phys. Stat. Sol. (b)* **136** 181
[2] Shrivastava U C, Nigam H L 1973 *Coord. Chem. Rev.* **9** 267
[3] Shah M C, Shrivastava B D, Pandeya K B, Shrivastava K, Prasad J. 1983 *J. Phys. C* **16** 3601
[4] Koul P N, Padalia B D 1983 *X-ray Spectrom.* **12** 128
[5] Levy R M 1965 *J. chem. Phys.* **43** 1846
[6] Lytle F W, Sayers D E and Stern E A 1975 *Phys. Rev.* B. **11,** 2795

Interpretation of optical conductivity in normal state of Iron-Based Superconductors CeOFeAs

K. K. Choudhary[1, 3] **and N. Kaurav**[2]
[1]Department Physics, Shri Vaishnav Institute of Technology and Science, Baroli, Sanwer Road, Indore -453331, India.
[2]Department of Physics, Govt. Holkar Science college, A. B. Road, Indore (MP)-452001 India.

[3]E-mail: kkchoudhary1@yahoo.com

Abstract. Quantitative analysis of the optical conductivity σ(ω) for the normal state of Iron-Based superconductors CeOFeAs have been made within the two-component scheme: one is the coherent Drude free carrier excitations and other is incoherent motion of carriers leading to a polaron formation, originated from inter and intra layer transitions of charge carriers. The model successfully accounts for the anomalies reported in the optical measurements for metallic state of the superconductors. The frequency dependent relaxation rates are expressed in terms of memory functions and the coherent Drude carriers from the effective interaction potential leads to a sharp peak at zero frequency which is an indication of metallic conduction and a long tail at higher frequencies, i.e. in the infrared region. While to that the hopping of carriers from Fe to Fe in the FeAs layer and from FeAs layer to CeO layer (incoherent motion of carriers) yields two-peak value around 100 cm^{-1} and 425 cm^{-1} respectively in the optical conductivity centred at mid-infrared region. Both the Drude and hopping carriers contribute to the optical process of conduction in the iron-based superconductors and shows similar results on optical conductivity in the mid-infrared as well as infrared frequency regions as those revealed from experiments.

1. Introduction

The discovery of non-copper-based superconductors has attracted wide attention to elucidate the mechanism of superconductivity and to explore higher T_c materials. It is believed that strong electron correlation and layered structures may play an important role. However, the T_c of the non-copper-based superconductors is still lower than 40 K as predicted by BCS theory and experimental evidence. Recently, superconductivity in iron and nickel-based layered quaternary compounds has been reported: LaOFeP (T_c = 4 K) [1], LaONiP (T_c = 3 K) [2], Later on Kamihara et al. [3] found that by substituting P with As, and partially O with F in LaOFeP, the resultant material $La(O_{1-x}F_x)FeAs$ (x=0.05–0.12) became superconductive at 26K. The Fe_2As_2 layer, which is sandwiched between the La_2O_2 layers, serves as a carrier conduction path [1, 2] Thus, conduction carriers are two-dimensionally confined in the Fe_2As_2 layer, causing strong interactions among the electrons.

Chen et al. has presented the reflectance and conductivity spectra in a far-infrared region for CeOFeAs and LaOFeAs[4]. The reflectance below 400 cm^{-1} is strongly suppressed at low frequency below the phase transition temperature, which is a strong indication for the formation of an energy gap. However, the low-frequency reflectance still increases fast towards unity at zero frequency, indicating a metallic behavior. Further more, the spin-density wave (SDW) partial gap is observed for

EuFeAs and LaOFeAs compounds [5]. The gap-like absorption peaks in the real part of conductivity for both the compounds scale with their respective T_{SDW}. The studies of reflectance and real part of optical conductivity of a single crystal of BaKFeAs [6] has shown that the absorption spectrum of this system consists of a noticeable Drude peak at low frequencies and a wide absorption band with maximum at 0.7 eV, which is attributed to carrier scattering by collective excitations with energies of the order of 25 meV and strongly temperature dependent coupling constant with carriers.

The aim of this paper is to explain theoretically some of the gross features observed in the experimental data on frequency dependent optical conductivity of iron-based superconductor, using a simple phenomenological model. The electronic dielectric function is evaluated following Random Phase approximation of polarizabilities. The first channel to the conductivity is the coherent Drude i. e., the intralayer component with temperature dependent damping. Secondly the hopping of carriers from one site to another (either Fe to Fe inter-layer hopping or from FeAs layer to CeO layer), small polaron as well as charge transfer.

2. The Model

The interaction potential in terms of dielectric function reads

$$V(q;\omega)=\frac{2\pi e^2}{q\varepsilon(q,\omega)} \tag{1}$$

$$\varepsilon(q,\omega)=\varepsilon_\infty+P(q,\omega). \tag{2}$$

The $\varepsilon(q,\omega)$ represents ε_{ab} for interaction confined to *a-b* plane. The charged quasi particles are being scattered by the optical phonons, the charge fluctuations as well impurities and posses a finite damping rate Σ. We can then calculate the optical conductivity as

$$\sigma(\omega)=\lim_{q\to 0}[\omega\,\mathrm{Im}\,\varepsilon(q,\omega)]/[4\pi] \tag{3}$$

We shall now proceed to include polarizability of carriers in the long wavelength limit.

The electronic polarizability within random phase approximation for carriers follows

$$P(q,\omega)=-2\pi e^2Z^2n_cq/[m_\mu^*\omega(\omega+i\Sigma(\omega)] \tag{4}$$

Ze being the sum of the ionic charge and the free electron charge, n_c is the 2D density and m_μ* is the effective mass. The frequency dependent damping function is denoted as $\Sigma(\omega)$.

The model longitudinal dielectric response function for in-plane interactions as

$$\varepsilon(q,\omega)=\varepsilon_\infty-\frac{2\pi Z^2e^2n_cq}{m_\mu^*\omega[\omega+i\Sigma(\omega)]}. \tag{5}$$

The frequency dependent damping function of the quasi particles are expressed as

$$\Sigma_\mu(\omega)=\Sigma_\mu(0)+[\Sigma_\mu(\infty)-\Sigma_\mu(0)][-i\omega\Gamma_\mu(\omega)] \tag{6}$$

in terms of memory functions [7] $\Gamma_\mu(\omega)$ and μ is either '0' or '1'. Furthermore, we denote $\Sigma_\mu(0)=\gamma_{\mu 0}$ and $\Sigma_\mu(\infty)=\gamma_{\mu\infty}$ as the low and high frequency limits of the damping function of the quasi particles, respectively. In order to satisfy the requirements of causality, the memory functions are

$$\Gamma_\mu(\omega)=\int_0^\infty\Gamma'_\mu(t)e^{i\omega t}dt, \tag{7}$$

where $\Gamma'_{\mu}(t=0)=1$ and $\Gamma'_{\mu}(t=\infty)=0$. Using the Gaussian forms of Γ for memory function,

$$\Gamma'_{\mu}(t)=\exp(-t^2/\gamma_{\mu c}^2) \tag{8}$$

$\gamma_{\mu c}$ being the characteristic relaxation rates. Physically, $\gamma_{\mu 0}$ and $\gamma_{\mu\infty}$ are the inverse of life times of quasi particles, due to the scattering by external fields. While to that, $\gamma_{\mu c}$ are regarded as those associated with the motion of scatterers.

3. Results and Discussion

For the numerical calculation of the optical response of iron-based superconductors, we have used the realistic values of some physical parameters as follows: The effective mass of the electron along the conducting FeAs plane is $m^* = 17$ [8] which is consistent with the value deduced from electronic specific heat coefficient $\gamma = 11$ mJ mol^{-1} K^{-2}. The behaviour of the multilayer systems critically depends on the electron density; we have used the charge carrier density $n_c \cong 1.7 \times 10^{21}$ electrons/cm^3 which is consistent with the Hall data [8]. The value of effective charge for conducting FeAs layer is used as Ze = -2e. The optical conductivity sensitively depends on the values of relaxation rates. The quasi particle relaxation rates γ_{00}, $\gamma_{0\infty}$ for the Drude component dominates the low frequency behaviour of the optical conductivity. They represent the inverse of the lifetime of the carriers having a well-defined Fermi surface. The relaxation rates are temperature dependent and for $T = 170$ K, we take $\gamma_{00} = 500$ cm^{-1}, $\gamma_{0\infty} = 1000$ cm^{-1} and $\gamma_{0c} = 2200$ cm^{-1} for CeOFeAs superconductors, respectively.

Physically, the choice of the relaxation rates corresponds to the scattering process well described by phenomenological Drude model. However, the time $1/\gamma_{00}$ and $1/\gamma_{0\infty}$ for which the carriers can move freely without being scattered is much larger than the time $1/\gamma_{0c}$ in which the scatterers return to the initial state, the scatterers can respond to the motion of the carriers quickly enough. With these presumptions, we show in Figure 1, the results for the optical conductivity that corresponds to simple Drude model. We find from the present calculations that a peak value of $\sigma(\omega)$ at zero frequency is observed which develops from the free carrier plasmon modes and a tail at higher frequencies. The magnitude of the Drude peak is about 240 Ω^{-1}cm^{-1}, which consistent with the experimental data [4] at $T = 170$ K in the CeOFeAs. Appearance of Drude peak at low frequency is an indication of metallic conduction in normal state of superconductor.

Switching to the second mode of conduction, i.e., the transfer of carriers (electrons) from site to site through hopping. We have considered first here the small polaron, originated from intra-layer transitions from Fe to Fe atom in FeAs layer which develops a peak value in optical conductivity at about 100 cm^{-1}. It is worth to comment that the effective mass m^* for the hopping component is difficult to determine through the experiments due to shortness of the mean free path. The relaxation rates γ_{10} and $\gamma_{1\infty}$, for the hopping component dominating the mid infrared behaviour of the optical conductivity. The characteristic rates γ_{0c} are regarded as the motion of the excitations, which scatter the free carriers. The excitations may be substantiated by the phonon, the magnon or the hopping component. While to that the localised spin system or the lattice system are to be considered to interact with the hopping component with characteristic rates γ_{1c}. The relaxation rates at $T = 170$ K are chosen as $\gamma_{10} = 600$ cm^{-1}, $\gamma_{1\infty} = 1000$ cm^{-1} and $\gamma_{1c} = 66000$ cm^{-1}, respectively. Let us begin our discussion with the pronounced peak centered at 425 cm^{-1} within the polaronic picture.

Another aspect of our analysis as shown in Figure 1 in the range from 300 cm^{-1} to 500 cm^{-1}, the conductivity is dominated by the strong absorption at $\omega = 425$ cm^{-1}. Incorporating an additional oscillator in the optical conductivity expression with $\gamma_{20} = 950$ cm^{-1}, $\gamma_{2\infty} = 1000$ cm^{-1} and $\gamma_{2c} = 6200$ cm^{-1}, respectively. The optical conductivity shows a prominent feature at about 425 cm^{-1} and is associated with the interlayer charge-transfer transition from the CeO layer to FeAs layer. Both contributions are than clubbed and the resultant optical conductivity is plotted in Figure 2 along with the reported data on CeOFeAs superconductor [4] at $T = 170$ K. We thus argued here that the low-frequency response should be coherent Drude peak and an incoherent mid-infrared absorption band, which takes account

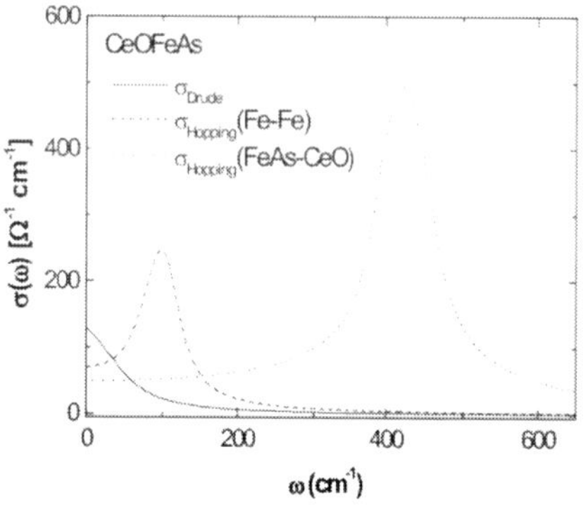

Figure 1. Variation of optical conductivity with frequency in CeOFeAs.

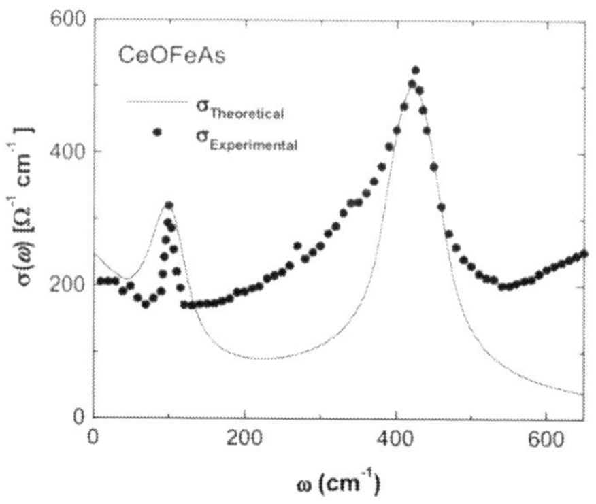

Figure 2. Theoretically estimated frequency dependent optical conductivity (solid line) is compared with experimental data (solid circles), taken from Chen et al. 2008 [4] for CeOFeAs.

of the two-component contributions, one is in between Fe-Fe and another feature is that the peak due to the CeO-FeAs charge-transfer excitations. We note that the relaxation rate γ_{10} satisfy a condition that $\gamma_{10} \geq \varepsilon_F$ reflects that the motion of the carriers is nodoubt hopping like. It is thus anticipated that Fermi surface of the hopping component almost loses its sense.

4. Conclusion

The two-component model of optical conductivity presents two channels: the first is coherent Drude component with a temperature dependent damping function, leads to a sharp peak at zero frequency which is an indication of metallic conduction and the second is the incoherent hopping component. The hopping of carriers from Fe to Fe in the FeAs layer and from FeAs layer to CeO layer i.e. the incoherent motion of carriers yields two-peak value around 100 cm^{-1} and 425 cm^{-1} respectively in the optical conductivity centred at mid-infrared region.

Acknowledgements:
Financial assistance from Department of Science and Technology, New Delhi and Madhya Pradesh Council of Science and Technology, Bhopal, India is gratefully acknowledged.

References

[1] Kamihara Y, Hiramatsu H, Hirano M, Kawamura R, Yanagi H, Kamiya T and Hosono H 2006 *J. Am. Chem. Soc.* **128** 10012.
[2] Watanabe T, Yanagi H, Kamiya T, Kamihara Y, Hiramatsu H, Hirano M and Hosono H 2007 *Inorg. Chem.* **46** 7719
[3] Kamihara Y, Watanabe T, Hirano M and Hosono H 2008 *J. Am. Chem. Soc.* **130** 3296
[4] Chen G F, Li Z, Wu D, Li G, Hu W Z, Dong J, Zheng P, Luo J L and Wang N L 2008 *Phys. Rev. Lett.* **100** 247002
[5] Wan-zheng HU, Dong J, Li G, Li Z, Zheng P, Chen G, Luo J, Wang N 2009 *Front. Phys. China* **4** 459
[6] Yang J, Huvonen D, Nagel U, Room T, Ni N, Canfield P C, Bud'ko S L, Carbotte J P and Timusk T, arXiv: 0807.1040
[7] Varshney D, Patel G S, and Singh R K 2002 *Supercond. Sci. Tech.* **15** 1617
[8] Sefat A S, McGuire M A, Sales B C, Jin R, Howe J Y and Mandrus D 2008 *Phys. Rev. B* **77** 174503.

EXAFS study of tetrakis(imidazole) dinitratocopper (II) complex

Davood Ah. Dar[a], Ajita Johari[a], B. D. Shrivastava[a], J. Prasad[b], K. Srivastava[b], S. N. Jha[c], D. Bhattacharyya[c], A. Poswal[c]
[a]School of Studies in Physics, Vikram University, Ujjain-452001, India
[b]Department of Chemistry, University of Allahabad, Allahabad-211002, India.
[c]Applied Spectroscopy Division, Bhabha Atomic Research Centre, Mumbai-400085, India

Email: rashmibasant@gmail.com

Abstract. The extended X-ray absorption fine structure (EXAFS) spectra at the K-edge of copper in tetrakis(imidazole)dinitratocopper(II) complex $[Cu(im)_4(NO_3)_2]$ (im= imidazole), has been recorded using the dispersive EXAFS beamline at the 2 GeV Indus-2 synchrotron source at RRCAT, Indore, India. The EXAFS data analysis has been done using the computer programs *Athena* and *Artemis*. Theoretical model has been generated using the available crystal structure of the complex and fitted to its experimental EXAFS data. The structural parameters thus obtained, which include bond-lengths, coordination numbers and thermal disorders, have been reported. The results obtained have been found to be in good agreement with crystallographic results. The coordination geometry around the copper site has been determined and depicted. Further, copper has been found to be in +2 oxidation state in this complex.

1. Introduction

The EXAFS (extended X-ray absorption fine structure) in the X-ray absorption spectrum is the normalized oscillatory part of the absorption coefficient above the absorption edge to approximately 1000 eV or higher energy. The EXAFS oscillations are of interest as they contain structural information specific to the scattering atom. In the present work, EXAFS of tetrakis (imidazole)dinitratocopper(II) complex $[Cu(im)_4(NO_3)_2]$ (im = imidazole) at the K-edge of copper has been recorded using synchrotron source. Theoretical model has been generated and fitted to the experimental EXAFS data. The obtained structural parameters, which include bond-lengths, coordination numbers and thermal disorders, have been reported.

2. Experimental details

The complex $[Cu(im)_4(NO_3)_2]$ has been prepared and characterized by standard methods [1]. The K-absorption spectra at the K-edge of copper in the complex has been recorded at the BL-8 dispersive EXAFS beamline at Indus-2 synchrotron, RRCAT, Indore [2-4]. The beamline has 460mm long Si(111) crystal having 2d value equal to 6.2709Å mounted on an elliptical bender, which can bend the crystal to take the shape of an ellipse (the elliptical optics offers minimum aberration). The radiation transmitted through the sample has been detected by a position sensitive CCD detector having 2048

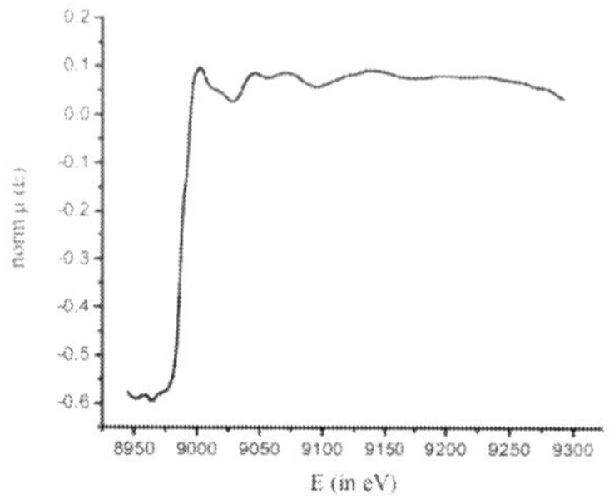

Figure 1.(a). Normalizesd μ(E) spectrum of complex at the K-edge of copper.

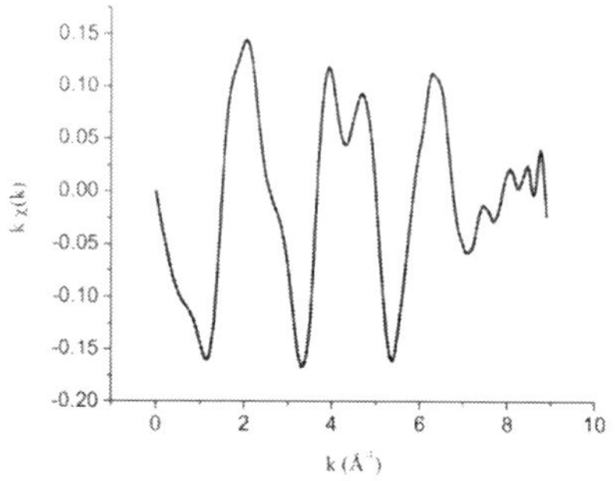

Figure1(b). χ(k) spectrum obtained from (a)

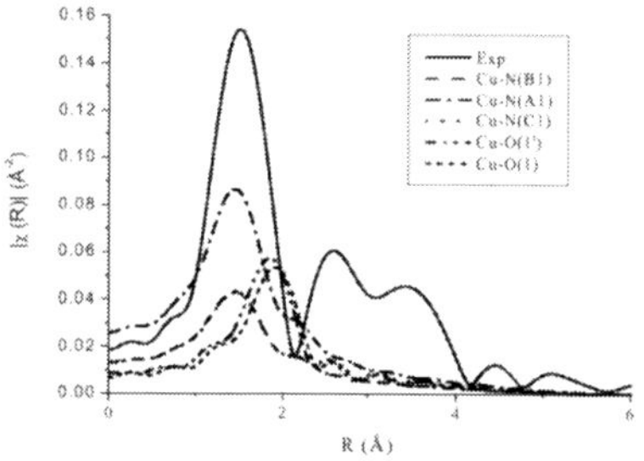

Figure 1.(c). The theoretically calculated contributions to χ(R) by single scattering paths. The experimental χ(R) is also given.

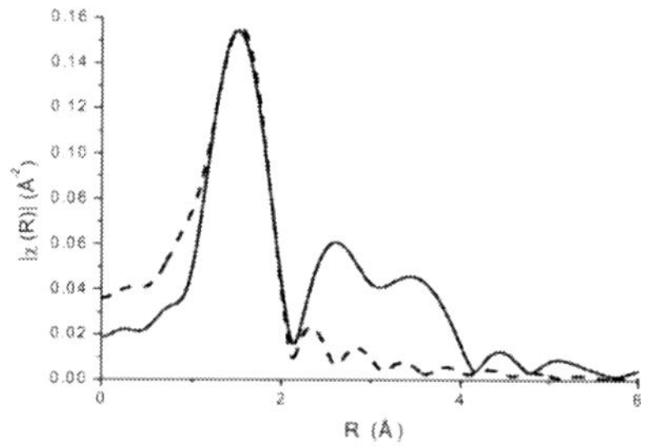

Figure1. (d) Fourier transformed EXAFS data of the complex. Solid line is experimental data and dashed line is modelled fit.

2048 pixels. The whole absorption spectrum can be recorded simultaneously in a short duration of even a few micro seconds in case of a rich sample. The beamline has a resolution of 1eV at the photon energies of 10 KeV. The plot of absorption versus photon energy is obtained by recording the intensities Io and It, as the CCD outputs, without and with the sample, respectively and using the relation, $I_t = I_0e^{-\mu x}$, where μ is the absorption coefficient and x is the thickness of the absorber. For energy calibration of that particular setting, absorption spectra at Cu K-edge of Cu metal foil and at Lu L_3-edge of Lu_2O_3 powder have been recorded under the same setting of crystal bender and the goniometer. Taking the values of the energies of Cu K-edge in metal and Lu L_3-edge in Lu_2O_3 as 8980.5 and 9249 eV, respectively, the CCD channels have been calibrated [5].

3. Data analysis

The EXAFS data has been analyzed by using the computer programs *Athena* version 0.8.056 and *Artemis* version 0.8.012 [6]. These programs include *FEFF6L* for generation of the theoretical EXAFS models and *FEFFIT* for parameter optimization of the model. The analysis procedure can be divided into two stages: reduction of the measured absorption spectra to EXAFS and analysis of the EXAFS data to obtain the structural parameters.

The spectrum is first normalized by regressing a linear function to the pre-edge region and by regressing a linear or quadratic function to the post-edge region. Normalized μ(E) spectrum is

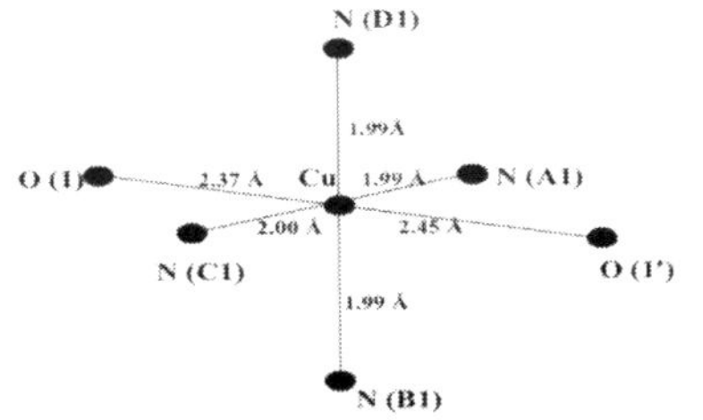

Figure 2. Coordination geometry about the copper metal in the complex $[Cu(im)_4(NO_3)_2]$, based on reference [1].

produced by subtracting the pre-edge line from the entire data spectrum and then dividing the spectrum by the step height [7]. Normalized $\mu(E)$ verses E (energy) curve for the complex under investigation is shown in figure1(a). Normalized $\mu(E)$ data is then converted into the EXAFS signal $\chi(k)$ data (figure 1(b)), which is then Fourier transformed (figure 1(c)). The Fourier transformed data is then fitted with the theoretical model in R-space to obtain the different fitting parameters.

The theoretical model is written as a sum of paths of the contribution from all scattering paths of photoelectron that travels from the absorbing atom and then scattered from one or more neighbouring atoms and finally returns to the absorbing atom [7]

$$\chi(k) = \sum_i [N_i S_0^2 F_{eff_i}(k)/kR_i^2]\cdot \sin[2kR_i + \phi_i(k)]\exp(-2\sigma_i^2 k^2)\exp[-2R_i/\lambda(k)], \quad (1)$$

with $R_i = R_{0i} + \Delta R_i$ and $k^2 = [2m_e(E - E_0 + \Delta E_0)/\hbar]$

In the above equations, the term F_{eff}, $\phi_i(k)$ and $\lambda(k)$ are the effective scattering amplitude, the phase shift and the mean free path of the photoelectron respectively, all of which can be calculated by FEFF program. The term R_i is the half path length of the photoelectron (the distance between the absorber and the coordinating atom for a single scattering event). The value of R_{0i} is the half path length used in the theoretical calculation which can be modified by ΔR_i. Each path has several adjustable parameters optimized by the computer code FEFFIT to fit the data. These parameters include the passive electron-reduction factor (S_0^2), the number of identical paths (N_i), the relative mean square displacement of the atoms included in the path (σ_i^2), an energy shift for each path (ΔE_0) and a change in the path length (ΔR_i).

The following parameters have been used as input in Artemis for calculating the theoretical model:
Space group: Pna21
Cell constants: a = 13.85Å, b = 9.83Å, c = 13.39Å
Cluster size = 8 Å

The remaining input parameters are listed in table1. The theoretically calculated contributions to $|\chi(R)|$ by the single scattering paths (1) three Cu-N and (2) two Cu-O paths along with experimental $|\chi(R)|$ are shown in figure 1(c). One of the Cu-N paths has degeneracy 2. For generating the theoretical models, we have followed the procedure outlined by Kelly et al. [7]. Accordingly, a single value of S_0^2 and ΔE_0 is used for all the paths in the fitting but different values ΔR and σ_i^2 can be used for different paths. Hence for the theoretical model, same ΔR and σ_i^2 values have been used for the three Cu-N paths and another values (i.e., different from the values used for the first three paths) for the remaining two Cu-O paths. The resulted fitting curve is shown in figure 1(d).

4. Results and discussion

The energy of the K-edge of copper in the given complex is found to be 8986.7 eV. Taking the energy of copper metal K-edge as 8980.5 eV, the chemical shift is 6.2 which indicates that copper is in +2 oxidation state in the complex [8].

Table 1. The input positional parameters for the complex [Cu(im)4(NO3)2] from reference [1].

S.No	Element	X	Y	Z
1	Cu	0.1244	0.0064	0.2500
2	N(A1)	0.1251	0.1302	0.3695
3	N(B1)	0.2686	-0.0108	0.2601
4	N(C1)	0.1240	-0.1196	0.1311
5	N(D1)	-0.0202	0.0188	0.2415
6	O(1)	0.1016	-0.2071	0.3628
7	O(1')	0.1501	0.2156	0.1436

Table 2. The EXAFS fitting results for [Cu(im)4(NO3)2] and their comparison with XRD results.

EXAFS results					XRD results
Atomic pair	N	R(Å)	ΔR(Å)	σ^2 (Å^{-2})	R(Å)
Cu-N(A1)	2	1.99	-0.020670	0.0099	2.010
Cu-N(B1)	1	1.99	-0.020670	0.0099	2.008
Cu-N(C1)	1	2.00	-0.020670	0.0099	2.018
Cu-O(1)	1	2.37	-0.153954	0.0001	2.606
Cu-O(1')	1	2.45	-0.153954	0.0001	2.526

For analysis of the EXAFS data of the given complex, the input parameter, R_{bkg}, that determines the maximum frequency of the background was set to 1.10 Å. Fourier transform was performed over k-range: k_{min}=2.553Å-1, k_{max} = 8.860 Å^{-1}. Theoretically modelled data were fitted in the R-space to the experimental data using k_w = 1. Fitting was performed for coordination shells in the R range of 1.10 - 4.0 Å. In the fitting procedure, we used the first five paths obtained from the FEFF calculations. The results obtained from fitting are given in table 2, which gives the local structural parameters obtained from the analysis. The S_0^2 value obtained is 0.9, ΔE_0 is also reasonable, i.e., 4.19 eV. Our results for the bond lengths, with ΔR ranging from 0.02 to - 0.15 Å, are given in table 2 and are in good agreement with the crystallographic bond distances reported by McFadden et al. [1]. The distances of the four coplanar imidazole nitrogen atoms which define the basal plane and two unidentate nitrato-oxygen atoms of the tetragonally distorted octahedral at copper (figure 2) i.e., Cu-N(A1), Cu-N(B1), Cu-N(C1), Cu-N(D1), Cu-O(1') and Cu-O(1) are respectively 1.99 Å, 1.99 Å, 2.00 Å, 1.99 Å, 2.45 Å and 2.37 Å. Based on the structure reported in reference [1], the coordinate geometry around copper atom has been depicted in figure 2, showing these distances.

5. Conclusion

The aim of the present work was to study the coordinate geometry of the complex $[Cu(im)_4(NO_3)_2]$ about the copper metal using the EXAFS spectroscopy. The structural parameters for the complex, obtained by fitting the theoretical model to the experimental EXAFS data, have been found to be in good agreement with the crystallographic data. Further, the copper metal in the complex has been found to be in +2 oxidation state.

References

[1] McFadden D L, McPhail A T, Garner C D and Mabbs F E 1976 *J. Chem. Soc. Dalton Trans.* **47**

[2] Das N C, Jha S N, Bhattacharyya D, Poswal A K, Sinha A K and Mishra V K 2004 *Sadhana* **29** 545

[3] Bhattacharyya D, Poswal A K, Jha S N, Sangeeta and Sabharwal S C 2009 *Bull. Mater. Sci.* **32** 103

[4] Bhattacharyya D, Poswal A K, Jha S N, Sangeeta and Sabharwal S C 2009 *Nucl. Instrum. Methods* A **609** 286

[5] Gaur A, Johari A, Shrivastava B D, Gaur D C, Jha S N, Bhattacharyya D, Poswal A and Deb, S K 2011 *Sadhana* **36** 339

[6] Ravel B and Newville M 2005 *J. Synchrotron Rad.* **12** 537

[7] Kelly S D, Hesterberg D, Ravel B 2008 *Chapter 14 in Methods of Soil Analysis, Part 5, Mineralogical Methods*, (Medison, Soil Science Society of America) 387

[8] Kau L S, Spira-Solomon D J, Penner-Hahn J E, Hodgson K O and Solomon E I, 1987 *J. Am. Chem. Soc.* **109** 6433

X-ray absorption studies of copper (II) mixed ligand complexes with ethylenediamine (en) as one of the ligands

V K Hinge[a], S K Joshi[b], B D Shrivastava[c], J Prasad[d] and K Srivastava[d]
[a]Physics Department, Ujjain Engineering College, Ujjain 456010, M.P.
[b]Government Arts and Science College, Ratlam 457001, M.P.
[c]School of Studies in Physics, Vikram University, Ujjain 456010, M.P.
[d]Chemistry Department, University of Allahabad, Allahabad 211002, U.P.

E-mail: vijayhinge1@rediffmail.com

Abstract. X-ray K-absorption spectra of six biologically important copper (II) complexes having ethylenediamine (en) as primary ligand and nitrate, thio-cynate, perchlorate, bromide, chloride and sulphate as secondary ligands have been studied using a bent crystal 0.4 m Cauchois-type transmission spectrograph. Various parameters, e.g., chemical shift, energy position of the principal absorption maximum, effective nuclear charge and edge-width have been determined. The K-absorption edges have been found to split into two components, i.e., K_1 and K_2 in all of these complexes. All the complexes have been found to have copper in oxidation state +2. The relative ionic character of the complexes has also been determined and correlated with spectrochemical series.

1. Introduction

The X-ray absorption fine structure analysis can give details about the electronic structure and the local arrangement of atoms around the absorbing atom. X-ray absorption near edge structure (XANES) is a powerful and promising tool for determination of the local structure of samples. Ethylenediamine is a well-known chelating agent. A chelating agent is a substance whose molecules can form several bonds to a single metal ion. In other words, a chelating agent is a multidentate ligand. A single molecule of ethylenediamine can form two bonds to a transition-metal ion such as Cu^{2+}. The bonds are formed between the metal ion and the nitrogen atoms of ethylenediamine. In the present work XANES spectra of six mixed ligands complexes of Cu(II) having ethylenediamine (en) as one of the ligand, has been studied.

2. Experimental

The six complexes studied in the present investigations are $Cu(en)_2(ClO_4)_2$, $Cu(en)_2(ONO_2)_2$, $Cu(en)_2(SCN)_2$, $Cu(en)_2Cl_2.H_2O$, $Cu(en)_2Br_2.H_2O$, $Cu(en)_2SO_4$ where en = ethylenediamine. All the complexes were prepared according to the standard methods reported in literature and their purity was checked [1]. A sealed Machlett tube with a tungsten target was employed as X-ray radiation source and was operated at 17 kV and 15 mA. The X-ray absorption spectra at the K-edge of copper were recorded on a bent crystal 0.4 m Cauchois-type transmission spectrograph employing a mica crystal, oriented to reflect from (100) planes, for dispersion. The absorption screens were prepared by spreading about 30 mg of substance on 1 cm^2 area of cellophane tape and covering it by another piece of cellophane tape. The exposure time was 2-3 hrs on Kodak X-ray films. For calibrating the spectra, the tungsten emission lines $WL\alpha_2$, $WL\beta_4$ and $WL\beta_6$ were used as reference lines and their energies

Published under licence by IOP Publishing Ltd

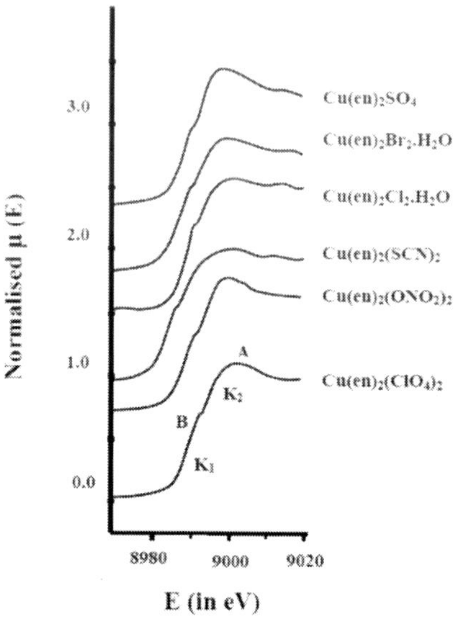

Figure 1. Copper K-edge XANES spectra for the copper complexes. The different spectra have been shifted vertically for better presentation

were taken from the tables of Deslattes et al [2]. The dispersion of the spectrograph was about 0.8 eV/10^{-2} mm of the X-ray film. The resolution of the spectrograph was tested by recording the Mo Kβ1,3 doublet as two separate lines, which was adequate for this study. The measurements at the K absorption edge were made on digital records obtained using a Carl-Zeiss G II microphotometer. The digital records were analyzed using computer programs Origin and Athena [3] were used for data analysis. Derivative spectra were used for obtaining the K-edge position.

3. Results and Discussion

The shapes of the copper K-absorption discontinuity and the associated near edge structure (XANES) for all the complexes are shown in Fig 1. The curves in this figure represent the normalized K absorption spectra. The energies of the copper K-edge (E_{K1} and E_{K2}) and the principal absorption maximum (E_A) along with the values of the edge-width (E_A-E_{K1}) and the chemical shift ΔE_{K1} are given in Table 1.

3.1 Chemical shift

The shift of the X-ray absorption edge i (i = K, L, M....) of an element in a compound/complex with respect to that of the pure element is written as:

$$\Delta E_i = E_i(\text{complex}) - E_i(\text{metal})$$

It is well known that the position of the K-edge depends on the valence state of the absorbing ion. In general, the chemical shift is towards high energy side of the metal edge , it increases progressively with increase of the valency of the cation, unless the shift is either suppressed by the covalent character of the bond or enhanced by the formation of the metal-metal bonding [4]. In the present work, the K-absorption edge for copper is found to be shifted towards the high energy side in all the six complexes studied. It is well established that on compound formation, usually the interatomic distance increases and dehybridization sets in. Consequently, the 4p band becomes narrower and no more admixes with 3d and 4s bands before it finds an empty 4p state. Transition from 1s to 4p levels thus, gives rise to the shift of K-edge in the complexes under present study towards the high energy side. The values of chemical shifts obtained for all the complexes are tabulated in Table 1. For the present complexes, the chemical shift values lie in the range 4.1 to 9.7 eV. Hence on the basis of values of the chemical shifts, all our complexes are found to have copper in oxidation state +2.

The chemical shift values of the $Cu(en)_2Cl_2.H_2O$ and $Cu(en)_2Br_2.H_2O$ complexes of the present work are 8.5 and 7.8 eV respectively. Shift for the chloro complex is greater than the shift for the

Table 1. Copper K-edge XANES data for the copper complexes

S.NO.	Complex	E_{K1} (eV)	E_{K2} (eV)	E_A (eV)	Chemical Shift (E_{K1}-8980.5) (eV)	Edge width (E_A-E_{K1}) (eV)	Z_{eff}
	Copper metal	8980.5	8988.2	8992.5	-	12.0	-
1	$Cu(en)_2(ClO_4)_2$	8989.1	8994.7	9000.3	8.6	11.2	1.00
2	$Cu(en)_2(ONO_2)_2$	8990.2	8994.9	9000.6	9.7	10.4	1.10
3	$Cu(en)_2(SCN)_2$	8984.6	8998.4	9002.3	4.1	17.7	0.52
4	$Cu(en)_2Cl_2.H_2O$	8989.0	8994.4	9003.0	8.5	14.0	0.99
5	$Cu(en)_2Br_2.H_2O$	8988.3	8993.9	9002.5	7.8	14.2	0.92
6	$Cu(en)_2SO_4$	8988.6	8992.3	8998.4	8.1	9.8	0.95

bromo complex. This has been attributed to the large polarizability of bromine as compared to the chlorine. Further on the Pauling's electronegative scale, the values for chlorine and bromine are 3.0 and 2.8 respectively and hence the metal-chlorine bond is more ionic as compared to metal- bromine bond. Thus, the contribution due to such a bonding towards chemical shift in the chloro complex is more than that of bromo complex.

The chemical shift values of the $Cu(en)_2(ONO_2)_2$ and $Cu(en)_2(ClO_4)_2$ complexes of the present work are 9.7 and 8.6 eV respectively. Earlier workers have reported the chemical shift values of $Cu(Ø_3PO)_2$ $(ONO_2)_2$ and $Cu(Ø_3PO)_2$ $(ClO_4)_2$ copper(II) complexes as 12.6 and 9.7 eV respectively. Similarly for the complexes $Co(oatsc)_2$ $(NO_3)_2$ to $Co(oatsc)_2$ $(ClO_4)_2$ the shift values are 7.1 and 6.5 eV respectively. From above discussion, it is obvious that $(ONO_2)_2$ and $(ClO_4)_2$ ligands follow the same trend for chemical shift variation irrespective of the fact that the other ligand in the mixed ligand complexes may be $(Ø_3PO)_2$, $(oatsc)_2$ or even $(en)_2$ ligand. The sequence of variation of the chemical shift remains the same for both the transition metals, i.e., copper and cobalt.

The chemical shift in the three complexes $Cu(en)_2Cl_2.H_2O$,$Cu(en)_2Br_2.H_2O$ and $Cu(en)_2(ONO_2)_2$ follows the order $Br^- < Cl^- < ONO_2^-$. This order can be explained on the basis of the order of these ligands in the spectrochemical series, which is also the same, i.e, $Br^- < Cl^- < ONO_2^-$. Thus, the variation of the chemical shifts follows the same sequence as in the spectrochemical series for the three ligands Br^-, Cl^- and ONO_2^-.

The chemical shift of 8.1 eV for $Cu(en)_2SO_4$ and 4.1 eV for $Cu(en)_2(SCN)_2$ are less than those for nitro and perchlorato complexes. It should be noted that the sulphato group is coordinated through two of its oxygen atoms in the complex $Cu(en)_2SO_4$ and in the complex $Cu(en)_2(SCN)_2$ the copper atom has four short bonds and two long bonds frequently encountered with Cu^{2+} ions. The sulfur atom of the thiocyanate group completes a distorted octahedron, which suppresses the value of chemical shift.

The chemical shift values for all the complexes studied in the present investigation are lower than those of the corresponding simple salt. The lowering of the chemical shift values occurs due to the covalent character of the metal-ligand bond, as also expected for the N and S donor atoms of the ligand. The linkage to Cu(II) in ON_2O and ClO_4 is with one oxygen, in SO_4 it is with two oxygen, where as for Cl_2 and Br_2 complexes it is through chlorine and bromine respectively. For the SCN complex the linkage with Cu(II) is through S.

For the complexes under study, the order in which the ligands contribute to the chemical shift is: $SCN < Br < SO_4 < Cl < ClO_4 < NO_3$. As is well known, an ionic bonding enhances the chemical shift, whereas a covalent bonding suppresses it. Hence, the above order may also be taken as representative of the relative ionic character of the bonding in these complexes.

3.2 Effective nuclear charge Z_{eff}

One can define effective nuclear charge as a total charge within a certain molecular volume around the nucleus [5]. Various theoretical and phenomenological methods have been proposed for the estimation of the effective nuclear charge. In the present work, Z_{eff} has been obtained from the measured chemical shift by using the semi-experimental method by employing the procedure suggested by Nigam and Gupta [5]. The effective nuclear charge on the copper in the complexes under present study varies between 0.52 – 1.10 electrons/atom.

3.3 Splitting of the main edge

The K-edges, as shown in Fig. 1, have been found to split into two components, i.e., K_1 and K_2 in all of the complexes. In an extensive XANES study of Cu(I) and Cu(II) compounds Kau *et al.* [6] have found that a peak about 8984 eV is diagnostic of Cu(I) and has been assigned to the Cu(I) 1s→4p transition. Further, no Cu(II) complex exhibits a significant peak below 8985 eV. Berry *et al.* [7] have reported that in Cu(II) compounds a shoulder B in the absorption edge has been observed which is strongly z-polarized and has been assigned Cu(II) 1s→4p transition simultaneous with ligand to metal charge transfer. Another characteristic feature of Cu(II) is the maximum A at about 9000 eV which also corresponds to 1s→4p transition as well as to the transitions to continuum states. In the spectra of presently studied complexes, the peak like feature at 8984eV, characteristic of Cu(I), is absent. Further, in the absorption edge the shoulder B is present and the feature A at about 9000eV is also present, which are characteristic of Cu(II). Thus, copper has been found to have oxidation state +2 in the complexes, on this basis also.

3.4 Edge-width

In Table 1, we have reported the values of the edge-width (E_A-E_{K1}). In general, edge-width of the K-absorption edge increases with the increase of covalent character of the bonds provided other factors like molecular symmetry etc., remain the same [8]. In chloro, nitro, bromo, sulphato, perchlorato and thio-cynate complexes, the edge-widths values are 14.0, 10.4, 14.2, 9.8, 11.2 and 17.7 eV, respectively.

3.5 Principal absorption maximum

In Table 1, we have also included the data for the principal absorption maximum E_A in the complexes and in the metal. It has been observed that for copper metal, the value of E_A is 8992.5 eV and for all the complexes it is shifted towards the higher energy side. The shift of principal absorption maximum depends on the type of overlap between metal and ligand orbitals. The greater the overlap of the metal and the ligand orbitals, more stable are the bonding molecular orbitals and hence the corresponding antibonding molecular orbitals are more unstable. Since, the principal absorption maximum occurs due to the transitions from 1s orbital to the unoccupied antibonding molecular orbitals (1s-5t_{1u*} in octahedral), the principal absorption maximum, therefore, shifts to the higher energy side [9].

4 Conclusions

XANES study at the K-edge of copper in the six complexes studied in the present work, suggests that copper has valence state +2 in all the complexes. The effective nuclear charge on the copper in the complexes under present study varies between 0.52 – 1.10 electrons/atom. The relative ionic character has been found to follow the sequence:

$Cu(en)_2(SCN)_2 < Cu(en)_2Br_2.H_2O < Cu(en)_2SO_4 < Cu(en)_2Cl_2.H_2O < Cu(en)_2(ClO_4)_2 < Cu(en)_2(ONO_2)_2$

References

[1] Goodgame M and Haines L I B 1966 *J.Chem. Soc.A* 174
[2] Deslattes R D, Kessler E G Jr., Indelicato P, de Billy L, Lindroth E and Anton J 2003 *Rev. Mod. Phys.* **75** 35
[3] Ravel B and Newville M, *J. Synchrotron Rad.* 2005 **12** 537
[4] Agarwal B K and Verma L P 1970 *J. Phys. C: Solid State Phys.* **3** 533
[5] Nigam A K and Gupta M K 1974 *J. Phys. F: Metal Physics* **4** 1084
[6] Kau L S, Spira-Solomon D J, Penner-Hahn J E, Hodgson K O and Solomon E I 1987 *J. Am. Chem. Soc.* **109** 6433
[7] Berry A J, Hack A C Mavrogenes J A, Newville M and Sutton S R 2006 *Am. Mineral.* **91** 1773
[8] Kumar A, Nigam A N and Shrivastava B D 1981 *X-ray Spectrom.* **10** 25
[9] Rajput O P, Nigam A N and Shrivastava B D, 1984 *X-ray Spectrom.* **13** 156

Exafs data analysis of some cobalt complexes

A.Mishra[1], Vishakha Vibhute[1], Garima Jain[1], Samrath Ninama[1]
[1]School of Physics, Vigyan Bhavan, Devi Ahilya University, Khandwa Road Campus, Indore [MP] - 452001, India

Email: mishraa1960@yahoo.in, truptivibhute@ yahoo.co.in

Abstract. X-ray spectroscopy has assumed great significance with its increasing application in the different fields of physics like solid state physics, chemical physics, atomic physics, Plasma Physics and Astro physics etc.in present paper we obtained bond lengths of cobalt metal complexes using Levy's, LSS and Lytle methods and are compared by FEFFIT (Fitting of EXAFS by Fast Fourier Inverse Transform) method. Bond length basically known as radial distance between ligand and metal complexes. FEFFIT is computer program for calculating bond length obtained theoretical data of EXAFS.

1.Introduction

Present paper deals with analysis of bond length calculation by EXAFS study In the absorption edges, EXAFS ranging from 50 to about 1000eV above the edge. Lytle, Sayers and Stern related these fluctuations of the absorption coefficient to the atomic arrangement surrounding of the absorbing atom. Structural information is obtained from EXAFS either by a comparison of an unknown compound with a known model With help of EXAFS using IFEFFIT method we are find the bond lengths and structural characteristics of Cobalt complexes [1] We have used Cobalt amino acid complexes for X-ray absorption fine structure studies are- 1.Cu-Aspartic acid, and 2.Cu-Glutamic acid, Amino acids are critical to life, and have many functions in metabolism. One particularly important function is to serve as the building blocks of proteins, which are linear chains of amino acids, Cobalt is a member of first transition series elements that have been of particular interest, for the possibility of various valance states is always there due 3d incomplete shell. The electronic configuration of cobalt is (A) 3d7 4s2. Cobalt compounds are widely used to make coloring materials the transition metals are capable of showing several oxidation states. The members of the first transition series form more stable complexes with ligands congaing nitrogen and oxygen while the members of the complexes. One of the most important properties of transition elements from s and p-block elements is that they show variable oxidation states. When an X-ray beam passes through a material, its intensity falls off according to the relation

$$I(x) = I_0 e^{-\mu T}$$

where I_0 is the intensity of X-rays incident on the material, I is the intensity at a distance T in the material and μ is the linear absorption coefficient. [2]
The physics of EXAFS, was originally explained by the short range order theory of Kronig in 1932 [3-4], and confirmed experimentally by Sayers, Lytle, and Stern in 1971 [5].In this region photo-electrons have sufficient energy to effectively propagate through the sample as free particles. Extended X-ray absorption fine structure (EXAFS) spectroscopy has been a valuable probe of local structure in molecules, surfaces, and solids.

2. Experimental details

The X-ray generator manufactured by Messrs.Rich Seifert and Co.Germany,was used for the production of X-rays. Its principle components are:-X-ray tube housing type SN60,.High voltage generator and control module basic equipment n.ID 3000,&Water cooling system no.KKS3.

X-ray absorption edge studies were carried out using a conventional Siefert sealed X-ray tube with Tungsten target operating at 20 kV and 40 mA.After this process, the scanning of the X-ray films were completed on Carl-Ziess Microdensitometer coupled with a computer to convert the data into IFEFFIT analysis

3. Results and Discussion

3.1. Extended X-ray Absorption fine structure

The bond lengths of cobalt complexes were calculated with using of IFEFFIT method, after this process, it has been compared with these three different methods.

1. Levy's method.
2. LSS method
3. Lytle method

According to Levy's method bond length given by R = (151/ ΔE) 1/2 where, ΔE is the energy separation between first maxima and first minima. The LSS method is a graphical method and according to this method, the simplified equation for wave vector is given by k = (0.263E) 1/2 where, E is energy position of different maxima and minima in fine structure curve measured from point in K-absorption edge. The total phase shift δj (k) parameter was also determined using the expression δj (k) = - αjk + βj- 1/2Π, where notations have their usual meaning. Lytle method is modification of LSS method is given by the expression R= (37.60/M)1/2 Where M is slope Between E verses Q plots. The absorption energy maxima are E=A, B, C, D and Q=2.4, 6.4, 12, 20 are constant.

Table 1. The bond length in (Å) of Cobalt complexes is calculated by FEFFIT programming and it compared by LSS, Levy's, and Lytle methods.

Cobalt complex	R_{LSS}	$R_{Levy's}$	R_{Lytle}	$R_{IFEFFIT}$
Complexe1	1.47	1.40	1.58	1.21
Complexe2	1.28	1.44	1.51	1.57

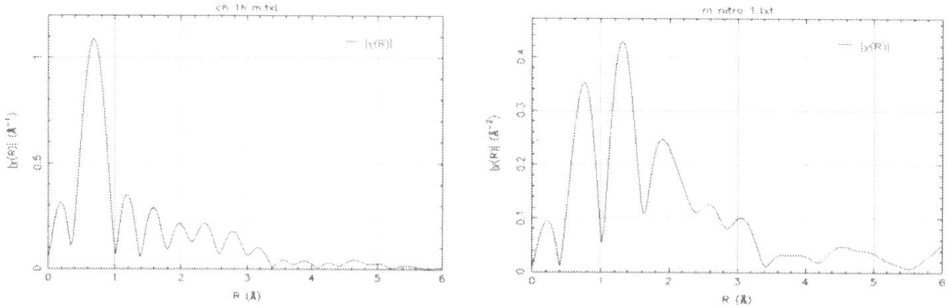

Figure1: Magnitude of Fourier transform for Cobalt complex obtained from ATHENA program.

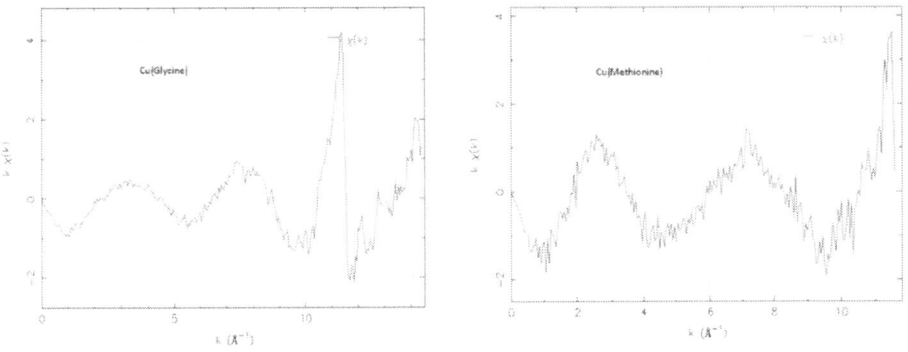

Figure2: EXAFS function for Cobalt complex obtained from ATHENA program.

4. Conclusion

Here we have compared the bond lengths of Cobalt complex experimentally by LSS, Levy's, and Lytle methods and we have also calculated theoretically by FEFFIT (ATHENA) programming. The calculated EXAFS curves for Co complexes are in good agreement with experimental EXAFS curves for all Cobalt complex studied within experimental error.

5. Acknowledgements

The authors are grateful to Dr. Pratibha Sharma, School of chemical Science, DAVV Indore, for their support in using laboratory.

References

[1] X-Ray Absorption. Principles, Applications, Techniques of EXAFS, SEXAFS and XANES, edited by Koningsberger D. C. and Prins R. (Wiley, New York,1988)
[2] 1. Nadzhakov E.G. and Barinskii R.L., 1960 *Sov. Phys.Doki* **4** 1319
[3] Bianconi A, in " EXAFS for inorganic system" Garner C D et al., 1981 *Daresbury laboratory report No. SERC/DL/ SCI/RIT.*, **13**
[4] Hernandez W, Spodine E, Richter R, Hallmeier K H, Schroder U, Beyer L Platinum(II) mixed ligand complexes with thiourea derivatives, dimethyl sulphoxid and chloride: Syntheses, molecular structures, and ESCA data Z.Anorg. Allg. Chem.629, (2003) 2559-2565.
[5] De L, Kronig Z 1931 *Phys* **70** 317

Effect of annealing on ZnO thin films grown on quartz substrate by RF magnetron sputtering

G.Anil Kumar[1], M.V.Ramana Reddy[1] and Katta Narasimha Reddy[2]
[1]Department of Physics, Osmania University, Hyderabad,A.P. India.
[2]Department of Physics, Mahatma Gandhi University, Nalgonda,A.P. India.

E-mail: anilhcu@gmail.com

Abstract: Zinc oxide thin films were grown on quartz substrates by RF magnetron sputtering technique in an Argon atmosphere with sputtering power of 50W and sputtering pressure of 2x10-2 Torr and studied the effect of annealing on the structural and optical properties. Crystalline properties of ZnO films as a function of annealing temperature were investigated using X-ray diffraction. XRD analysis revealed that the deposited films were polycrystalline in nature with strong preferential orientation of grains along the c-axis. The micro structural parameters, such as the lattice constant, crystallite size, stress and strain are calculated. The effect of annealing on the deposited films was discussed. All the films present a high transmittance of above 90% in the wavelength range of the visible spectrum and sharp absorption edge near 380 nm.

1. Introduction

Zinc oxide (ZnO), a II–VI compound semiconductor with a direct band gap of 3.3 eV, has recently received great interest as a promising candidate for the fabrication of ultraviolet (UV) light-emitting devices [1]. Its large exciton binding energy of 60 meV enables the realization of efficient exciton-related UV lasing at room temperature [2,3]. High quality ZnO thin films are therefore desirable to achieve such applications. So far, various preparation methods have been used to deposit the ZnO films, such as: RF magnetron sputtering [4], spray pyrolysis [5], chemical vapor deposition (CVD) [6], pulsed laser deposition [7], and sol–gel process [8]. Among these means, RF Magnetron sputtering is preferred due its reproducibility and for getting highly oriented films. This physical vapor deposition is used to produce films with the desired optoelectronic and nanostructure properties by precisely controlling the parameters like Ar-O2 flux deposition pressure, target-substrate distance, RF power and substrate temperature. In the present study, we have made an attempt to produce highly textured ZnO nanocrystalline films in the absence of oxygen environment on quartz substrates using RF magnetron sputtering and studied the effect of annealing treatment on the structural and optical properties.

2. Experimental details

2.1.Substrate cleaning

The substrate cleaning is very important in the deposition of thin films. The quartz substrates were first cleaned by a mild soap solution then degreased with acetone, etched with 5% of HCl for 30 min., ultrasonically cleaned by de-ionized water and finally dried in the air.

2.2.ZnO thin film deposition

ZnO thin films were grown on quartz substrates by using RF magnetron sputtering. The sputtering system is capable of creating an ultimate vacuum of 5 x 10^{-5} Torr. A power supply operated at a crystal-controlled frequency of 13.56 MHz. The sputtering chamber was pumped with turbo molecular pump and rotary pump combination. The pressure in the sputter chamber was measured using digital Pirani and Penning gauge combination. The magnetron target assembly was mounted on top of the sputter chamber such that the sputtering could be done by sputter down configuration. A high purity (99.99%) ZnO target of 2-in. diameter and 3 mm thickness was used for the sputtering. Pure argon was used as sputtering gas .The flow rate of argon gas was controlled by Tylan mass flow controller. The sputtering chamber was pumped down to the vacuum of 5 x 10-5 Torr before the introduction of Argon gas. The target was pre-sputtered for 15 min to remove any impurity on the surface of the target. A shutter was placed between the target and substrate to control period of coating, The target was water-cooled and the substrate was kept at room temperature initially. During the film deposition, the substrate temperature was increased due to the bombardments of the sputtering ions and the final substrate temperature was nearly 40–60 0C. To measure the deposition rate and film thickness, a vibrating quartz crystal was used. The sputtering conditions maintained during the growth of ZnO films are given in Table 1. ZnO films deposited at room temperature on quartz substrate were subjected to annealing treatment in air for 2 h at temperature of 200,400,600 and 800^0C, respectively.

The crystalline structure of the films was studied by x-ray diffraction measurements obtained with a Philips X'Pert X-Ray Diffractometer (XRD) in the 2θ range of 20°- 80° using CuKα radiation of wavelength λ=1.5406 Å at room temperature. X-ray tube was operated at 40 kV (Voltage) and 30 mA (current) with scanning speed of 0.5°/min. Optical transmittance spectra of films were recorded with a SHIMAZDU 3100 UV-Vis spectrometer over wavelength range from 200 to 800 nm.

3.Results and discussion

3.1.Structural properties

XRD patterns of ZnO films annealed at different temperatures are shown in Fig. 1. It is found that XRD patterns of ZnO thin films annealed at different temperatures show the same form, that is only a (002) diffraction peak is detected in the films. This indicates that ZnO films prepared by sputtering deposition show a good c-axis orientation perpendicular to the substrate and also means that orientation of prepared ZnO films is not destroyed after further annealing treatment. The diffraction peak angle of the films annealed at RT, 200, 400,600 and 800 ^{0}C appears at 33.84^0, 34.59^0,

Table 1:

Deposition parameters maintained during the deposition of ZnO films by RF magnetron sputtering .

Sputtering target	ZnO (99.99%) 2-in. diameter and 3 mm thickness
Target to substrate distance	70 mm
Substrates	quartz
Ar gas flow rate	80 SCCM
Ultimate pressure	5x10^{-5}Torr
Sputtering pressure	1.2x10^{-2} Torr
Substrate temperature	273K
Sputtering power	50 W
Thickness	100nm

34.80^0, 34.92^0 and 35^0, respectively. Compared with the (002) peak position of ZnO powder (2θ = 34.420), the diffraction angle of the films grown at RT decreases, which results in the increase of c-axis value. This indicates that ZnO films grown at RT suffer compressive stress along the interfaces. The diffraction angle of the films annealed at higher temperature increases in comparison with bulk ZnO, which indicates that c-axis values of ZnO films are shorted. So the stress in ZnO film shows tensile along the interfaces. The stress is related to the defects in the films, i.e., the compressive stress may be due to zinc interstitials, but the tensile stress may be due to the oxygen vacancies in the lattice of ZnO crystallites in the film. In addition to intrinsic stress originating from the defects (zinc interstitials or oxygen vacancies) in the ZnO lattice, the stress also originates from thermal stress, which results from the difference in coefficient of thermal expansion (CTE) of ZnO thin films and quartz substrate [9]. As the annealing tempareture increases from 200 to 800 ^{0}C, the FWHM decreases from 0.33 to 0.14 and the crystallite sizes increases from 24 to 57 nm.
The average crystallite sizes of the films annealed at different temperature have been calculated using the Scherrer's formula:

$$D = 0.94\lambda/(\beta \cos \theta) \qquad (1)$$

Where λ and β are X-ray wavelength, the Bragg's diffraction angle and the full width at half maxima (FWHM) of the peak corresponding to the "θ" value.
Dislocation density (δ) and Number of crystallites per unit area (N) have been calculated using the crystallite size and thickness of the film [10-12].
The average uniform strain (ezz) in the lattice along the c-axis in the randomly oriented ZnO films annealed at different temperature have been estimated from the lattice parameters using the following expression [13],

$$e_{zz}=(d-d_0)/d_0 \qquad (2)$$

Where "d" is the lattice parameter of the ZnO film calculated from (0 0 2) peak of XRD pattern and the "d_0" is the lattice parameter for the ZnO bulk.
For hexagonal crystals, the stress (σ) in the plane of the film can be calculated using the biaxial strain model [14]:

$$\sigma = (2C_{13} - (C_{11}+C_{12})\, C_{33}/C_{13}).e_{zz} \qquad (3)$$

here C_{ij} are elastic stiffness constants (C_{11} = 2.1 x 10^{11} N/m^2 , C_{33} = 2.1x 10^{11}N/m^2, C_{12} = 1.2 x 10^{11} N/m^2, and C_{13} = 1.05 x10^{11} N/m^2). This yields the following numerical relation for the stress derived from the change in the 'd' lattice parameter:

$$\sigma\ (N\ m^{-2}) = -\ 4.5\times10^{11}.e_{zz} \qquad (4)$$

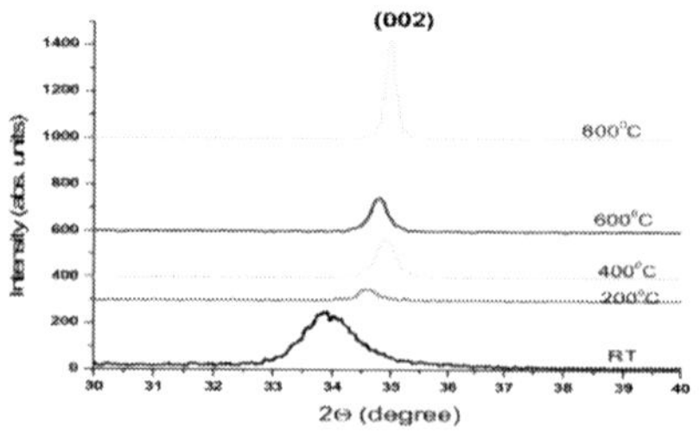

Figure.1. XRD patterns of ZnO films annealed films at different tempartures.

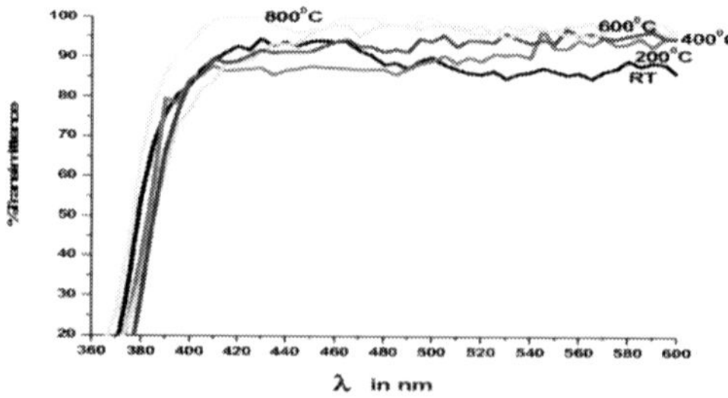

Figure. 2. Optical transmittance spectra of ZnO annealed at different tempartures.

Table 2: Structural information of ZnO thin films annealed at different temperature.

annealing temperature (^{0}C)	Lattice parameter (nm)	Grain size (nm)	Dislocation density (10^{15} linesm^{-2})	Number of grains/unit area (10^{16} m^{-2})	Strain (10^{-3})	Stress (GPa)
RT	0.5292	22.21	2.027	0.9127	17.09	-7.69
200	0.5180	24.77	1.629	0.6579	5.18	2.33
400	0.5151	26.7	1.402	0.5253	10.75	4.83
600	0.5134	21.7	2.123	0.9786	14.01	6.30
800	0.5123	57.87	0.298	0.0515	16.13	7.25

The various structural parameters for ZnO thin films annealed at different temperature were calculated and are represented in Table 2.
From the Table 2. it has been observed that the lattice constant decreases with increasing annealing temperature. The change in lattice constant for the thin film deposited over the bulk clearly suggests that the film's grains are under stress and that may be due to a change in the nature and concentration of the native imperfections. This causes either elongation (tensile) or compression of the lattice parameters. The density of film is, therefore, expected to change in accordance with changes in the lattice constant. The film crystallite size increases with increasing annealing temperature.The average internal stress for the ZnO films deposited at RT is found to be compressional in nature. The negative sign indicates that the films are in a state of compressive stress.As annealing tempareture increases the tensile stress increases. The microstrain (e) and the dislocation density (δ) are observed to exhibit a slow decreasing trend. The decrease in e and δ at higher annealing tempareture may be due to the movement of interstitial Zn atoms from inside the crystallites to the grain boundary, which leads to a reduction in the concentration of lattice imperfections. Further, the number of crystallites decreases with increasing annealing tempareture .

3.2.Optical properties

Fig. 2 shows the optical transmittance spectra of ZnO films annealed at different temperatures.All the films present a high transmittance of above 90% in the wavelength range of the visible spectrum and sharp absorption edge near 380 nm.The optical transmittance of the films increased with increase of annealing tempareture.The absorption edge was shifted towards lower wavelength with the increase of annealing tempareture.The increase in transmittance of the filmswas related to an increase in grain size of the films.

4.Conclusion

C-axis oriented ZnO films on quartz substrate have been deposited by RF magnetron sputtering technique and have been thermally annealed. Based on the result of XRD measurement, the increase of annealing temperature increases the crystalline quality, grain size and relaxes residual compressive strain. The optical transmittance of the films increased and the absorption edge was shifted towards lower wavelength with increase of annealing temperature.

Acknowledgments

The authors thank UGC-SAP(DSA-II) for providing necessary facilities. One of the authors (GAK) thank UGC,New Delhi for awarding the JRF under the UGC scheme of RFSMS.One of us (MVRR) thank OU-DST-PURSE Programme,Osmania University, Hyderabad, for providing financial assistance to carry out this work.

References

[1] Webb J B, Williams D F, Buchanan M 1981 *Appl.Phys.Lett.* **39** 640
[2] Brett M J, McMahon R W, Affinito J, Parsons R R 1983 *J. Vac.Sci. Technol.* **A1** 352
[3] Sato H , Minami T, Tamura Y, Takata S, Mori T, Ogawa N 1994 *Thin Solid Films* **246** 86
[4] Pizzini S, Butta N, Narducci D, Palladino M 1989 *J. Electrochem.Soc.* **136** 1945
[5] Kino G S , Wagner R S 1973 *J. Appl. Phys.* **44** 1480
[6] Aranovich J A , Golmayo D , Fahrenbruch A L, Bube R H 1980 *J.Appl. Phys.* **51** 4260
[7] Jin Z C, Hamberg I, Granqvist C G, Sernelius B E, Berggren K F 1988 *Thin Solid Films* **164** 381
[8] Thareja R K, Mitra A 2000 *Appl. Phys. B: Lasers Opt.* **71** 181
[9] Zhu B L, Sun X H, Zhao X Z, Su F H, Li G H, Wu X G 2008 *Vacuum* **82** 495
[10] Zeng J N, Low J K, Ren Z M, Liew T, Lu Y F 2002 *Appl. Surf.Sci.* **197–198** 362
[11] Natsume Y, Sakata H 2002 *Mater. Chem. Phys.* **78** 170
[12] Kim Y J, Kim Y T, Yang H K, Park J C, Han J I, Lee Y E, Kim H J 1997 *Vac. Sci. Technol.A* **15** 1103
[13] Minami T, Sato H, Takata S, Ogawa N, Mouri T 1992 *Jpn. J. Appl. Phys.* **3** L1106
[14] Gupta V, Mansingh A 1996 *J. Appl. Phys.* **80** 1063

International Conference on Recent Trends in Physics (ICRTP 2012) IOP Publishing
Journal of Physics: Conference Series **365** (2012) 012032 doi:10.1088/1742-6596/365/1/012032

Origin of the Lβ2 satellites in the X-ray emission spectra of the middle-Z elements from $_{40}$Zr to $_{52}$Te

Renuka Kendurkar and B. D. Shrivastava
School of Studies in Physics, Vikram University, Ujjain-452001, India

Abstract. In this communication, the origin of the $L\beta_2$ satellites $L\beta_2^{I}$, $L\beta_2^{(b)}$, $L\beta_2^{II}$ and $L\beta_2^{(c)}$, in the L emission spectra of the middle-Z elements from $_{40}$Zr to $_{52}$Te, has been explained on the basis of multiple ionization theory. As done in earlier reports, it has been assumed that the various transitions, which may give rise to these satellites, belong to the transition scheme L_3M_x-$M_xN_{4,5}$ (where x = 1-5). The energies and the intensities of the 41 possible transitions corresponding to this transition scheme have been calculated theoretically. The energies have been calculated using the available Hartree-Fock-Slater data on K-LM and L-MN Auger transition energies. The intensities of the various transitions have been estimated by considering cross sections for Coster-Kronig transitions as well as for M-shell shake-off process. Theoretical $L\beta_2$ satellite spectra have been computed. There are four intense peaks in the theoretical $L\beta_2$ satellite spectra which have been identified as the four satellites $L\beta_2^{I}$, $L\beta_2^{(b)}$, $L\beta_2^{II}$ and $L\beta_2^{(c)}$. Consequently, the satellites have been assigned those intense transitions which have been found to give rise to the corresponding four peaks in the theoretical $L\beta_2$ satellite spectra. The present work is an improvement over the earlier reports.

1. Introduction

In the elements $_{40}$Zr to $_{52}$Te, the intense diagram line $L\beta_2$ is accompanied with its satellites on the higher energy side which have been designated as β_2^{I}, $\beta_2^{(b)}$, β_2^{II} and $\beta_2^{(c)}$ [1]. Soni and Poonia [2] have theoretically investigated the origin of these satellites. They have suggested that all these satellites arise mainly due to the transition array $2p_{3/2}^{-1}3x^{-1}$ - $3x^{-1}$ $4d_{3/2,5/2}^{-1}$ (x = s, p, d), i.e., L_3 M_x - M_x $N_{4,5}$ (where x = 1-5). These authors have calculated relative intensities of the transitions, corresponding to this transition scheme, using McGuire's tables for Coster-Kronig transition probabilities [3]. McGuire's tables have been superseded by the tables of Chen *et al.* [4]. While McGuire's tables are based on non-relativistic calculations, the tables of Chen *et al.* are based on relativistic calculations. Hence, we thought it proper to do fresh calculations of intensities of transitions using Chen *et al.*'s tables. The values of relative intensities calculated in the present work are different than those calculated by Soni and Poonia [2]. The result is that the assignments of transitions to the satellites in the present work are slightly different than those of these authors. Also, Soni and Poonia [2] have studied selected elements only, leaving several elements between. Hence, we thought that it would be proper to study all the elements from Z = 40 to 52.

2. Theoretical calculations

2.1 Initial and final states for $L\beta_2$ satellites

As assumed by Soni and Poonia [2], we also assume that the origin of the $L\beta_2$ satellites, can be ascribed to the transition scheme $2p_{3/2}^{-1}3x^{-1}$ - $3x^{-1}$ $4d_{3/2,5/2}^{-1}$ (x = s, p, d), i.e., L_3M_x - $M_xN_{4,5}$, (x = 1-5).

Table 1. Assignments of transitions to the satellites $L\beta_2^{I}$, $L\beta_2^{(b)}$, $L\beta_2^{II}$ and $L\beta_2^{(c)}$

S. No.	Atomic number (Z)	Satellite $L\beta_2^{I}$			Satellite $L\beta_2^{(b)}$		
		Experimental value of energy [1] (eV)	Theoretical values: Energy (eV)	Theoretical values: Relative intensity	Experimental value of energy [1] (eV)	Theoretical values: Energy (eV)	Theoretical values: Relative intensity
1	40	2241.3	-	-	2243.8	2248.0	289.80
2	41	2397.3	2396.8	31.52	-	2399.0	271.20
3	42	2543.2	2549.2	28.6	2547.8	2552.9	265.50
4	43	-	2706.8	36.1	-	2710.0	279.01
5	44	2860.1	2866.9	34.27	2865.3	2870.8	237.64
6	45	3026.6	3034.8	39.79	3030.7	3038.8	260.92
7	46	3197.9	3204.6	95.54	3202.7	3208.7	227.17
8	47	3377.1	3381.1	88.88	3381.7	3385.1	192.44
9	48	3558.1	3561.1	104.80	3562.9	3565.9	188.40
10	49	3744.6	3747.1	113.68	3749.6	3753.1	179.27
11	50	3937.4	3938.0	35.83	3941.7	3944.1	180.98
12	51	4133.9	4134.1	122.66	-	4141.0	175.92
13	52	4337.1	4335.2	125.71	-	4342.0	232.99
		Satellite $L\beta_2^{II}$			Satellite $L\beta_2^{(c)}$		
1	40		2251.0	190.60	-	2255.0	212.98
2	41	2247.6	2403.0	184.97	-	2406.0	158.84
3	42	-	2556.8	173.70	2554.5	2560.1	178.90
4	43	2550.6	2714.0	185.03	-	2718.0	182.10
5	44	-	2874.8	154.01	2874.5	2878.0	141.76
6	45	2869.7	3043.1	160.08	3041.5	3047.9	116.60
7	46	3035.8	3212.1	120.93	3213.5	3215.1	88.81
8	47	3207.1	3388.0	141.28	3392.9	3391.4	153.70
9	48	3386.1	3569.1	131.29	3573.6	3573.1	192.70
10	49	3567.4	3756.1	163.12	3760.1	3759.1	131.53
11	50	3753.8	3948.1	200.50	3953.0	3951.1	171.81
12	51	3947.1	4144.0	157.90	-	4148.0	133.60
13	52	4144.3	4348.0	112.50	-	4355.2	111.20

Considering the dipole selection rules, the 41 possible transitions have been found out corresponding to this transition scheme. The energies and intensities of these transitions have been calculated as follows.

2.2 Calculation of energies of transitions in doubly ionized atoms

The formula for calculation of transition energy of say, L_3M_x - $M_xN_{4,5}$ transition, has been written as –

$$E\,(L_3M_x - M_xN_{4,5}) = E\,(K\alpha_1) + E\,(L_3 - M_xN_{4,5}) - E\,(K - L_3M_x)$$

The energy of $K\alpha_1$ line, i.e., E (K - L_3) has been taken from the tables of Cauchois and Senemaud [1] instead of the earlier tables of Bearden and Burr [5] used by Soni and Poonia [2]. The energies of the Auger transitions L_3 - $M_xN_{4,5}$ and K - L_3M_x have been taken from the tables of Larkins [6].

2.3 Calculation of intensities of transitions in doubly ionized atoms

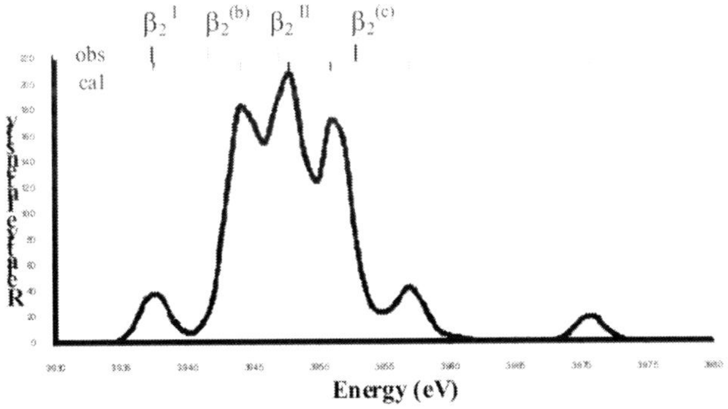

Figure 1. The computed L_3M_x - M_x $N_{4,5}$ spectrum of $_{50}Sn$. The experimentally observed and theoretically calculated satellite positions in the region of $L\beta_2$ satellites are shown at the top.

The probabilities of creation of a single hole states, $1s^{-1}$ and $2p_{3/2}^{-1}$ (K^{-1} and L_3^{-1}), have been calculated by using Moore *et al.*'s method [7]. Once the singly ionized state has been created, it is the probability of a particular subsequent process that will lead to the formation of two-hole state. The single hole so created may get converted through a Coster-Kronig transition to a double hole state. In the present work, the value of the probabilities of formation of double hole states via this process have been taken from the tables of Chen *et al.* [4].

The singly ionized state may get converted to doubly ionized state also by an associated shake-off of an electron. We have determined the shake-off transition probabilities by interpolation method from the values reported by Carlson and Nestor [8].

Finally, the total probability of creation of doubly ionized initial state, say, L_3M_x has been determined by adding both the cross sections calculated above for the two processes, i.e., by Coster-Kronig transition and by shake-off process. For calculating the relative intensities of the transitions of each multiplet, this probability has then been statistically distributed among all the allowed transitions from this set of levels, considering first of all the multiplets of super multiplets from various $^{(2S+1)}L$ levels of the set and then using the tables of White and Eliason as given by Condon and Shortley [9].

2.4 Synthesis of satellite spectra

A composite spectrum formed by spectral lines emitted by the 41 transitions for the elements has been computed by taking each as a Gaussian line. For this we have taken energy on x-axis and intensity on y-axis. The peak height of each line is taken equal to transition probability and peak position on x-axis is taken at the energy of the transition. The theoretical satellite spectrum obtained for the element $_{50}Sn$ is given in the figure 1 as a representative spectrum for all these elements under consideration.

3. Results and discussion

The $L\beta_2$ satellite spectra as computed in the present study for the middle-Z elements have four peaks, which have been identified by us as the four observed satellites: $L\beta_2^{I}$, $L\beta_2^{(b)}$, $L\beta_2^{II}$ and $L\beta_2^{(c)}$. From the analysis of the transitions which give rise to these four peaks, the assignments of the transitions, which can be responsible for the origin of these four satellites, have been made. Table 1 gives the theoretical energies and relative intensities of the satellites along with experimental values of the energy.

3.1 $L\beta_2^{I}$ satellite

The first peak in the computed spectra has been identified as the $L\beta_2^{I}$ satellite. Considering the various transitions which give rise to this first peak, we have found that the transitions $^3D_3$ - $^3F_4$ (3p) and $^3D_3$ - $^3D_3$ (3p) are the common transitions in all the elements under consideration. As a result, we have assigned these transitions for the origin of the satellite $L\beta_2^{I}$ in all the elements. The transition $^3F_4$ - $^3F_4$

(3d) has also been found to contribute to the intensity of the satellite $L\beta_2^I$ in the elements with Z = 46 to 52.

3.2 $L\beta_2^{(b)}$ satellite

The second peak in the computed spectra has been identified as the $L\beta_2^{(b)}$ satellite. We have found that the intense transitions $^3F_4$ - $^3G_5$ (3d) and $^3F_4$ - $^3D_3$ (3d) along with the weak transitions $^1D_2$ - $^1F_3$ (3d), $^1D_2$ - $^1P_1$(3d), $^3F_4$ - $^3G_4$ (3d) and $^3P_2$ -$^3D_3$ (3s) , i.e., $^3D_3$ - $^3F_4$ (3p) and $^3D_3$ - $^3D_3$ (3p) are the common transitions in all the elements under consideration, which give rise to this peak. Hence, the $L\beta_2^{(b)}$ satellite has been assigned mainly to these two intense transitions and the six weak transitions in all the elements. The transition $^3F_4$ - $^3F_4$ (3d) has also been found to contribute to the intensity of the satellite $L\beta_2^{(b)}$ in the elements with Z = 40 to 45.

Transition $^3F_4$ - $^3F_4$ (3d) deserves special mention because it contributes to the intensity of the satellite $L\beta_2^{(b)}$ in the elements from Z = 40 to 45, but in the elements with Z > 45, it contributes to the intensity of the satellite $L\beta_2^I$.

3.3 $L\beta_2^{II}$ satellite

The third peak in the computed spectra has been identified as the $L\beta_2^{II}$ satellite. Considering the various transitions which give rise to this third peak, we have found that the transitions $^3D_3$ - $^3F_4$ (3d), $^3D_3$ - $^3P_2$ (3d) and $^3D_2$ - $^3F_3$ (3d) with the weak transitions $^3D_2$ - $^3D_1$ (3d), $^3D_3$ - $^3D_2$ (3d) and $^3D_2$ - $^3F_2$ (3d) are the common transitions for all the elements which give rise to this peak. Thus, according to the present study, these three intense transitions are responsible for the origin of $L\beta_2^{II}$ satellite along with the three weak transitions.

3.4 $L\beta_2^{(c)}$ satellite

The fourth peak in the computed spectra has been identified as the $L\beta_2^{(c)}$ satellite. Considering the various transitions which give rise to this fourth peak, we have found that the intense transitions $^1F_3$ - $^1G_4$ (3d) and $^1F_3$ - $^1D_2$ (3d) along with weak transitions transitions $^3P_1$ - $^3D_2$ (3d), $^3P_1$ - $^3P_2$ (3d), $^3P_1$ - $^3S_1$ (3d), $^3P_1$ - $^3P_0$ (3d) and $^3P_0$-$^3D_1$ (3d) can be attributed to the origin of the $L\beta_2^{(c)}$ satellite in all the elements under consideration with Z = 40 to 52.

Finally, whereas, Soni and Poonia [2] have studied selected elements only, leaving several elements between, we have studied all the elements from Z = 40 to 52.

4. Conclusions

The $L\beta_2$ theoretical satellite spectra have been computed for the middle-Z elements from $_{40}$Zr to $_{52}$Te. The theoretical spectra show four peaks which have been identified as the four satellites $L\beta_2^I$, $L\beta_2^{(b)}$, $L\beta_2^{II}$ and $L\beta_2^{(c)}$. From the analysis of the transitions which give rise to these four peaks the assignments of the transitions to these four satellites have been done. Our assignments are slightly different than those reported by Soni and Poonia [2].

5. References

[1] Cauchois Y and Senemaud C 1978 *Tables Internationnal de constantes selectionnees. 18.Longeurs d'onde des emissions X et des discontinuites d'absorption X* (London: Pergamon Press)
[2] Soni S N and Poonia S 2000 *J. Phys.Chem. Solids* **61** 1509
[3] McGuire E J 1971 *L-Shell Auger, Coster-Kronig and radiative matrix elements, and Auger and Coster-Kronig transition rates in j-j coupling* (Report No. Sc-RR-710075) (New Mexico: Sandia Laboratories)
[4] Chen M H, Crasmann B and Mark H 1979 *At. Data Nucl. Data Tables* **24** 13.
[5] Bearden J A and Burr A F 1965, *Atomic energy levels* (Oak Ridge, Tennesse, USA: U.S. Atomic Energy Commission)

[6] Larkins F P 1977 *At. Data Nucl. Data Tables* **20** 311
[7] Moores D L, Golden L B and Sampson D H 1980 *J. Phys.* B **13** 385
[8] Carlson T A and Nestor C W Jr. 1973 *Phys. Rev.* A **8** 2887
[9] Condon E U and Shortley G H 1967 *The theory of atomic spectra* (Cambridge: Cambridge University Press) 241

Coordination geometry around copper in a Schiff-base trinuclear copper complex using EXAFS spectroscopy

Abhijeet Gaur[a], B. D. Shrivastava[a], D. C. Gaur[b], J. Prasad[c], K. Srivastava[c], S. N. Jha[d], D. Bhattacharyya[d] and A. Poswal[d]
[a]School of Studies in Physics, Vikram University, Ujjain-452001, India
[b]Government P. G. College, Rajgarh (Biaora)-465611, India
[c]Department of Chemistry, University of Allahabad, Allahabad 211002, India.
[d]Applied Spectroscopy Division, Bhabha Atomic Research Centre, Mumbai-400085, India

Email: abhijeetgaur9@gmail.com

Abstract. In the present investigation, we have studied extended X-ray absorption fine structure (EXAFS) spectra of a trinuclear Schiff-base copper complex tetraaqua-di-μ3-(N-salicylidene-DL-glutamato)-tricopper(II)heptahydrate, $[Cu_3(C_{12}H_{10}NO_5)_2\ (H_2O)_4].\ 7H_2O$, in which three metal sites are present. One metal site is square-pyramidal (4+1) and other two similar metal sites are tetragonally distorted octahedral (4+2). EXAFS has been recorded at the K-edge of copper in the complex at the dispersive EXAFS beamline at 2 GeV Indus-2 synchrotron source at RRCAT, Indore, India. The analysis of EXAFS spectra of multinuclear metal complexes pose some problems due to the presence of many absorbing atoms, even when the absorbing atoms may be of the same element. Hence, using the available crystal structure of the complex, theoretical models have been generated for the different copper sites separately, which are then fitted to the experimental EXAFS data. The two coordination geometries around the copper sites have been determined. The contributions of the different copper sites to the experimental spectrum have been estimated. The structural parameters, which include bond-lengths, coordination numbers and thermal disorders, for the two types of copper sites have been reported. Further, copper has been found to be in +2 oxidation state at these metal sites.

1. Introduction

The extended X-ray absorption fine structure (EXAFS) spectroscopy is an effective technique for selectively investigating the local coordination environment around the metal active site of complexes. However, the analysis of EXAFS spectra of multinuclear metal complexes pose some problems due to the presence of many absorbing atoms, even when the absorbing atoms may be of the same element [1]. In the present work, EXAFS spectra of a Schiff-base copper complex catena-tetraaqua-di-μ_3-(N-salicylidene-DL-glutamato)-tricopper(II)heptahydrate, $[Cu_3(C_{12}H_{10}NO_5)_2(H_2O)_4].7H_2O$, formed from amino acid and salicylaldehyde have been investigated. In this complex, each Schiff-base ligand is hexadentate, coordinated through one N and four O atoms to two Cu atoms in the trimeric unit and through one carboxyl O atom to Cu in the adjacent trimeric unit, thus forming a polymeric structure [2]. This trinuclear Schiff-base copper complex in which there are three metal sites, one site being square-pyramidal (4+1) and other two similar sites being tetragonally distorted octahedral (4+2) is an interesting sample from the point of view of analysis of its EXAFS spectral data.

Table 1. The EXAFS fitting results for square -pyramidal structure

Atomic pair	EXAFS results				XRD results
	N	R(Å)	delr(Å)	σ^2(Å^{-2})	R
Cu1-O3	1	1.90	-0.008	0.0045 ± 0.0040	1.91
Cu1-N1	1	1.92	-0.008	0.0045 ± 0.0040	1.93
Cu1-O1	1	1.95	-0.008	0.0045 ± 0.0040	1.96
Cu1-O4	1	1.97	-0.008	0.0045 ± 0.0040	1.98
Cu1-O5	1	2.26	-0.221	0.0081 ± 0.0045	2.48

Table 2. The EXAFS fitting results for tetragonally distorted octahedral structure

Atomic pair	EXAFS results				XRD results
	N	R(Å)	delr(Å)	σ^2(Å^{-2})	R
Cu1-O6	2	1.89	-0.052	0.0061 ± 0.0017	1.94
Cu1-O2	2	1.90	-0.052	0.0061 ± 0.0017	1.95
Cu1-O7	2	2.70	-0.004	0.0174 ± 0.0167	2.71

2. Experimental

Copper complex has been prepared by following the standard methods. The prepared complexes were characterized by standard chemical methods [2]. The complex was finely powdered and then absorption screens were prepared by carefully spreading calculated amount of the powder on 1 cm^2 area of kapton tape. The copper K-edge EXAFS spectra of these absorption screens have been recorded at the BL-8 dispersive EXAFS beamline at 2 GeV Indus-2 synchrotron source at Raja Ramanna Center for Advanced Technology (RRCAT), Indore, India. The details of this beamline and the procedure for recording the spectra have already been described [3].

3. Data analysis

The EXAFS data of the complex has been analyzed using the available computer software packages *Athena* version 0.8.061 and *Artemis* version 0.8.013 [4]. The spectrum is first normalized and the normalized $\mu(E)$ vs. E (energy) curve for the complex is shown in figure 1. Normalized $\mu(E)$ data is converted to the EXAFS signal $\chi(k)$ data, which is then Fourier transformed. $\chi(k)$ vs. k spectrum for complex is shown in figure 2. The Fourier transformed data is then fitted with theoretical model in R-space to obtain the different fitting parameters. As the complex has three copper sites, one site having square-pyramidal structure and other two sites having distorted octahedral structure, two theoretical models have been generated, one for square-pyramidal structure and another for distorted octahedral structure. In the present work we have used the following parameters as input in Artemis for generating the two theoretical models: Space group: P -1, Cell constants: a = 7.874 Å, b = 8.702 Å, c = 13.278 Å, $\alpha = 97.32^\circ$, $\beta = 104.06^\circ$, $\gamma = 97.83^\circ$. Cluster size = 10 Å. The remaining input positional parameters are taken from ref [2].

For square-pyramidal structure, the theoretically calculated contribution to $|\chi(R)|$ by the single scattering three axial Cu-O paths, one axial Cu-N path and one apical Cu-O path are shown in figure 3. Similarly, for distorted octahedral structure of the complex, the theoretically calculated contribution to $|\chi(R)|$ by the single scattering four Cu-O axial paths and two longer apical Cu-O paths are shown in figure 4. This figure shows only three paths as each path has a degeneracy of 2.

Table 3. The EXAFS fitting in complex with 30% square-pyramidal site and 70% octahedral site contributions.

Atomic pair	EXAFS results				XRD results	Atomic pair	EXAFS results				XRD results
	N	R(Å)	delr(Å)	σ^2(Å^{-2})	R		N	R(Å)	delr(Å)	σ^2(Å^{-2})	R
Cu1-O3	1	1.83	-0.080	0.0054 ± 0.0017	1.91	Cu1-O6	2	1.93	-0.014	0.0101±0.0035	1.94
Cu1-N1	1	1.85	-0.080	0.0054 ± 0.0017	1.93	Cu1-O2	2	1.94	-0.014	0.0101±0.0035	1.95
Cu1-O1	1	1.88	-0.080	0.0054 ± 0.0017	1.96	Cu1-O7	2	2.69	-0.014	0.0101±0.0035	2.71
Cu1-O4	1	1.90	-0.080	0.0054 ± 0.0017	1.98						
Cu1-O5	1	2.40	-0.080	0.0054 ± 0.0017	2.48						

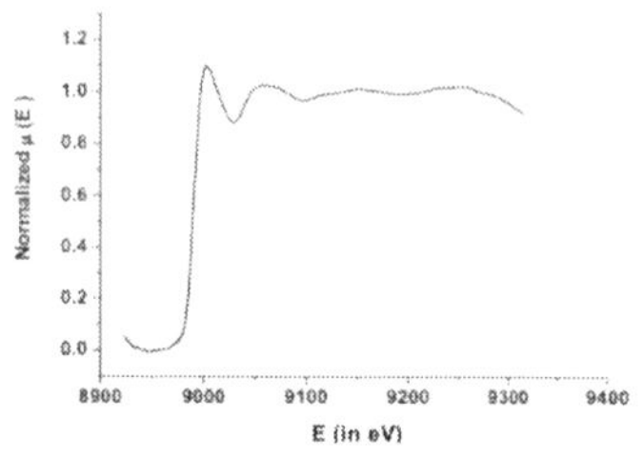

Figure 1. . Normalized EXAFS spectra at the K-edge of copper in complex

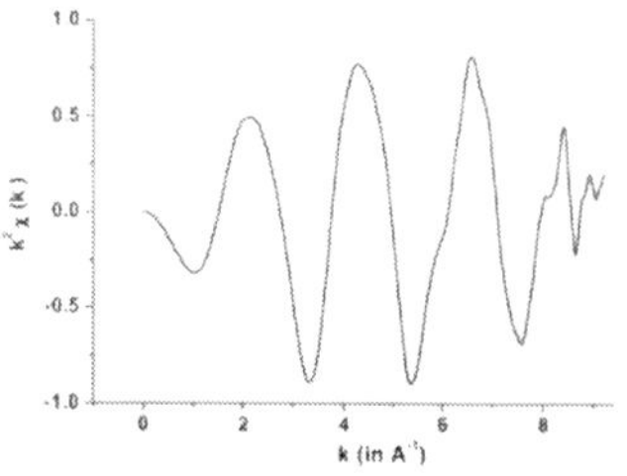

Figure 2. $\chi(k)$ vs. k spectra for complexes.

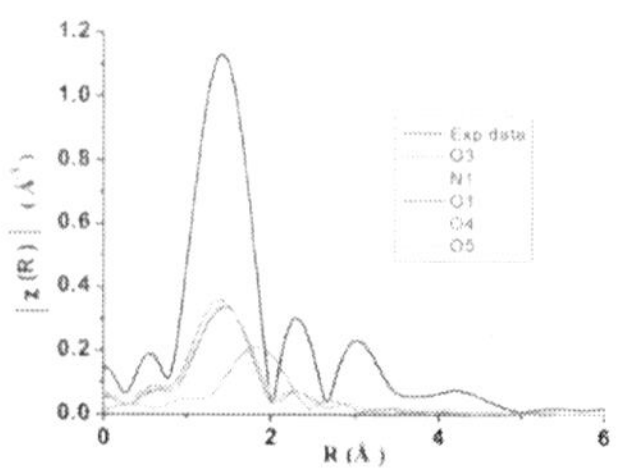

Figure 3. Contribution to $|\chi(R)|$ by single scattering paths for square pyramidal structure

For generating the two theoretical models we have followed the procedure outlined by Kelly *et al.* [5]. Accordingly, a single value of S_0^2 and ΔE_0 is used for all the paths in the fitting but different values of ΔR can be defined for different paths. Unique Debye-Waller factor (σ^2) values are given to each of the scattering paths in the model. As multiple scattering paths also have important contribution, they have also been included in modeling the data. The resulting fitted curves are shown in figure 5 for square-pyramidal structure of complex and in figure 6 for distorted octahedral structure of the complex. After doing the above mentioned two types of fittings, a third type of fitting has also been done (not shown here). This fitting has been done using the theoretical models of complex for both the square-pyramidal and octahedral structure, simultaneously. Using this type of fitting we have tried to estimate the contributions of the different copper sites to the experimental EXAFS spectra.

For this third type of fitting we have set a new parameter x1 which can be varied from 0.1 to 0.9. Using this parameter, S_0^2 for paths of octahedral structure have been defined as amp*x1 and for paths of square pyramidal structure as amp*(1–x1). Same value of E_0 has been defined for all the paths of both structures. One value of ΔR has been defined for paths of octahedral structure and another value of ΔR has been defined for paths of square pyramidal structure. Unique Debye-Waller factor (σ^2) values are given to each of the scattering paths in the model. The value of parameter x1 has been varied during different fittings and the values of statistical factor reduced χ_v^2 have been noted each time. The value of parameter x1 for which minimum value of reduced χ_v^2 is obtained has been taken as contribution of octahedral copper site to the experimental EXAFS spectrum.

4. Results

For the complex, the energy of the K-edge of copper is found to be 8988.03 eV which suggests that copper is in +2 oxidation state in this complex [6, 7].

The EXAFS data of complex has been fitted in three different ways: firstly, by its own octahedral structure, secondly, by its own square pyramidal structure and lastly, by both of its structures simultaneously. These three fittings are discussed below. In all these fittings, Fourier transform was performed over k-range: k_{min}= 2.67 Å^{-1}, k_{max}= 7.97 Å^{-1}. Theoretically modeled data was fitted in the R-space to the experimental data using k_w= 2 and in the R ranges of 1.0 -5.0 Å.

4.1 Octahedral structure fit

For the analysis of the EXAFS data of complex, the first type of fitting has been done using the theoretical model for its own octahedral structure. In the fitting procedure, we have used first eleven scattering paths obtained from Feff calculations. The value of goodness-of-fit parameter, i.e., reduced chi-square (χ_v^2) obtained is 676.12. The results obtained from fitting are given in table 2, which are in agreement with the crystallographic bond distances (also given in table). The four axial Cu-O distances are found to be 1.89 Å, 1.89 Å, 1.90 Å and 1.90 Å. Two apical Cu-O distances are found to be 2.70 Å. Hence, this theoretical model generated by us appears to be quite reasonable.

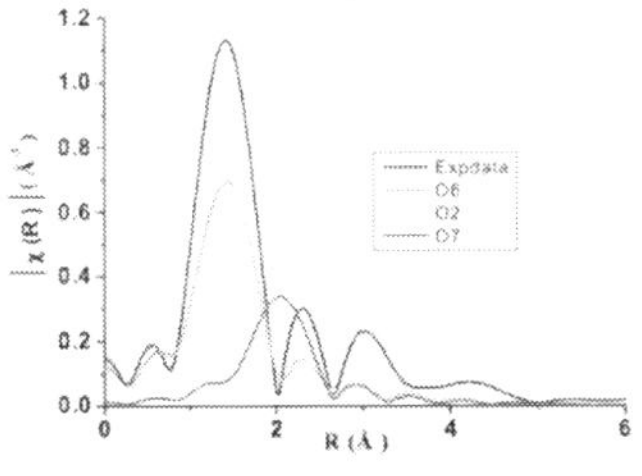

Figure 4. Contribution to |χ(R)| by single scattering paths for octahedral structure.

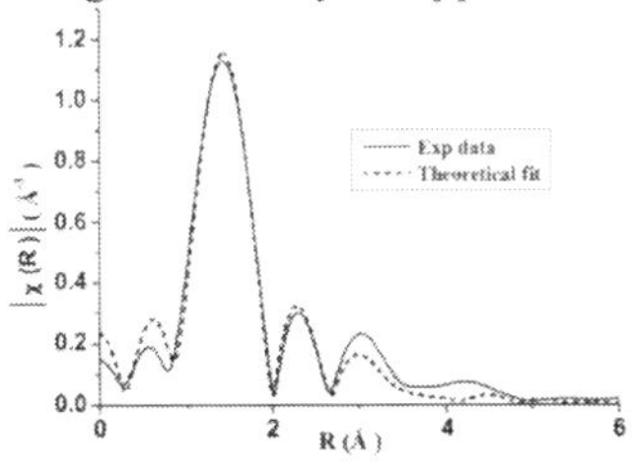

Figure 5. Fourier transformed experimental data and the theoretical fit for square pyramidal structure.

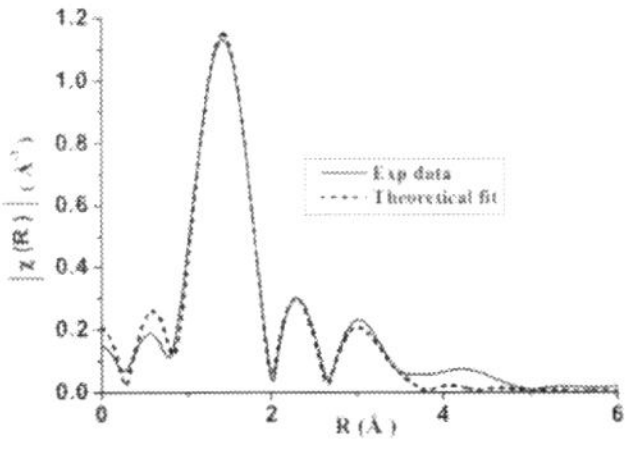

Figure 6. Fourier transformed experimental data and the theoretical fit for octahedral structure.

4.2 Square pyramidal structure fit

For the analysis of the EXAFS data of complex, the second type of fitting has been done using the theoretical model for its own square pyramidal structure. In the fitting procedure, we have used ten scattering paths obtained from Feff calculations. The value of goodness-of-fit parameter, i.e., reduced chi-square (χ_v^2) obtained is 326.06. The results obtained from fitting are given in table 1, which are in agreement with the crystallographic bond distances (also given in table). The three axial Cu-O distances are found to be 1.90 Å, 1.95 Å and 1.97 Å and one axial Cu-N distance is found to be 1.92 Å. Hence, this theoretical model generated by us appears to be quite reasonable.

4.3 Fitting both the structures simultaneously

Third type of fitting has been done by using the theoretical models for both of its own octahedral and the square pyramidal structures simultaneously. In the fitting procedure, we have used first three scattering paths of octahedral structure and first five scattering paths of square pyramidal structure, obtained from Feff calculations.

As already explained in data analysis a new variable x1 has been defined for this fitting procedure, which gives the contribution of octahedral Cu site to the experimental EXAFS data. By varying x1 from 0.1 to 0.9 during consecutive fittings, the values of reduced χ_v^2 have been determined for each fitting. The value of x1 for which minimum value of reduced χ_v^2 is obtained has been taken as contribution of octahedral copper site to the experimental EXAFS spectrum. In the present case, the minimum value of χ_v^2 has been obtained at x1 = 0.7. Thus, it can be said that octahedral Cu site has 70% and square pyramidal Cu site has 30% contributions to the experimental EXAFS spectrum. The results obtained from fitting are given in table 3, which are in agreement with the crystallographic bond distances (also given in table). These results are nearly the same as obtained from fitting the two structures separately (as mentioned above).

5. Conclusions

The aim of the present work is to show how EXAFS spectra of the complexes, having different types of coordination environment, can be analyzed to yield the geometry around the central metal ion. In the present case, the EXAFS data of complex has been fitted in three different ways: firstly, by theoretical model of its own octahedral structure, secondly, by theoretical model of its own square pyramidal structure and lastly, by theoretical models of both of its structures simultaneously. The results obtained by fitting two structures simultaneously are nearly the same as obtained from fitting the two structures separately as well as those obtained from the crystallography. Using this kind of fitting, the contributions of the different copper sites to the experimental spectrum have been found to be 70% from the two octahedral copper sites and 30% from the one square pyramidal copper site. The obtained value of the chemical shift suggest that copper is in +2 oxidation state in this complex.

6. Acknowledgement
Thanks are due to Madhya Pradesh Council of Science & Technology (MPCST), Bhopal (India) for a research grant.

7. References

[1] Riggs-Gelasco P J, Stemmler T L, Penner-Hahn J E 1995 *Coord. Chem. Rev.* **144** 245
[2] Korhonen K, Hamalainen R and Turpeinen U 1984 *Acta Cryst.* **C40** 1175
[3] Gaur A, Johari A, Shrivastava B D, Gaur D C, Jha S N, Bhattacharyya D, Poswal A and Deb, S K 2011 *Sadhana* **36** 339
[4] Ravel B and Newville M 2005 *J. Synchroton Rad.* **12** 537
[5] Kelly S D, Hesterberg D and Ravel B 2008 *Chapter 14 in Methods of Soil Analysis, Part 5, Mineralogical Methods* (Medison, Soil Science Society of America) pp 387
[6] Kau L S, Spira-Solomon D J, Penner-Hahn J E, Hodgson K O, Solomon E I, 1987 *J. Am. Chem. Soc.* **109** 6433
[7] Fulton J L *et al.*, 2000 *J. Phys. Chem. A* **104** 11651

Effect of Mixed Glass Former on Ionic Conductivity of Silver Boro Tungstate glass system x[0.75AgI:0.25AgCl]: (1-x) [Ag_2O-{B_2O_3:WO_3}]

Harsha Dehariya, R Kumar* and A R Polu
Dept. of Physics, Dr. H. S. Gour University, Sagar

E-mail: ranveerssi@yahoo.com

Abstract. The idea to explore new 'Superionic Electrolytes', " Fast ionic conductors" is due to their tremendous potential applications in solid state electrochemical devices viz. solid state batteries , fuel cells, sensors, super capacitors. Superionic glasses have attracted great deal of attention due to their several advantageous over their crystalline counterparts such as high ionic conductivity, easy preparation, wide selection of compositions, isotropic properties and high stability etc [4-7]. Large numbers of silver ion based glasses have been reported in the literature for the glassy system of AgI:Ag2O: MxOy (MxOy = B2O3, SiO2, P2O5, GeO2, V2O5, As2O5, CrO3, SeO2, MoO3 & TeO3 etc many of them shows high silver ion conductivity [8]. Ion transport behavior of Silver Boro Tungstate glass system x[0.75AgI:0.25AgCl]: (1-x) [Ag2O{B2O3:WO3}], where $0 \leq x \leq 1$ in molar wt% prepared by melt quench technique were reported. The new host [0.75AgI:0.25AgCl] was used as a better alternate in place of conventional host salt AgI. Conductivity measurement were carried out on this glass system as a function of frequency from 50 Hz to 5 MHz, over a temperature range of 27°C to 200°C, for different compositions by Impedance spectroscopy. The composition 0.7[0.75AgI:0.25AgCl]: 0.3[Ag2O{B2O3:WO3}] shows the highest conductivity of the order of $\sigma rt \sim 2.76 \times 10^{-2}$ S/cm, referred to as the Optimum Conducting Composition (OCC). The enhancement in the conductivity has been obtained by mixed former effect. XRD result shows that the system is completely amorphous. Temperature dependence of conductivity of all compositions were studied & reported. Activation energies (Ea) were also evaluated from the slope of .Log(σ) vs 1000/T, Arrhenius plots.

1. Introduction

Ever since the discovery of MAg4I5 (M = Rb, K, NH4)[1-2] and fast ion conduction in Na-β alumina [3] a remarkable progress in the field of Solid State Ionics was made. This discovery had opened up the pathways to explore new superionic materials with different cations and anion conduction. Amongst these solids a great deal of effort has been made for the development of novel glassy electrolyte with innovative transport properties using various silver - oxy salts for their electrochemical device applications [4-7]. Since than a large number of Ag+ ion based oxide glasses with variety of glass former / glass modifier ratios with different doping salts were studied and reported [5,10-11]. In silver borate glasses, incorporation of Ag2O in B2O3 modifies the Boron-Oxygen network by converting some of the BO3 unit to BO4 units, which permits the dissolution of AgI salt in glass matrix providing Ag+ for the ion transport [9]. Conductivity of glasses can be enhanced by modification of the network by an alkali oxides or alkali sulphide by addition of dopant salt AgX (X=

Cl, B, I) and by mixed glass former in the network. The enhancement in conductivity due to partial substitution of the network former by another, while network modifier is kept constant is called "mixed former effect" 'MFE'[10].

In the present paper, we report here the effect of two glass formers [B2O3-WO3] on conductivity of new Silver Boro Tungstate system x[0.75AgI:0.25AgCl] : (1-x)[Ag2O:{B2O3- WO3}],where $0 \leq x \leq 1$. An alternate dopant salt a quenched [0.75AgI:0.25AgCl] mixed system solid solution has been used in place of conventional salt AgI [11]. Both the host exhibits identical transport properties, however new host salt yield better superionic glass/ composite electrolytes as compared to those prepared with AgI [12]. Earlier attempts has been made to optimize the room temperature conductivity using single glass former such as B2O3[13], WO3[14], MoO3[15], CrO3[16] using a new host [0.75AgI:0.25AgCl] in place of conventional AgI salt. XRD of OCC of quenched system was performed to verify its amorphousity. Temperature dependence of conductivity of all compositions were studied and activation energy values (Ea) were evaluated from the slope of Log (σ) vs 1000/T Arrhenius plots.

2. Experimental

Solid electrolytes with the general composition x[0.75AgI:0.25AgCl]:(1-x) [Ag2O{B2O3:WO3}] for $0 \leq x \leq 1$ have been prepared from analar grade chemicals (CDH) AgI, AgCl, Ag2O, H3BO3 and WO3. The appropriate amounts of the chemicals of the desired composition were weighed and grinded thoroughly by mortar and pestle. This powder was placed in a borosilica test tube and melted at 900°C in an electric muffle furnace. The melt was rapidly quenched by dropping the melt directly into liquid bath.

For ionic conductivity measurement the electrolytes were pulverized into fine powder and then pressed into pellets at a pressure of 2-3 ton/ cm-2 using perkin elmer hand press. The impedance studies in the present work were carried out using a HIOKI LCR HITESTER (3532-50) in the frequency range 50 Hz to 5MHz at different temperatures varying from 27°C to 200°C. Colloidal silver paint was painted on both sides of the electrolyte pellet as a reversible electrode and the true bulk resistance was calculated from the complex impedance plot.

3. Results and Discussion

Figure 1 shows the compositional dependence of ionic conductivity at room temperature for the system x[0.75AgI:0.25AgCl]:(1-x) [Ag2O{B2O3:WO3}] and x [AgI]:(1-x) [Ag2O{B2O3:WO3}] where $0 \leq x \leq 1$. The ionic conductivity of all the compositions are ranging in order of 10-3 to 10-2 S/cm. We observe from figure that the conductivity increases as x increases, attains a peak value at x = 0.7, then decreases as x is increased further from 0.8 and 0.9. The system with composition 0.7[0.75AgI:0.25AgCl]: 0.3 [Ag2O{B2O3:WO3}] exhibited highest ionic conductivity of the order of σrt = 2.76 x 10-2 S/cm and referred to as "Optimum Conducting Composition" (OCC). Conductivity values were compared with glassy electrolytes prepared with AgI salts also for comparison which shows similar trends but have lower conductivity value. This increase in conductivity 'σ' has been attributed due to the mixed former effect. The reason for the decrease in conductivity for x = 0.8 & 0.9 may be due to blocked pathways for Ag+ ion migration created by interstitial borons. Boron may occupy interstitial positions of WO3 networks may cause hindrance to Ag+ ions which otherwise would have passed through those of WO3 interstitial windows. The ionic conductivities are relatively higher than those obtained for single former electrolyte [14]. This feature can be accounted for , on the basis of thermodynamic approach to the weak electrolyte theory [17] where GM/GFs ratios i.e. [Ag2O/B2O3-WO3] can be treated as 'solvent' and the Ag+ ion conduction occurs through a dissociation of the solute of the type AgI → Ag+ + I-.

The solvent of the amorphous solid electrolytes with an additional glass former can be characterized by a relatively high dielectric constant which increases the solubility and polarisability of the solute [0.75AgI:0.25AgCl], which thereby increases the silver ion conductivity over the ternary solid electrolytes with only single glass former.

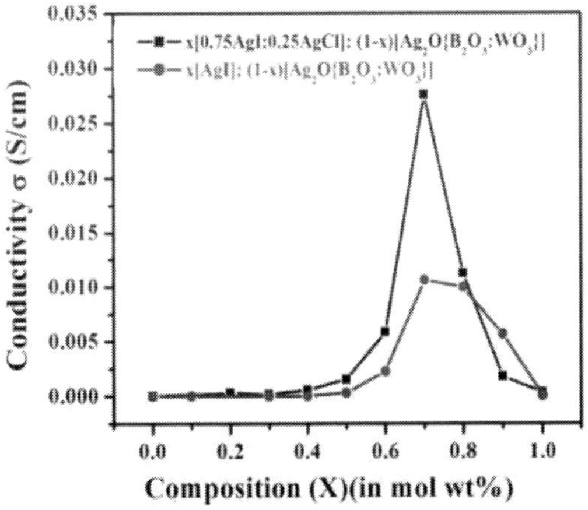

Figure 1. Composition dependence of room temperature conductivity of Silver Boro Tungustate glass system (-■-) with [0.75AgI:0.25AgCl] and (-●-) with [AgI]

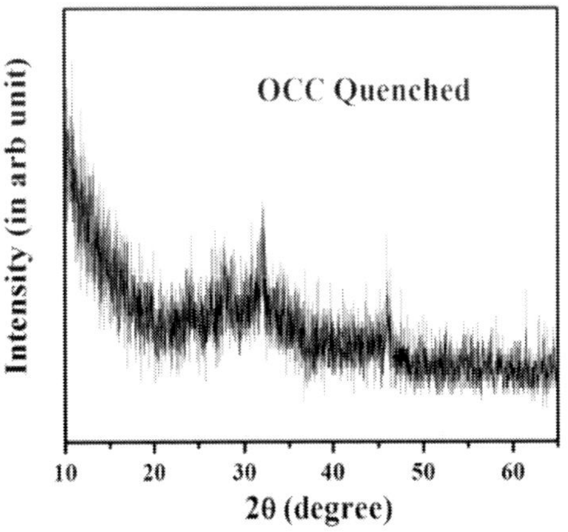

Figure 2. XRD pattern of OCC 0.7[0.75AgI:0.25AgCl]: 0.3 [Ag_2O-{B_2O_3:WO_3}] quenched system

Figure 2 shows the XRD results of OCC of quenched system. It is clearly observed from the diffractogram that the quenched composition (x = 0.7) have diffused shallow peak which confirms the formation of completely glassy/ amorphous phase of the system.

Figure 3 shows the temperatures dependence of conductivity for different compositions of x[0.75AgI:0.25AgCl]:(1-x) [Ag2O{B2O3:WO3}] where mol fraction x varies in range from 0 to 1. Conductivity plots for the host compound [0.75AgI:0.25AgCl] is also drawn for comparison. One can

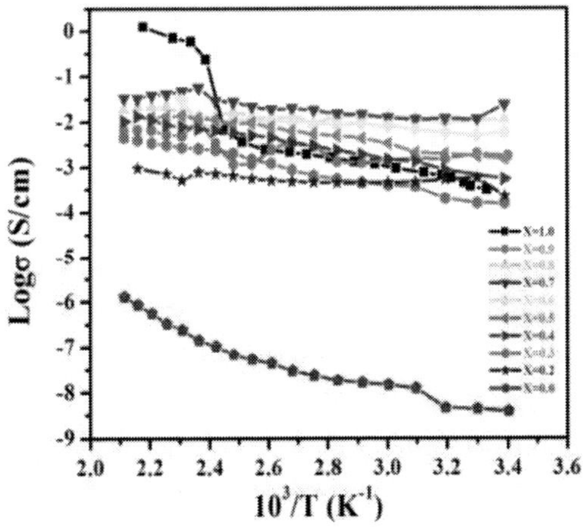

Figure 3. Logσ vs 1/T plots of Silver Boro Tungstate glass system x[0.75AgI:0.25AgCl]: (1-x) [Ag_2O{B_2O_3:WO_3}]for different compositions along with the host salt [0.75AgI:0.25AgCl].

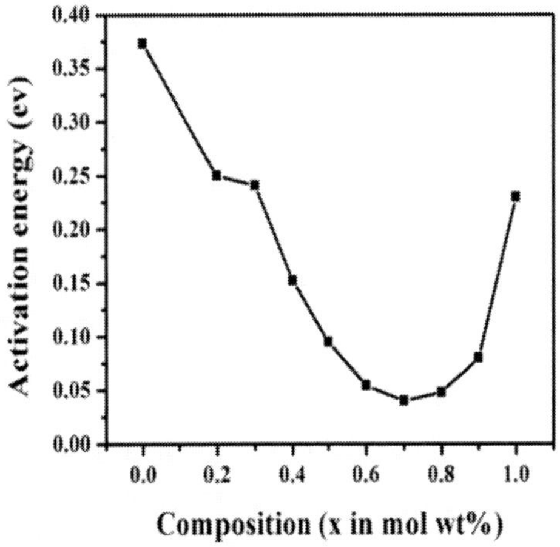

Figure 4. Composition dependence of activation energy of Silver Boro Tungstate glass system x[0.75AgI:0.25AgCl]: (1-x) [Ag_2O{B_2O_3:WO_3}].

note from the figure that although the high conducting α – phase of the host material appeared to be stabilized completely in the glass former/network modifier, however optimum conductivity enhancement from the host were achieved for the composition range x = 0.7 as observed from figure. This is in accordance with the behaviour generally observed in AgI- based glasses in which conductivity increases with AgI content.
Fig. 4 shows activation energy (Ea) verses compositional dependence of x. The activation energy (Ea) values were computed from the slope of Logσ vs 1/T Arrhenius plots. It is observed that the composition x = 0.7 have minimum activation energy Ea ~ 0.04 ev. The Arrhenius equation governing the variation of the quenched (OCC) 0.7[0.75AgI:0.25AgCl]: 0.3 [Ag2O{B2O3:WO3}] can be expressed as: $\sigma = \sigma 0$ x exp (-Ea/kT), where Ea is the activation energy which is very small in comparison with that of host compound Ea ~ 0.234 eV. It can be noted that Ag+ ion transport in the glass system takes place with relative ease.

4. Conclusion

The effect of mixed glass formers on ionic conductivity of a new Ag+ ion conducting Silver Boro Tungstate glass system x[0.75AgI:0.25AgCl]:(1-x) [Ag2O{B2O3:WO3}] has been studied. An alternate host a quenched [0.75AgI:0.25AgCl] mixed system / solid solution has been used in place of conventional salt AgI. The ratio of two glass former B2O3 and WO3 was kept constant while salt concentration of Ag+ was varied. An enhancement of the order of ~10-2 S/cm was achieved at room temperature for composition 0.7[0.75AgI:0.25AgCl]: 0.3[Ag2O{B2O3:WO3}] which is referred as OCC. This has been attributed due to MFE. Glassy/amorphous phase of OCC was confirmed by XRD. Temperature dependence of conductivity shows that α - like phase of the host salt is completely stabilized in the glass matrix and mechanism of ion transport in OCC can be explained by weak electrolyte model suggested for glasses. Minimum activation energy value (Ea ~ 0.04eV) was obtained for OCC suggest the easy ion transport in the system.

5. Reference

[1] Bradley J N and Greene P D 1976 *Trans Faraday Soc.* **63** 424
[2] Owens B B and Argue G R 1967 *Science* **157** 308
[3] Yao Y F T and Kumer J T 1967 *Inorg. Nucl. Chem.* **29** 2453
[4] Chandra S 1981 Superionic solids-Principals and Applications (Amesterdam; North Holland), 336
[5] Takahashi T 1995 Handbook of Solid State Batteries and Capacitors (MZA) Munshi Ed world Scientific Singapore 79
[6] Gool W V 1973 Fast Ion Transport in Solids: Solid State Batteries and Devices (North Holland Amsterdam) 593
[7] Mohan G D and Roth W L 1976 Superionic Conductors, (Plenum Press, New York)-
[8] Hosono M, Kawamura J, Itoigawa H, Kuwata N, Kamiyama T and Nakamura J 1999 *J. of Non Cryst. Solids* **244** 81
[9] Kumar A and Shaju K M, Chandra S 1995 *J. Phys.* **73** 369
[10] Chowdari B V R and Kumari P P 1996, *J. of Non Cryst. Solids* **197** 31
[11] Agrawal R C, Gupta R K, Kumar R and Kumar A 1994 *J. of Mater Sci.* **29** 3673
[12] Agrawal R C, Kumar R, Gupta R K and Saleem M 1995, *J. Non cryst.solids* **181** 110
[13] Agrawal R C and Kumar R 1994, *J. Phys. D. Appl. Phys.* **27** 2431
[14] Agrawal R C, Verma M L and Gupta R K 2004 *Solid State Ionics* **171** 199
[15] Agrawal R C, Verma M L, Gupta R K, and Kumar R 2002 *J. Phys. D. Appl. Phys* **35** 810
[16] Agrawal R C and Kumar R 1996, *J. Phys. D. App. Phys.* **29** 156
[17] Kone A and Souquet J L 1986 *Solid State Ionics* **18-19** 454

Structural induced changes in properties and crystallization behaviour of $Co_{80}Zr_{10}V_{10}$ alloy

S. N. Kane[a, 1], **S. Panchal**[a], **E. Fleury**[b]
[a]School of Physics, D. A. University, Khandwa Road Campus, Indore – 452001, India
[b]Center for High Temperature Energy Materials, KIST, Seoul 136-791, South Korea

E-mail: kane_sn@yahoo.com (S. N. Kane)

Abstract. X-ray diffraction (XRD), differential scanning calorimetery (DSC) and magnetic measurements have been used to study the structure-induced modification in the properties and crystallization behaviour of $Co_{80}Zr_{10}V_{10}$ alloy. DSC studies reveal a single step crystallization with activation energy of 3.84 ± 0.05 eV. The best magnetic properties were obtained after annealing the studied alloy at 425 °C for 1 hour with following values: coercivity = 5.0 A/m, saturation induction = 0.7 Tesla and watt loss 0.4 watt/kg (at 50 Hz and at 0.7 Tesla). The average grain diameter ***D*** of the precipitated nanograins was found to vary between 10.0 – 13.5 nm with moderate dependency on the annealing temperature and time, for a volume fraction ranging between 30.5 – 42.0 %. XRD measurements reveal the presence of $ZrCo_{2.9}$, $Co_{23}Zr_6$, Co_2Zr and ZrV_2O_7 phase.

1. Introduction

Co-based amorphous alloys are an important class of metallic glass particularly for magnetic applications, such as magnetic core, electronic surveillance articles and electromagnetic wave attenuator. Their advantages in comparison to Fe-based amorphous alloys are the zero-magnetostriction and lower value of coercivity [1]. Amorphous alloys are metastable solids with a non-crystalline disordered atomic-scale structure which upon heating relaxes towards a more stable structure before reaching the final crystallization [2].

In the present work, the magnetic measurements, x-ray diffraction (XRD) and differential scanning calorimetery (DSC) have been used to study the influence of the thermally induced changes in structure and its effect on the properties and crystallization behaviour of a metalloid-free Co-based $Co_{80}Zr_{10}V_{10}$ alloy developed by Nose et al. [1].

2. Experimental details

Ribbon (about 30 μm thick and 8 mm wide) of nominal composition $Co_{80}Zr_{10}V_{10}$ was prepared using the planer flow casting method. DSC analyses, performed at a heating rate of 5, 10, 20 and 40 °C / minute, were used to determine the activation energy using the Kissinger's method. The crystallization peak temperature, T_x, as obtained by DSC measurements performed at 20 °C / minute was found to be 580 °C. In order to have the lower grain size leading to better exchange coupling between residual amorphous matrix and the growing nano-grains that would in-turn lead to better soft magnetic properties, the maximum annealing temperature was chosen to be 450 °C, which is 130 °C below T_x,

[1] To whom any correspondence should be addressed.

for a maximum annealing time of 1 hour. The following thermal treatments were thus chosen : i) annealing the samples at 350, 370, 400, 425 and 450 ^{0}C for 60 minutes, and ii) samples annealed at 425 ^{0}C for 10, 20, 30, 50 and 60 minutes. Bruker D8 Advance X-ray diffractometer was used to carry out XRD measurements with the Cu-K$_\alpha$ radiation (λ = 0.154 nm). The XRD patterns were analyzed by fitting a crystalline and an amorphous component using pseudo-Voigt profiles. The average grain size ***D*** was obtained by Scherrer's formula using integral width of the (110) line and the lattice parameter ***a*** was calculated using Nelson-Taylor-Sinclair correction in order to take into account the peak shift due to sample offset. For the amorphous phase, the first nearest neighbour distance X_m was calculated using the expression: $X_m = 1.227\ \lambda/\ 2\mathrm{Sin}\theta$ [3 - 4]. The hysteresis loops of as-cast and annealed samples were measured at 50 Hz (H_{max} = ± 1000 A/m), to get information on thermal annealing induced changes in the structure and its effect on the magnetic properties, i.e., the coercive field (H_c) and saturation induction (B_{1000}) were determined by analyzing raw data using a Matlab® program, described elsewhere [5].

3. Results and discussions

By performing DSC analyses at various heating rates between 5 and 40°C/min, the activation energy of crystallization, calculated using the Kissinger's method, was estimated to be 3.84 ± 0.05 eV.

Figure 1 depicts the temperature dependence of the hysteresis loops of the studied samples annealed at various temperatures for 1 hour. Noticeable variation of the magnetic properties after thermal annealing can clearly be observed. The table 1 and table 2 depict the variation of H_c and B_{1000} as a function of the annealing temperature and time, respectively. Perusal observation of the table 1 reveals that under these annealing conditions, the H_c and B_{1000} range between 5.0 – 29.9 A/m and 0.7 Tesla, respectively. The best values for H_c : 5.0 A/m and B_{1000}: 0.7 Tesla were obtained after annealing at 425 °C/1hour. The low B_{1000} values can be ascribed to the presence of Zr and V in the first near-neighbour shell of Co, which reduce its magnetic moment. The lowest loss value of 0.4 W/kg (at 50 Hz and at 0.7 Tesla) was obtained after annealing at 425 °C for 1 hour.

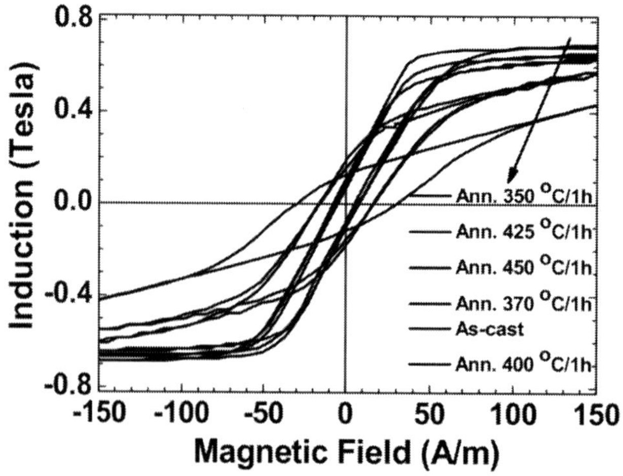

Figure 1. Temperature dependence of hysteresis loops.

Table 1. Annealing temperature dependence of H_c, B_{1000} and losses.

Sample details	H_c (A/m)	B_{1000} (T)	Losses (W/kg) at 50 Hz
As-cast	16.5	0.7	0.7
Ann. 350 °C/1h	7.2	0.7	0.4
Ann. 370 °C/1h	15.9	0.7	0.7
Ann. 400 °C/1h	29.9	0.7	1.2
Ann. 425°C/1h	5.0	0.7	0.4
Ann. 450 °C/1h	6.9	0.7	0.4

Table 2. Annealing time dependence of H_c, B_{1000} and losses.

Sample details	H_c (A/m)	B_{1000} (T)	Losses (W/kg) at 50 Hz
Ann. 425°C / 10 min.	6.2	0.7	0.5
Ann. 425°C / 20 min.	5.8	0.7	0.5
Ann. 425°C / 30 min.	7.9	0.7	0.5
Ann. 425°C / 50 min.	10.0	0.7	0.7
Ann. 425°C / 60 min.	5.0	0.7	0.4

In order to further improve the magnetic properties, the dependency of magnetic properties with the annealing time was investigated at 425 °C and the results have been reported in the table 2. The value of H_c varies between 5.0 and 10.0 A/m and B_{1000} is 0.7 Tesla. It is worth noting that after 30. min. annealing at 425 °C, best H_c and B_{1000} values can be obtained.

The XRD patterns (not shown) revealed the co-existence of crystalline and residual amorphous phases. Table 3 depicts the parameters for samples annealed as a function of temperature and time, obtained by analyzing XRD data. For comparison, H_c values (obtained from magnetic measurements) are also shown. The variation of H_c with D and V_x suggests that there is a counter-balance between these two parameters. Within the experimental errors the first nearest-neighbour distance in the residual amorphous matrix (0.248 nm ± 0.002) almost remains constant for the studied specimens, showing that the studied samples have similar mass densities. From the table 3 it is suggested that the value of D varies between 10 – 13.0 nm and does not change much with the annealing temperature and time. The volume fraction of nanograins V_x varies between 30.5 – 42.0 %. Within experimental error, the D values are around 10 nm, exhibiting better exchange coupling between growing nanograins and hence displaying lower H_c values. Values of the lattice parameter, a, suggest the precipitation of bcc Co phase containing Zr (Zr atomic radius : 155 pm, in comparison to that of Co : 135 pm). XRD measurements also revealed the presence of $ZrCo_{2.9}$, $Co_{23}Zr_6$, Co_2Zr, and ZrV_2O_7 phase, as confirmed from the JCPDS database.

Table 3. Temperature time transformation behaviour of grain size < ***D*** >, lattice parameter ***a***, volume fraction of nanograins V_x for nanocrystalline specimens, and X_m – the first-near neighbour distance in the residual amorphous matrix, as obtained after analyzing XRD measurements. For comparision, H_c values are also given, as obtained from magnetic measurements.

Sample details	< ***D*** > (nm) (± 2)	***a*** (nm) (± 0.0002)	V_x (%) (± 2)	X_m (nm) (± 0.002)	H_c (A/m) (± 0.5)
Temperature transformation					
As-cast	–	–	–	0.248	16.5
Ann. 350 °C/1h	10.0	0.2882	30.5	0.248	07.2
Ann. 370 °C/1h	10.0	0.2883	33.0	0.250	15.9
Ann. 400°C/1h	13.0	0.2873	38.0	0.248	29.9
Ann. 425°C/1h	12.0	0.2871	40.0	0.248	05.0
Ann. 450 °C/1h	13.0	0.2871	42.0	0.248	06.9
Time transformation					
Ann. 425°C / 10 min.	12.0	0.2875	38.0	0.249	06.2
Ann. 425°C / 20 min.	12.0	0.2879	39.0	0.248	05.8
Ann. 425°C / 30 min.	12.0	0.2882	39.0	0.249	07.9
Ann. 425°C / 50 min.	10.0	0.2876	40.0	0.249	10.0
Ann. 425°C / 60 min.	12.0	0.2871	40.0	0.248	05.0

4. Summary

The investigation of the structure and magnetic properties of the metalloid-free $Co_{80}Zr_{10}V_{10}$ rapidly quenched ribbon indicated that the best magnetic properties were obtained after annealing at 425 °C for 1 hour: H_c = 5.0 A/m, B_{1000} = 0.7 Tesla and watt loss 0.4 watt/kg (at 50 Hz and at 0.7 Tesla). The low B_{1000} value was attributed to the presence of Zr and V in the first near- neighbour shell of Co, reducing its magnetic moment. The size of the precipitated grains varied between 10.0 – 13.0 nm, with a moderate dependency with the annealing temperature and time. The volume fraction V_x varies between 30.5 – 42.0 % with a tendency of increasing as the annealing temperature and time increased. The large value of the lattice parameter suggested the presence of Zr in the precipitated bcc Co phase. XRD measurements also show the presence of $ZrCo_{2.9}$, $Co_{23}Zr_6$, Co_2Zr and ZrV_2O_7 phase.

Acknowledgement

Eric Fleury acknowledges the financial support by the by 21st Frontier Program CNMT (#2011K000165) of the Korean Ministry of Education Science and Technology.

References

[1] Nose M, Kanehira J, Ohnuma S, Shrokawa K and Masumoto T 1981 *J. Appl. Phys.* **52** 1911
[2] Masumoto T and Egami T 1981 *Mater. Sci. Eng.* **48** 147
[3] Guinier A 1963 *X-ray Diffraction in Crystals, Imperfect Crystals and Amorphous Bodies.* (Freeman, San Fransisco) p 72
[4] Kane S N , Coisson M, Tiberto P, Vinai F and Mazaleyrat F 2011 *Current Appl. Phys.* **11** 981
[5] Kane S N , Gercsi Z, Ghodke N, Khinchi S S and Gupta A 2005 A digital hysteresis loop tracer for studying soft magnetic materials (Proc. of the DAE Solid State Physics Symposium vol. 50) ed. V K Aswal, K G Bhushan and J V Yakhmi, Prime (Time Education, Mumbai, India) (ISBN: 81-8372-019-6) p. 287-288

Spin Relaxation in Mn doped CdTe/ZnTe QDs

Shivani Rana[1*], **Pratima Sen**[1] **and P. K. Sen**[2]
[1]School of Physics, Devi Ahilya University, Indore-452001, M.P, India
[2]Department of Applied Physics, Shri S. G. S. Institute of Science & Technology, Indore-452003, M.P., India

*E-mail: shanshi.rana@gmail.com

Abstract. The spin relaxation times of the electron have been calculated for phonon assisted transitions between the Zeeman sublevels arising as a result of magnetic impurity doping and applied magnetic field in Mn doped CdTe/ZnTe quantum dots. Taking into account effective mass approximation with four band k.p perturbation theory and incorporating the sp-d exchange interaction, the energies of the split states have been obtained and analyzed with magnetic field and effective concentration. The spin relaxation times have been found to be considerably longer with higher dopant concentration in high magnetic fields at very low temperature.

1. Introduction

In the recent years, spin dynamics in semiconductor nanostructures has drawn considerable attention due to the enormous potential for spintronic devices.[1] Semiconductor quantum dots (QDs) doped with transition metal (TM) ions form a new class of materials called diluted magnetic semiconductors (DMS), which appear to be promising candidates for the spintronic devices because of their ability to control single spin carriers. The DMSs exhibit interesting magnetic, optical and magneto-optical properties such as large Faraday rotation, giant Zeeman splitting etc with many practical applications in spin transport, spin filtering, charge-controlled magnetism, etc [1, 2]. Such magneto-electronic phenomenon in the DMS QDs is attributed to the exchange interaction between the magnetic moments of the dopant TM ions and electron or hole spins of the host semiconductor [2, 3]. Moreover, quantum confinement effects in semiconductors significantly enhance the sp-d exchange interactions. This enhancement can be ascribed to the strong overlapping of the carrier wave functions in the TM ion doped QDs. As a consequence, the spin-splitting is enhanced manifold in DMS QDs in the presence of magnetic field.

In II-VI semiconductor QDs, the optical transitions between the electron and heavy hole bands are polarization sensitive. The spin-up and spin-down electrons in the conduction band can be created by exciting the QD with right and left circularly polarized light, respectively. The spin dephasing from the Zeeman sublevels generates circularly polarized light with distinct phase shifts. In this article, we have theoretically calculated the spin relaxation times in CdTe/ZnTe QDs doped with Mn. The spin flipping is considered to be longitudinal optical (LO) phonon mediated transition between the spin-split Zeeman states. The numerical analysis has been carried out for Mn doped CdTe/ZnTe QDs.

2. Theoretical Formulation

The DMS QDs are considered to experience parabolic axial confinement $V_{\perp}^{i}$ and a step like V_{II}^{i} confinement along the growth axis such that the total confinement potential is given by $V_i(\rho_i, z_i) = V_{II}^{i} + V_{\perp}^{i}$ where

$$V_{II}^{i}(\rho_i) = \frac{1}{2} m_i \omega_i^2 \rho_i^2 \tag{1}$$

and
$$V_{\perp}^{i}(z_i)=\begin{cases}\Delta V^{i}, & |z_i|\ge w/2,\\ 0, & |z_i|< w/2,\end{cases}\tag{2}$$

The superscript i represents electron (e) or hole (h). ρ_i and z_i are the cylindrical coordinates, m_i represents the effective masses, ω_i corresponds to the oscillator frequencies and ΔV^{i} refers to the conduction ($i=e$) and valence ($i=h$) band offset. Also, w is the width of the potential well. Under the effective mass approximation, the electron and hole wave functions for the above mentioned confinement can be written as [4]

$$\Psi_{l}^{e}(\rho_e,z_e,\phi)=\sum_{n,s}c_{nl}\psi_{n,l}(\rho_e,\phi)f_{n'}(z_e)u_s(\rho_e,z_e)\tag{3a}$$

$$\Psi_{j}^{h}(\rho_h,z_h,\phi)=\sum_{j_z,n,s}c_{nsj_z}\psi_{n,j}(\rho_h,\phi)f_{n'}(z_h)u_{j_z}(\rho_h,z_h)\tag{3b}$$

Here, c_{nl} and c_{nsj_z} are the constants of normalization for the electron and hole, respectively with n and s being the principal and spin-quantum numbers, respectively. Also, j and l are the angular momentum quantum numbers for electron and hole, respectively and $j_z=\pm 3/2, \pm 1/2$ corresponds to the hole spin. $u_s(\rho_e,z_e)$ and $u_{j_z}(\rho_h,z_h)$ are the Bloch functions for electron and hole, respectively. In eq. (3) $f_{n'}(z_i)$ is the wave function arising due to the step like confinement potential of the form [7]

$$f_{n'}(z_i)=\sqrt{\frac{2}{L}}\sin\left[\frac{n'\pi}{2}\left(z_i+\frac{L}{2}\right)\right];\quad |z_i|\le L/2\tag{4}$$

In eq. (4), q (= 1,2..........) represents the band index number arising due to the z-confinement. The envelope function due to parabolic confinement potential slowly varying over the nanodimensions of the QD is defined as

$$\psi_{n,l}(\rho,\phi)=C_{nl}(i\rho)^{|l|}e^{-\rho^2/2a^2}e^{il\phi}L_n^{|l|}(\rho^2/a^2)\tag{5}$$

where $C_{nl}=\sqrt{\frac{n!}{\pi(n+|l|)!}}\left(\frac{1}{a_i}\right)^{|l|+1}$ is the normalization constant; $a_i\left(=\frac{\sqrt{2}\lambda_i a_c}{(\lambda_i^4+4a_c^4)^{1/4}}\right)$ is known as the effective confinement length parameter that incorporates the effects of both geometrical and magnetic confinements which influence the electronic spectra. $\lambda_i=(\hbar/m_i^{*}\omega_i)^{1/2}$ is the geometrical confinement length and $a_c=(\hbar/eB)^{1/2}$ is the cyclotron radius. Also, $L_n^{|l|}$ is the generalized Laguerre polynomial. The electron Hamiltonian in the TM impurity doped QD in presence of the magnetic field B (applied along the z-direction) is described by

$$H_e=\frac{(\vec{p}+e\vec{A})^2}{2m_e^{*}}+V_e(\rho_e,z)+J_{s-d}\vec{S}\cdot\sigma_e\pm\frac{1}{2}g_e^{*}\mu_B B\tag{6}$$

where $\vec{A}$ is the vector potential in the symmetric gauge, g_e^{*} and μ_B are the effective g-factor for the electron and Bohr magneton, respectively. $\vec{S}$ is the spin of the magnetic ion and σ_e is the pauli spin corresponding to the spin of the conduction band electron. The second term in Eq. (6) represents the geometrical confinement potential; the third term describes the exchange interaction between the conduction band electron and the magnetic impurity ions while the last term gives the Zeeman splitting.

Taking into account four-band k.p approximation in the Luttinger Hamiltonian H_L for the hole states, the hole Hamiltonian H_h in the presence of magnetic field and confinement potential for a DMS QD can be written as

$$H_h=H_L+V_h^{\perp}(\rho_h,z_h)+J_{p-d}\vec{S}\cdot\sigma_h-\frac{e\hbar B\kappa j_z}{m_0}\tag{7}$$

where m_0 being the rest mass of the electron. In Eq. (7), last term refers to the Zeeman splitting energy and the last term stands for the p-d exchange interaction between the carriers of p-subband and Mn2+ ions. This exchange interaction is expressed within the molecular field approximation for electron and hole as

$$H_{ex} = J_{s,p-d}\langle S_z(X)\rangle j_z \tag{8}$$

Here, $J_{s-d} = x_{eff}N_0\alpha/2$ and $J_{p-d} = x_{eff}N_0\beta/2$; N_0 being the number of cations per unit volume and x is the effective concentration of TM ions contributing to the exchange interaction in the quantum dot [5]. α and β are the exchange constants for s-d and p-d exchange interactions, respectively. $\langle S_z(X)\rangle = S_0 B_s(X)$ is the thermodynamical average of the spin component along the magnetic field B where $X = Sg\mu_B B/k_B(T+T_0)$; g_{Mn} being the effective Lande g-factor for transition metal ions and S is the spin of the localized d-electrons of the TM ions and BS (X) is the Brillouin function describing overall paramagnetic behavior of the magnetization at temperature T [5]. S0 and T0 are the fitting parameters accounting for magnetism arising due to TM-TM ionic interaction.

The spin relaxation rate of the electron has been calculated from the Fermi Golden Rule, considering electron spin-phonon coupling and is given by

$$\frac{1}{\tau_s} = Ws = \frac{8\pi^2\mu_0^{\,2}\Delta\omega^3}{3\hbar c_l^{\,3}}\left(\frac{1}{\varepsilon_\infty} - \frac{1}{\varepsilon_s}\right) \tag{9}$$

where c_l is the longitudinal sound velocity, $\Delta\omega$ is the transition frequency between the Zeeman sublevels defined as $\Delta\omega = E_e^{\pm} - E_{hh}^{\pm}$. $E_e^{\pm}$ and $E_{hh}^{\pm}$ are the electronic and heavy hole energies of the system, respectively. ε_∞ and ε_0 are the high frequency and static dielectric constants. Also, μ_0 is the induced dipole moment for the transitions between the spin split states given by

$$\mu_0 = \frac{ep_{cv}}{m_0\Delta\omega}\int \psi_l^e(\rho_e,\phi) f_{n'}(z_e) f_{n'}(z_h) \psi_j^h(\rho_h,\phi)\, d\tau \tag{10}$$

Here, $p_{cv} = \langle u_s(\rho_e, z_e)|\vec{p}|u_j(\rho_h, z_h)\rangle$ is the interband momentum matrix element.

3. Conclusions

In this section we have studied the effect of magnetic field and TM concentration on the transition frequency and the spin relaxation times between the spin-split Zeeman sublevels in Cd1-xMnxTe/ZnTe QD using the formulation developed in section 2. The material parameters used in the present numerical analysis are: m_e = 0.096m0 [7], the conduction (valence) band offset ΔV_c (ΔV_v) is 0.85 eV (0.05 eV) [6]. Luttinger parameters under spherical approximation are taken as γ_1 = 4.7, γ_2 = 1.45 and γ_3= 1.9 [6] while the exchange constants $N_0\alpha$ ($N_0\beta$) are 0.22 eV (− 0.88 eV) [2].

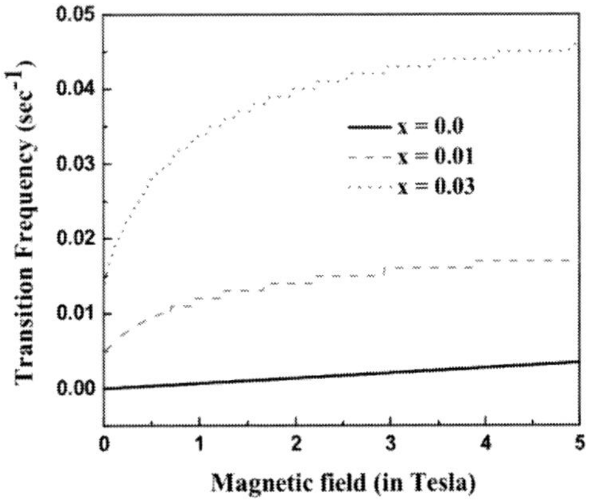

Figure 1. Transition frequency as a function of magnetic field at effective concentration (x = 0, 0.01, 0.02)..

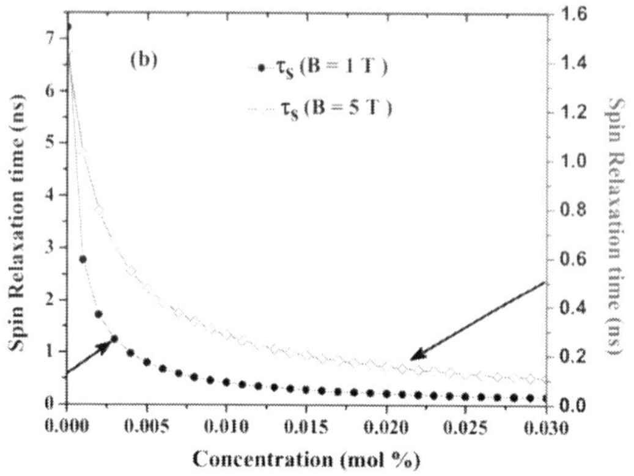

Figure 1. Variation of spin relaxation time with effective Mn concentration at B = 1 T and 5 T for Mn-doped CdTe/ZnTe quantum dots at 1.5 K.

Lande g-factor for electron g_e = 1.12 [2] and for Manganese g_{Mn} = 2.0, spin quantum number of 3d electrons S = 5/2. The speed of sound of LO phonons c_l = 3083 m/s [7] and high frequency and static dielectric constants are ε_∞ = 10.9 and ε_0 = 7.1. [7]

We have considered very small DMS QDs in the strong confinement regime with both magnetic field and geometrical confinements influencing the optical properties of the QDs. The dimensions of the cylindrical QDs are taken to be of 1.25 nm radius and 1.5nm height (under strong confinement limit). In Fig. 1, the transition frequency has been plotted as a function of magnetic field for undoped (x = 0.0) and Mn2+ doped (x = 0.01, 0.02) CdTe/ZnTe. It is clear from this figure that the transition frequency increases with increasing magnetic field and dopant concentration. Since the Zeeman splitting enhances with increasing magnetic field and Mn concentration due to the presence of strong sp-d exchange interaction. Hence, the transition energy between the Zeeman sublevels also increases.

The spin relaxation time has been plotted with increasing Mn concentration. Longer spin relaxation time has been observed in Mn doped QDs as compared to undoped QDs. Also, it is evident from Fig. 2 that the spin relaxation time increases with increasing magnetic field. The effect of dopant concentration is also apparent from the figure. We have obtained 1.24 ns of spin relaxation time for a small concentration of 0.003 mol % at 1T field and 1.5 K temperature which is in agreement with Worschech et al. [3]

In conclusion, the incorporation of TM ions into the semiconductor QDs enhances the spin relaxation time of the electron for the transition between the Zeeman sublevels.

4. Acknowledgement

Authors are thankful to DST, Govt. of India for financial assistance.

5. References

[1] Zutic I, Fabian J and Das Sarma S 2004 *Rev. Mod. Phys.* **76** 323
[2] Furdyna J K 1988 *J. Appl. Phys.* **64** R29
[3] Worschech L, Schmidt T, Forchel A, Slobodskyy T, Schmidt G and Molenkamp L W 2008 *Semicond. Sci. Technol.* **23** 114018
[4] Chang K, Li S S, Xia J B and Peeters F M 2004 *Phys. Rev. B* **69** 235203
[5] Ogawa T and Kanemitsu Y 1998 Optical Properties of Low-Dimensional Materials (Singapore: World Scientific) 181
[6] Bhattacharjee A K and Guillaume C Benoit ′ a la 1997 *Phys. Rev. B* **55** 10613
[7] Adachi S 2005 Properties of Group-IV, III-V and II-VI Semiconductors (Great Britain:John Wiley & Sons)

Journal of Physics: Conference Series **365** (2012) 012037 doi:10.1088/1742-6596/365/1/012037

Effect of Shell and Shell Thickness on Photoluminescence (PL) of a CdSe/ZnS Core – Shell Quantum Dot

Saikat Chattopadhyay[1], Pratima Sen[1], J. T. Andrews[2], P. K. Sen[2]
[1]Laser Bhawan, School of Physics, Devi Ahiliya University, Indore, India – 452001
[2]Department of Applied Physics, Sri G. S. Institute of Technology & Science, Indore, India – 452003

E-mail : the_saikat@yahoo.com

Abstract. The Photoluminescence (PL) intensity is theoretically calculated for a CdSe/ZnS Core – Shell Quantum Dot incorporating WKB approximation. Coating CdSe bare quantum dots with a ZnS layer indicates an enhancement in the PL intensity remarkably with a red-shift in spectra. The PL intensity increases with initial shell growth and then it starts to decrease if outer shell growths continue. Smaller Core – Shell Quantum Dots shows greater PL intensity with more red-sift in PL spectra.

1.Introduction

Modern technological developments have made it possible to prepare three dimensionally confined quantum dot (QD) structure. Self assembled QDs, colloidal QDs are examples of it. In each case the surface to volume ratio of the number of atoms in QDs is very large as compared to their bulk counterpart. The dangling bonds of the surface atoms in QDs can be passivated by either using suitable organic capping or by deposition of inorganic material of appropriate band gap energies and lattice matching on it. Deposition of an additional layer of inorganic material on QDs is called as Shell and such Quantum Dots are popularly known as Core – Shell QDs (CSQDs). CSQDs find variety of applications ranging from biological tagging to single photon emitters, entangled Q-bit registers etc. Some basic linear optical properties of QDs e.g. appearance of discrete absorption spectra reveal the evidence of excitonic state in a QD. Due to large confinement effect the separation of optically induced electron-hole pairs in QDs becomes appreciably small and gives rise to stronger Coulombic interaction as compared to bulk. In small QDs the confinement dominates over the Coulombic interaction energy resulting into the blue shifting of electron-hole states. Embedding the core of a QD in a medium of different dielectric constant can have profound effect on its electronic and optical properties. In CSQDs the electron and hole wave functions may penetrate in the shell region. The extent of penetration depends on the properties of the core and shell material. The electrons (holes) experience a barrier potential equivalent to the conduction (valence) band offset of a core and shell. A proper choice of core and shell material gives an additional parameter to control the electronic and optical properties of CSQDs. In general, deposition of shell causes red-shift in PL peak and improves the PL quantum efficiency due to the proper passivation of surface dangling bonds and nonradiative recombination sites with strong confinement of electrons and holes inside the core [1]. The basic requirements to determine the PL intensity are the excitation mechanism which can generate population in various excited states of the system and radiative as well as nonradiative recombination

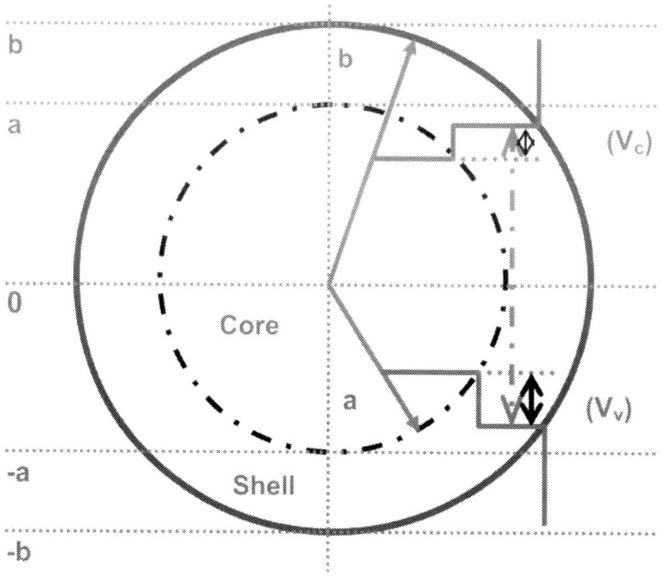

Figure 1. Schematic of a Core – Shell quantum dot (CSQD) and dimensions as assumed in the present calculations

processes. In this paper, we have theoretically investigated the linear optical properties like PL intensity of two different CSQDs by taking into account the temperature dependence of energy levels dephasing mechanisms and exciton-phonon interaction.

2. Theoretical Formulation:

We consider a type – I spherical core shell quantum dot with the band gap energy of the core material being smaller than that of the shell material. The geometry and the dimensions of the dot is illustrated in Fig.(1). Here, a is the radius of the core and b is the radius of the CSQD as a whole such that the annular shell thickness is d= (b-a). The two-dimensional electron and hole confinement potentials are given by [2, 3]

$$V_{e,h}(r) = \frac{V_{c,v}}{a^2}\left(r^2 - a^2\right) \tag{1}$$

r being the quasi-particle position satisfying the condition $a < r < b$. The subscripts e, h, c and v denote the electron, hole, conduction band and valence band, respectively. Vc and Vv are the conduction and valence band offsets between the core and the shell. The single particle wave functions under such situation can be described using WKB approx. as [2, 3, 4]

$$\phi_i = \begin{cases} \dfrac{A_S}{\sqrt{\kappa_i}} \exp\left[\displaystyle\int_{-b}^{-a} \kappa_i dx\right] & ; \quad -b < x < -a \\ \dfrac{A_c}{\sqrt{k_i}} \sin\left[\displaystyle\int_{-a}^{a} k_i dx + \dfrac{\pi}{4}\right] & ; \quad -a < x < a \\ \dfrac{A_S}{\sqrt{\kappa_i}} \exp\left[-\displaystyle\int_{b}^{a} \kappa_i dx\right] & ; \quad a < x < b \end{cases} \tag{2}$$

In order to describe the optical transitions in a three level system, we use the density matrix approach. It is appropriate to describe the optical properties of the quantum dot by taking into account the optical transitions among the ground, excitonic and biexcitonic states. The direct ground to biexciton state transitions is considered to be optically forbidden. The transition dipole moment operators corresponding to the transitions between the exciton to ground states (μ_{0ex}) and biexciton to exciton states (μ_{exbx}) are defined as [2, 5]

$$\mu_{0ex}(a,b) = \frac{ep_{cv}}{m_0\omega_{0ex}} \left|\phi_e^*(a,b).\phi_h(a,b)\right| \tag{3}$$

and

$$\mu_{exbx}(a,b) = \frac{-1}{2}\frac{ep_{cv}}{m_0\omega_{exbx}} \left|\phi_e^*(a,b).\phi_h(a,b)\right|^3 \tag{4}$$

Here, (ω_{0ex}) and (ω_{exbx}) are the transition frequencies corresponding to exciton and biexciton levels. They are temperature sensitive and taken as described in [2, 6]. p_{cv} is the interband transition momentum matrix element [7]. Using the density matrix formulation described in [3, 7] we have calculated theoretically the PL intensity for a CdSe/ZnS CSQD.

3. Conclusion

Photoluminescence (PL) spectra of CdSe/ZnS Core – shell quantum dot with varying ZnS shell thickness are presented in Fig. 2. The PL intensity arises due to the radiative decay from the exciton state and is proportional to the vacancy in the ground state and the occupation in excitonic state [2]. Following the similar theoretical formulation as described in [2, 3, 6] we calculated the PL intensity for two different CdSe/ZnS CSQDs having core radius 2 nm and 3 nm. The shell thickness (d) varies form 0.2a to 0.6a. Here in these two cases we have changed the core radius (a) keeping the core radius to shell thickness ratio (i.e. a: d) same. We find that for smaller core radius as the external shell growth starts, the PL intensity increases and also shows a red-shifted nature. For a shell thickness of 0.6a the intensity starts to decrease. For a bigger core size (a = 3 nm) the PL intensity increases for 0.2a thick ZnS shell layer and starts to decrease if we put another layer of same thickness (d = 0.4a). The red-shift in the PL spectra is also smaller than the first case as one can see from Fig. 3. The reason of decreasing PL intensity due to a thick external shell is described below.

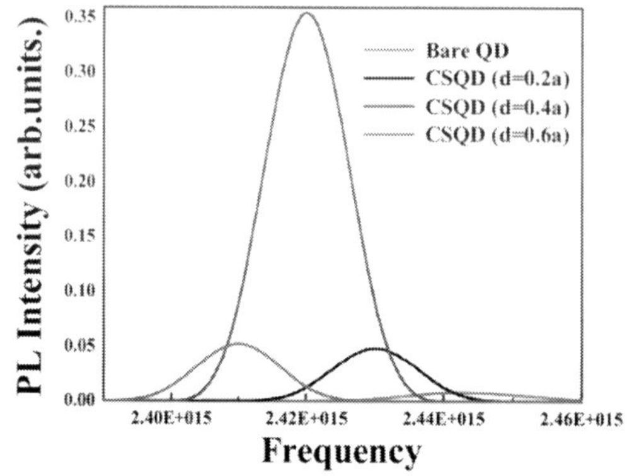

Figure 2. Photoluminescence (PL) Intensity for CdSe/ZnS CSQD with different shell thickness (d) with a core radius of a = 2 nm and shell thickness varies from 0.2a to 0.6a.

In small CSQDs, there is greater possibility that the carrier particles can tunnel and spread over the shell region also, which is the main reason behind the red-shift in PL spectra. Again in case of a CdSe/ZnS CSQD, primarily due to the shell layer, the carrier particles experience a strong confinement potential which causes a huge electron-hole Coulomb interaction. First-principle study

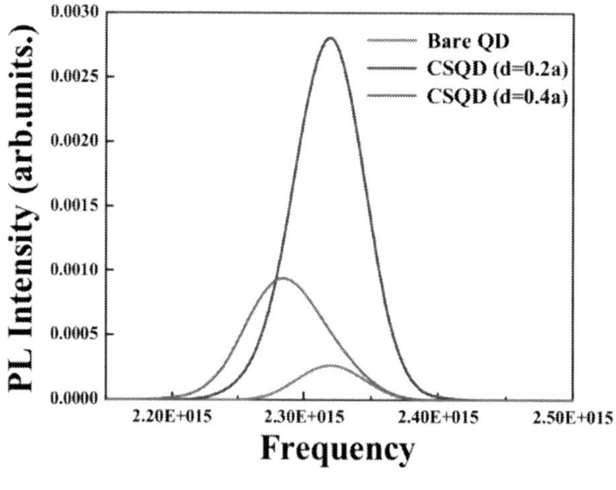

3. **Figure 3.** . Photoluminescence Intensity for CdSe/ZnS CSQD with different shell thickness (d) with a core radius of a = 3 nm and shell thickness varies from 0.2a to 0.4a.

also exhibit large electron-hole Coulomb interaction in different CSQDs due to strong confinement [8]. The localization of carrier wave functions in core region is a major factor responsible to increase the PL intensity. The reduction in the PL quantum efficiency is observed as shell thickness is increased further. The reason behind the decrease in PL intensity is due to the lattice mismatch and other radiation less relaxation processes which become important when a thick ZnS shell is grown around the CdSe core. To conclude, we can sum up that, the electron and hole wave functions get modified inside the core due to the presence of shell in conventional semiconductor core-shell QD. WKB approximation method is more appropriate to select the exact wave function for the electrons and holes. The knowledge of the effect of ZnS shell formation on CdSe QD will help to fabricate various optoelectronic devices.

4. Acknowledgement

Authors are thankful to DST, Govt. of India for financial assistance.

References

[1] Lad A D and Mahamuni S 2008 *Phys. Rev. B* **78** 125421
[2] Sen P, Chattopadhyay S, Andrews J T and Sen P K 2010 *J. Phys. Chem. Solids.* **71** 1201
[3] Chattopadhyay S, Sen P, Andrews J T and Sen P K 2012 *J Appl. Phys.* **111** 034310
[4] Ghatak A and Lokanathan S 2004 *Quantum Mechanics Theory and Applications*, 5th ed. (Macmillan India Ltd., New Delhi) pp. 380
[5] Andrews J T and Sen P 2002 *J. Appl.Phys.* **91** 2827
[6] Bafna M K, Sen P and Sen P K 2006 *J. Appl. Phys.* **100** 103515
[7] Sen P K 1982 *Solid State Commun.* **43** 141
[8] Li J J, Wang Y A, Guo W, Keay J C, Mishima T D, Johnson M B and Peng X 2003 J. Am. Chem. Soc. 125, 12567

Study Of $Co_{90}Fe_{10}$ Magnetic thin Film On MgO Substrate Using In-situ Moke Technique

Gagan Sharma[1], **D.Kumar**[1] , **Ajay Gupta**[1]
[1]UGC-DAE Consortium for Scientific Research, Khandwa Road, Indore-452001

E-mail: gags.phy@gmail.com

Abstract. Growth behavior of $Co_{90}Fe_{10}$ film on MgO substrate has been studied. In-situ magneto-optical Kerr effect measurements have been used to study evolution of the magnetic properties of the film with increasing film thickness. Extrapolating linear variation of Kerr signal with film thickness provides the evidence of a magnetic dead layer of thickness 2.2 nm at the interface with MgO substrate. X- ray reflectivity fitting confirms presence of an intermixed layer of about 2 nm.

Keywords: MOKE, Magnetic dead layer, Co90Fe10 film, XRR.

1. Introduction

Conventional MTJ with amorphous Aluminum Oxide barrier have been extensively studied for device applications but low magneto resistance limits the feasibility of such spintronics devices. MgO based MTJ is widely used device of this type and high tunnel magneto resistance (TMR) values are critical for high performance. Magnetic tunnel junctions involving MgO barrier, show promising application as non-volatile storage cells used in high performance magnetic random access memories (MRAM). Modification at the interfaces of thin film, significantly affect the performance of MTJ. Hence interfacial study and characterization of growth of magnetic thin film on MgO substrate provides relevant information for improving the performance. Moreover in high density MRAM, since lateral dimensions of magnetic cells are expected to be in nano scale range and to reduce critical current density for current induced Magnetization switching (CIMS), thickness of magnetic layer in free layer structure needs to be minimized. Therefore from application point of view, it is important to precisely determine the magnetic dead layer, because magnetic dead layer does not contribute to magnetic volume [1].In the present work, development of the magnetic properties of Co90Fe10 alloy film on MgO substrate as a function of film thickness has been done using in-situ MOKE measurements.

2. Experimental

Thin films of $Co_{90}Fe_{10}$ were deposited on MgO (100) substrate in a ultra high vacuum chamber using electron beam evaporation. Film thickness was measured in situ using a calibrated quartz crystal thickness monitor. Base pressure in the deposition chamber was 2x10-7 mbar. Deposition rate was

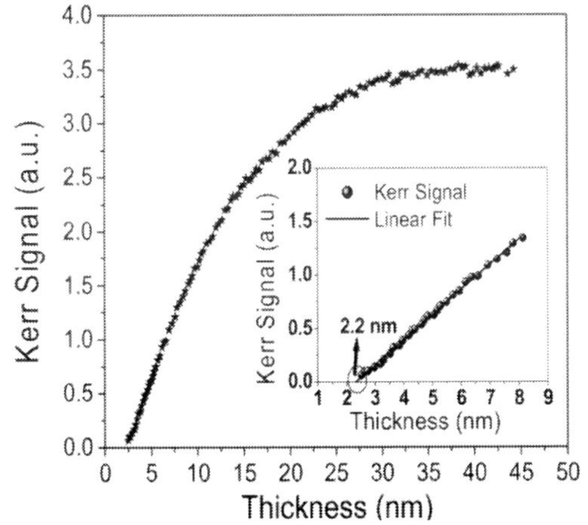

Figure 1. Kerr Signal as a function of film thickness. Inset of figure shows magnified curve in lower thickness region, which is fitted by straight line.

maintained at 0.004 nm/s. Thickness dependent Magneto-optical Kerr effect measurements were done simultaneously with deposition in Longitudnal geometry (L-MOKE) using He-Ne laser of wavelength 632.8 nm. Magnetic field inside chamber was applied using a pair of Helmholtz coils. Ex-situ x-ray reflectivity measurements were done on the deposited film using Bruker D8 diffractometer with a Gobbel mirror, so as to generate parallel monochromatic beam of Cu Kα radiation of energy 8.04 KeV.

3. Results and discussion

Figure 1 represents the dependence of the intensity of Kerr signal on film thickness. Initially Kerr signal increases linearly with film thickness and then saturates beyond a particular thickness. The linear

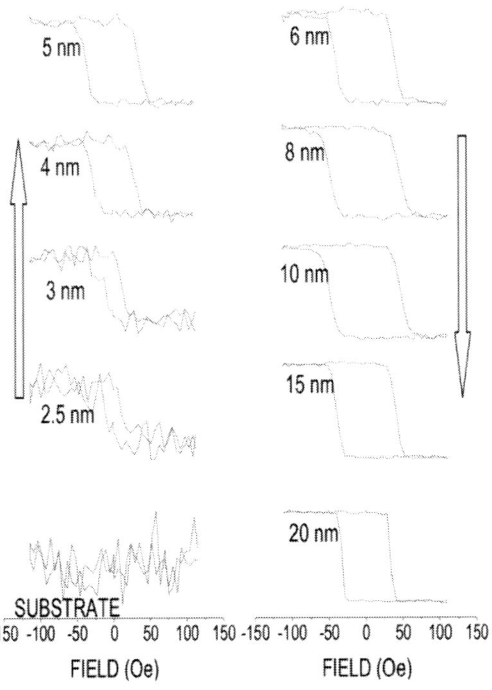

FIGURE 2. Few representative hysteresis loops of deposited film on MgO substrate at different film thickness

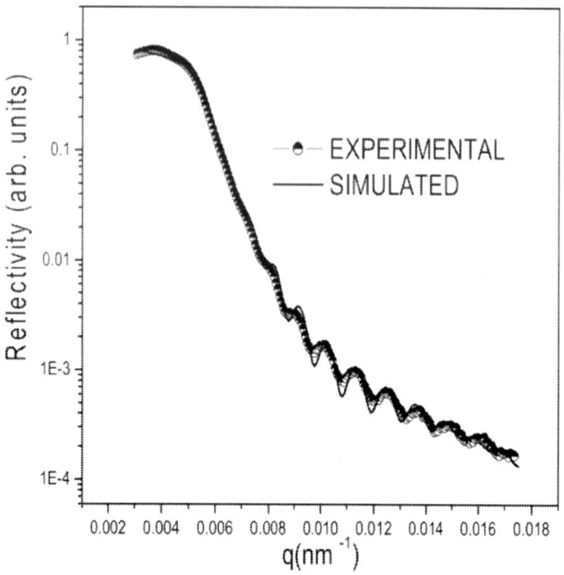

FIGURE 3. XRR of deposited film along with simulated curve.

dependence of Kerr Signal with film thickness up to 8 nm is shown in the inset of the figure. In the limit of the thickness of the magnetic layer being much smaller (typically of the order of 20 to 30 nm), the saturation Kerr signal is linearly proportional to the thickness of the magnetic layer [2]. Therefore, thickness dependence of saturation Kerr signal can yield information about a possible magnetic dead layer. The intercept of straight line on the x-axis, obtained by linear fitting, gives thickness of magnetic dead layer. In the present case of Co90Fe10 film on MgO substrate there is a magnetic dead layer of 2.2 nm.
Figure 2 depicts the development of MOKE hysteresis loops of deposited film with increasing film thickness. One can note that a faint signal starts appearing beyond 2 nm which improves with increasing film thickness.
Therefore, thickness dependence of saturation Kerr signal can yield information about a possible magnetic dead layer. The intercept of straight line on the x-axis, obtained by linear fitting, gives thickness of magnetic dead layer. In the present case of Co90Fe10 film on MgO substrate there is a magnetic dead layer of 2.2 nm.

Table1: Fitting parameters obtained from XRR

Material	Thickness (nm)	Roughness(nm)
Oxide layer	2.6±0.1	0.4±0.05
Co90Fe10	44.5±0.1	1.5±0.05
Intermixed layer	2.0 ±0.1	0.8±0.05
MgO substrate	------	0.6±0.05

Figure 3 shows X- ray reflectivity pattern of Film on MgO substrate taken after Deposition. The continuous curve represents the best fitting of the experimental data done using Parratt's formalism [3]. From the fitting, thickness of film comes out to be 44.5 nm.
Fitting parameters is shown in the Table 1.
It can be seen from XRR fitting parameters that there forms an intermixed layer at the interface with MgO substrate, of thickness around 2nm. This agrees very well with the thickness of the magnetic dead layer as obtained from In-situ MOKE measurement.

4. CONCLUSIONS:

Evolution of magnetic properties of ultrathin Co90Fe10 films on MgO(001) surface has been studied using in-situ MOKE measurements. The measurements provide clear evidence for the existence of a magnetic dead layer at MgO / Co90Fe10 interface, the thickness of which is found to be 2.2 nm. The presence of intermixed layer is also corroborated by X-ray reflectivity. The observed non-monotonous behavior of coercivity with film thickness can be understood in terms of a combined effect of superparamagnetic relaxation, and surface pinning of domain walls.

5. REFERENCES

1. Jang S.Y, Lim S. H. and Lee S. R. , 2010 *J. Appl. Phys.* **107**, 09C707
2. Xu Y.B., Kernohan E. T. M., Freeland D. J., Ercole A., Tselepi M., and Bland J. A. C. 1998 *Phys. Rev B* **58**, 2
3. Parratt L. G., 1954 *Phys. Rev.* **95**, 359

A newly Synthesis and characterization of metal complexes of 3-(N- phenyl) thiourea-pentanone-2 as ligand

A. Mishra*, S.Ninama*, P. Sharma$, N. Soni* and R. Awate*
*School of Physics & $School of Chemical Sciences, Devi Ahilya University, Indore-452001, (India)

Email:amishra1960@yahoo.co.in

Abstract: Present communication deals with the synthesis of copper and cobalt complexes derived from 3-(N- phenyl) thiourea-pentanone-2 as the ligand. Structure of all the compounds was characterized by elemental, XRD, FTIR, UV-VIS, SEM and ^{1}HNMR spectroanalytical studies. The X-ray diffraction studies indicate that 3-(N- phenyl) thiourea-pentanone-2 and their complexation with copper and cobalt metals are crystalline in nature. The FTIR spectroscopic data were used to assign the characteristic vibrational frequencies of various groups present in the compounds. Scanning electron Micrograms were used to assign the morphology and particle size of the compounds.

1. Introduction

Thiourea (Tu) derivatives have a long history as a ligand in coordination chemistry and coordinate to a metal via both sulfur and oxygen.[1] These hard and soft donor atoms provide a multitude of bonding possiblities.[2] Hydrogen bonding behaviors of some thioureas have been investigated and it is found that the intermolecular hydrogen bond between the sulphur and a hydrogen atom on N is common.[3] Thiourea derivatives and their transition metal complexes have been known for over a century and are easily synthesized in good yields. Recently metal complexes of thiourea and thiourea analogs have been investigated.[4] Metal complex of thiourea, commonly called semi organics, include the advantages of both organic and inorganic part of the complex. A number of thiourea derivatives has been reported to form complexes with copper and cobalt.[5] It is revealed from the literature that a number thiourea derivatives have been shown to posses antibacterial, antifungal, antitubercular, antithyroid and insecticidal properties.[6] Thus in order to observe the enhanced complexing behavior of thiourea a number of its extended derivatives coupled with the β-diketones have been prepared as the powerful complexion agents. These 3-(N- phenyl) thiourea-pentanone-2 have been noticed to possess a remarkable capacity to coordinate with the transition metals giving rise to highly colored cheated structure. Hence, in view of the aforementioned importance associated with the chelating behavior of 3-[N-phenyl thiourea-pentanone-2] vis- a- vis to get a deep insight into the complexation behavior, it was our endeavor to synthesize its copper and cobalt complexes followed by their characterization using XRD, SEM, FTIR, ^{1}HNMR and elemental analyses.

Published under licence by IOP Publishing Ltd

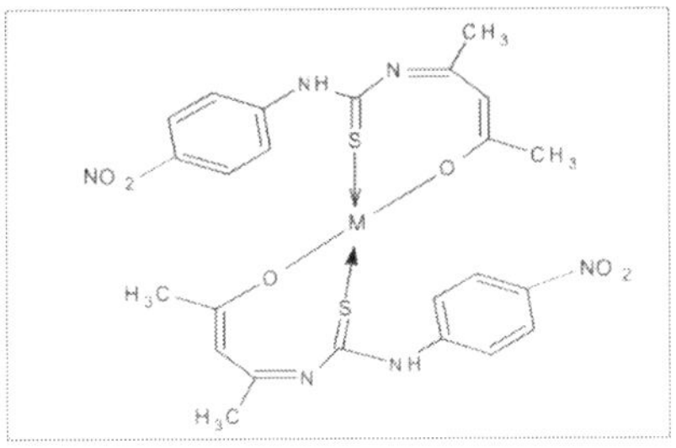

Figure 1. The structure of [3-(N-phenyl)-thiourea pentanone-2] M- (Cu and Co)

2. Experimental

Chemicals: All the chemicals and solvents were obtained from the Fluka Chemical CO., Germany and used as received. ^{1}H NMR labeled thiourea was obtained from the Isotec Co., USA; 5% H NMR enriched Tu was made by mixing 5% labeled 95% unlabelled Tu by Weight. The strcture of the complex is shown in Figure 1.

Synthesis: The preparation of ligand and the formation of complexes with metals have been carried out by chemical reaction with combination of compounds (aniline, ammonium thiocynate). The yield was 90%.

Table 1. Physical and Analytical data of the Metal (II) Complexes

Comolexes /Mol.formula	M.P. (^{0}C)	Yield (%)	Calc. (Found) % C H N S
[Ligand] [$C_{12}H_{14}N_3O_3S$]	250-251	65	51.4 5.03 14.9 11.4 (50.1)(4.9)(14.5)(11)
[Cu (L)(Cl)$_2$] [$C_{12}H_{14}N_3O_3SCuCl_2$]	230-232	58	34.7 3.4 17.1 7.7 (33.9)(3.1)(16.7)(7.1)
[Co (L)(Cl)$_2$] [$C_{12}H_{14}N_3O_3SCoCl_2$]	240-242	56	41.1 3.7 11.08 8.4 (40)(3.5)(11.04)(8.2)

Table 2 –XRD peaks and crystal thickness calculated from Scherer's formula

Compounds	hkl	Lattice parameters	Crystal size
Ligand	33, 37, 38,41	16.5, 14.9, 14.7, 14.1	463,334,330,290
Cu Complex	2,3,4,6	8.2,8.5,8.2,8.2	440,572,488,524
Co Complex	1,2,3,4	22.3,22.4,22.4,22.3	395,218,599,101

Table 3: IR, UV, and NMR spectroscopic Data of the ligand and Meta (II) Complexes

Ligand/complexes	IR (cm^{-1})	λ_{max} in nm (ε in M^{-1} cm^{-1})	^{1}H-NMR (DMSO-d_6) (ppm)
Ligand	3470(NH); 3480(=C-H, Sp2); 1598(C=S); 1615(C=O); 1468(C-C, ring structure); 836,749 (sub phenyl)	272(1103),	-CH_3 (2.19 -NH (6.80) –Aroma protons (7.95-8.21)
Cu Complex	3330(NH); 3281(=C-H, Sp2); 1600(C=S); 1630(C=O); 1517(C=C) 1490(C-C, ring Structure) 740(M→L); 921,824(sub phenyl)	241(9801), 294(9830), 364(1284)	-CH_3(3.10) -NH (6.60) Aromatic Protons (7.96-8.2)
Co Complex	3500(NH); 2944(=C-H, Sp2); 1616(C=S); 1630(C=O); 1533(C=C)1522(C-C, ring structure) 594(M→L); 927,782(sub phenyl)	(277), 382.4(227.5), ---------	-CH_3 (3.10) -NH (6.60) Aromat protons (7.96-8.2)

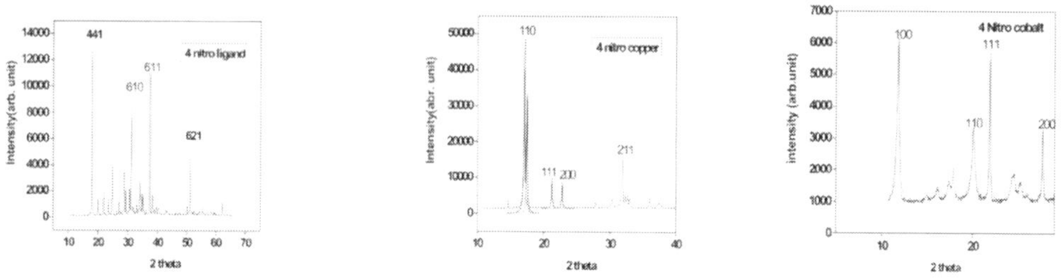

Figure2. shows XRD pattern of (a) ligand (b) complexation of ligand with copper metal (c) complexation of ligand with cobalt metal.

After this preparation of 3-(N-phenyl)-thiourea-pent-2-one was done using 1-2 pellets of NaOH added with pentone-2, 4-dione. Finally a pale yellow precipitate of 3-(N-phenyl)-thiourea-pentanone-2, was prepared, in which 85%yield was formed. The copper and cobalt metal chloride were added to 3-(N-phenyl)-thiourea-pentanone-2 were dissolved in 10ml of pure ethyl alcohol. After some time 60% yield of copper and cobalt complexes were formed.

Elemental analysis (C, H, N, S) was performed by using a Carlo Erba elemental analyzer. Melting points were determined in open capillary tubes using an electric melting point apparatus and are uncorrected. The data are reported in Table1.

X-ray diffraction: XRD patterns were recorded by Rikangu RINT-2000 diffractometer equipped with rotating anode with tube voltage of 40kV and current of 100mA on Philips PW3040/60 X-ray powder diffractometer. Cu target was used as the source of X-rays at λ=1.54A^{o} wavelength Results are summarized in Table 2.

^{1}HNMR :^{1}HNMR spectra (δ, ppm) were recorded on a 300 MHz Bruker spectrometer with TMS as an internal standard. (SeeTable3).

FTIR: The solid state FTIR spectrum of the ligand and their complex was recorded in the range 4000 to 400 cm-1 on perkin-Elmer 16 FPC FT-IR using KBr pellets. Results are summarized in Table 3.

Electronic Spectra: Electronic spectra of 1.862X10^{-4}M solution of the ligand and complex were recorded in methanol at different time intervals in the range 500-200nm on a UV-1700 SHIMADZU corporation spectrophotometer. The following instrument conditions were used; slit=2nm, scan speed=60nm/min, response=0.2s. The data are reported in Table 3.

SEM: The surface morphology of the sample was examined using scanning Electron Microscopy on a JEOL JSM 5600 SEM

3. Results and Discussion:

A thiourea derivative is capable of bonding to transition metals via sulfur or nitrogen. When thiourea derivatives are not attached as complex with metal, than XRD pattern is different those that of their complexes. Figure 2 (a, b, c) shows the XRD pattern of synthesized ligand i.e. [3-(N-phenyl) thiourea pentanone-2] and its complaxation with copper and cobalt metals.

These samples show the diffraction pattern with large number of peaks. The patterns show sharp intense peaks throughout the spectrum indicating the crystalline nature of the sample.[7] Upon observing Fig 2(a), we find that the XRD pattern of ligand shows sharp and highly intense peaks. Some of the sharp peaks are disappeared when this ligand is attached with copper metal as represented in fig 2(b). Moreover, to in fig 2(c) some additional sharp peaks are observed when ligand is attached to the cobalt metal. All the parameters calculated from XRD are computed in table 2. The crystal size, hkl and lattice parameter for ligand, copper complex and cobalt complex are showing that the samples are simple cubic structure. IR spectra of the complexes in comparison to the ligand 3-(N- phenyl) thiourea-pentenone-2 are listed in table 3. The medium peak (3470, 3330 and 3500cm^{-1} for ligand and their complexation with metals) is attributed to stretching of the

Figure 3.SEM image of a) ligand b) Cu metal with Ligand c)SEM image of Co metal with ligand

N-H group adjacent to the carbonyl group[8]. A strong absorption (1615cm^{-1} for ligand) is ascribed to the stretching vibrational of the carbonyl groups and shifts upon complexation.[8] Deprotonation induces delocalization of C=O stretching vibration (~180 cm^{-1} decrease).and confirming coordination through the oxygen. The C=S modes of thiourea was observed at a lower frequency (1598 cm^{-1}) compared to that of their complexation with copper and cobalt metals (1616, 1600 cm^{-1}). The characteristic absorption peaks in the IR spectra of all the complexes are similar, which indicate that the complexes have similar structures[9]. For free Tu the n-π* transition occurs at 272nm (λ) with molar extension coefficient (ε) value of 1103cm^{-1}M^{-1}. The value of λ_{max} for the complex in methanol at 264.5 nm with molar extension coefficient 1134 cm^{-1}M^{-1} has been observed, probably due to a π-π^* transition. Complex with copper metal, three prominent absorption at 241, 294 and 364nm with molar extension coefficients of 9801, 9830 and 1284cm^{-1}M^{-1} respectively, have been From the micrographs, it is observed that small fibers type structure for the ligand Co [3-(N-phenyl) thiourea pentanone-2], pyramidal type structure for the copper complex, while the needle type structure is observed for the Co complex. From the figure, we clearly estimated the grain size by SEM of the ligand is 4μm and their complex compounds are 5μm and 3μm respectively.
recorded and for complex with cobalt, two prominent absorptions at 277 and 371.5nm with molar extension coefficient of 9826 and12900 cm^{-1}M^{-1} respectively have also been recorded. Absorption at 364 and 371nm with a ε value of 9830 cm^{-1}M^{-1} indicates ligand to metal charge transfer (LMTC). Since sulfur is soft ligand and C=S and low lying π^* orbital and at the same time Co (II) 3d^7 and Co 3d^9 back bonding (MLCT) transition with an value of ε 9801, 9830 and 9826 cm^{-1}M^{-1}is also significant while both the peaks there is very slight change both in λ and ε value with time. In ^{1}H NMR a broad signals between (7.9 and 8.2) ppm. Indicate that both free thiourea and its complexation with metals cotaining Cu-Tu and Co-Tu are present. It is noted that the peaks for two transition metals Cu and Co bounded with thiourea at 7.66 and 7.96ppm.The signals 7.66 and 7.96 ppm indicates possible Cu-S and Co-S bond formations. Scanning electron microscopy is a convenient technique to study the microstructure of powdered samples[10]. Fig 3a, fig3b and fig3c show the surface morphology of ligand i.e. 3-(N-phenyl)-thiourea-pentanone-2, and their complexation with copper and cobalt metals respectively.

4. Conclusions

All the studied compounds were synthesized with the chemical reactions. The proposed general structure is also reported. The compounds were characterized by elemental, XRD, FTIR, UV-VIS, SEM and ^{1}HNMR spectro-analytical studies. The X-ray diffraction patterns indicate that the ligand 3-(N- phenyl) thiourea-pentanone-2 and their complexation with copper and cobalt metals are crystalline in nature and show simple cubic lattice structure. The IR spectroscopic data were used to assign the characteristic vibrational frequencies of various groups present in these compounds. Scanning electron Micrograms were used to assign the morphology and particle size of the compounds.

References

1. Burrows D A,Mare D C, Mahon F M 1999 *Polyhedron* **18** 2665
2. Hederson W, Nicholson B K, Dinger M B, Bennett R L 2002 *Inorg.Chim.Acta* **338** 210
3. Arslan H, Florke U, Kulcu N, Emen M F 2006 *Journal of coordination chem.* **59** 223
4. Hou W B, Xu D. Yuan D R, Liu M G 1994 *Cryst. Res. Technol.* **29** 939
5 Dominguez M,Antico E, Beyer L, Aguirre A, Garcia-Granda S, Salvado V 2002 *Polyhedron,*

21 1429
6. Madan V K, Taneja A D 1991 *J. Indian Chem. Soc.* **68** 471
7. Cullity B D Elements of X-Ray Diffraction, Second Edition, Addison-Wesley Publishing Company, 1978, chpt 5.
8. Yesilkaynak T, Binzet G, Emen F M, Florke U, Kulcu N, Arslan H 2010 *European journal of chem.* **1(1)** 1
9. Nakamoto K, "Infrared and Raman spectra of inorganic and coordination compounds", VIth Edn., Wiley Interscience, New York, 2009, chpt 3 pp 333-402.
10. Goldstein J, Newbury D E, Joy D C, Lyman C E, Echlin P, Lifschin E, Sawyer L, Michael J, Scanning electron microscopy and X-ray Microanalysis, 3rd Edition Kluwer Academic/Plenum Publishers, New York, 2003, pp 21-60

Jeans instability of self-gravitating magnetized strongly coupled plasma

R P Prajapati[1], P K Sharma[2*], R K Sanghvi[3] and R K Chhajlani[4]
[1]Dept. of Pure & Applied Physics, G. G. Central University, Bilaspur (C.G.), India
[2]BUIT, Barkatullah University, Bhopal (M. P.), India
[3]Govt. Holker Science College, Indore (M.P.), India
[4]S. S. in Physics, Vikram University, Ujjain (M. P.), India

E-mail: pks_buit30@yahoo.com

Abstract. We investigate the Jeans instability of self-gravitating magnetized strongly coupled plasma. The equations of the problem are formulated using the generalized hydrodynamic model and a general dispersion relation is obtained using the normal mode analysis. This dispersion relation is discussed for transverse and longitudinal mode of propagations. The modified condition of Jeans instability is obtained for magnetized strongly coupled plasma. We find that strong coupling of plasma particles modify the fundamental criterion of Jeans gravitational instability. In transverse mode it is found that Jeans instability criterion gets modified due to the presence of magnetic field, shear viscosity and fluid viscosity but in longitudinal mode it is unaffected due to the presence of magnetic field. From the curves we found that all these parameters have stabilizing influence on the growth rate of Jeans instability.

1. Introduction

In natural and laboratory situations many objects exist in strongly coupled state viz. neutron star, dwarf star, interior of planets, ultra cold neutral plasma and liquid plasma crystal [1]. In these objects the Coulomb potential energy is much greater that the average thermal energy of the plasma particles. The ratio of the Coulomb potential energy to the thermal energy is called the coupling parameter (Γ) thus for strongly coupled plasma $\Gamma >> 1$. In the presence of dust particles, plasma exhibits variety of rich phenomena. Under certain conditions, the charged dust grains may organize themselves in a regular crystal-like structure. For $\Gamma > \Gamma_c$ (where Γ_c is the critical value for crystallization) the dust component goes into an ordered crystalline phase which forms plasma crystal [2,3]. The mutual interaction among the dust particles can be represented by the Yukawa (Debye–Hückel) potential $\Phi(r) = \left(Q^2 / 4\pi\varepsilon_0 r\right)\exp\left(-r/\lambda_D\right)$, where λ_D is the Debye length of plasma and Q is the charge of dust particle [4]. The waves and instabilities in strongly coupled plasmas are studied by many authors using the generalized hydrodynamic model. Kaw and Sen [5] have discussed the low frequency modes in strongly coupled dusty plasmas. Rosenberg and Shukla [6] have shown the effect of strong ion correlations on the dispersion relation of possible ion-beam plasma instabilities. Many authors [7-9] have discussed the problem of strongly coupled plasma for the investigation of acoustic

* To whom any correspondence should be addressed.

wave and instabilities both analytically and numerically. However, most papers on collective effects in self-gravitating plasmas are concerned with the problem of the coupling between purely electrostatic and gravitational disturbances. Yaroshenko et al. [10] have discussed the linear coupling of electromagnetic and Jeans modes in self-gravitating plasma streams. Recently, Janaki et al. [11] have investigated the Jeans instability of viscoelastic fluid and found that the threshold for the onset of instability appears at higher wavelength in a viscoelastic medium. Looking to the importance of present problem in astrophysical situations, we investigate the Jeans instability of self-gravitating magnetized strongly coupled plasma.

2. Perturbation equations and dispersion relations

Let us consider a compressible, infinitely conducting viscoelastic strongly coupled plasma embedded in uniform magnetic field $\boldsymbol{B}_0(0,0,\mathrm{B})$. In the absence of a magnetic field, the mechanical equations can be described small amplitude, longitudinal sound (compressional) waves with speed c_s, the square of which is equal to the derivative of the pressure p with respect to the density ρ at constant entropy i.e. $c_s^2 = \gamma p_0/\rho_0$. The basic equations of the problems are linearized using the standard linearization process. We suppose all the physical quantities are sum of their equilibrium and perturbed parts i.e. $\boldsymbol{v} = \widetilde{\boldsymbol{v}}(\boldsymbol{x},\mathrm{t}), \boldsymbol{B} = \boldsymbol{B}_0 + \widetilde{\boldsymbol{B}}(\boldsymbol{x},\mathrm{t}),\ \phi = \phi_0 + \widetilde{\phi}(\boldsymbol{x},\mathrm{t})$ and $\rho = \rho_0 + \widetilde{\rho}(\boldsymbol{x},\mathrm{t})$.

The generalized equation of motion for magnetized strongly coupled plasma is given by

$$\left(1+\tau\frac{\partial}{\partial t}\right)\left[\rho_0\frac{\partial\widetilde{\boldsymbol{v}}}{\partial t}+\rho_0\nabla\widetilde{\phi}+c_s^2\nabla\widetilde{\rho}-\frac{1}{4\pi}(\nabla\times\widetilde{\boldsymbol{B}})\times\boldsymbol{B}_0\right]=\vartheta\nabla^2\widetilde{\boldsymbol{v}}+\left(\xi+\frac{\vartheta}{3}\right)\nabla(\nabla.\widetilde{\boldsymbol{v}}), \tag{1}$$

The continuity equation is

$$\frac{\partial\widetilde{\rho}}{\partial t}+\nabla\dot{}(\rho_0\,\widetilde{\boldsymbol{v}})=0, \tag{2}$$

The Poisson's equation for self-gravitational potential is written as

$$\nabla^2\widetilde{\phi}=4\pi G\widetilde{\rho}, \tag{3}$$

The equation for magnetic field is given by

$$\frac{\partial\widetilde{\boldsymbol{B}}}{\partial t}=\nabla\times(\boldsymbol{v}\times\mathbf{B}). \tag{4}$$

For linear instability analysis we now solve equations (1)-(4) using normal mode analysis subject to perturbation $\exp[i(k_x\hat{x}+k_z\hat{z})+\sigma t]$, (where σ is frequency of the harmonic disturbance and $k_{x,z}$ are the wave numbers in transverse and longitudinal directions to the magnetic field). We obtain three linear equations in terms of velocity components v_x, v_y, v_z which gives a 3 × 3 matrix whose vanishing solution yields a general dispersion relation. The general dispersion relation is discussed for transverse and longitudinal mode of propagations.

2.1 Transverse mode of propagations $\boldsymbol{k}\perp\boldsymbol{B}$:

$$\left\{(1+\sigma\tau)\sigma+\frac{k^2\vartheta}{\rho_0}\right\}^2\left\{(1+\sigma\tau)\left[\sigma^2+\left(c_s^2+V_A^2\right)k^2-4\pi G\rho_0\right]+\frac{k^2\sigma}{\rho_0}\left(\xi+\frac{4\vartheta}{3}\right)\right\}=0. \tag{5}$$

The above equation shows the dispersion relation of homogeneous magnetized, self-gravitating strongly coupled plasma. In classical limit $\sigma\tau \ll 1$ we recover the classical dispersion relation of self-gravitating, magnetized, viscous plasma. The above dispersion relation has two factors, in which the first factor equating zero gives the non-gravitating, damped viscous mode modified due to relaxation time for the strongly coupled plasma. In the kinetic limit $\sigma\tau \gg 1$, second factor of dispersion relation (5) gives

$$\sigma^2+\left[\left(c_s^2+V_A^2\right)+\frac{1}{\tau\rho_0}\left(\xi+\frac{4\vartheta}{3}\right)\right]k^2-4\pi G\rho_0=0. \tag{6}$$

The condition of Jeans instability can be obtained from the constant term of above dispersion relation which is given by

$$k<k_J=\left[\frac{4\pi G\rho_0}{\left(c_s^2+V_A^2\right)+\frac{1}{\tau\rho_0}\left(\xi+\frac{4\vartheta}{3}\right)}\right]^{1/2}. \tag{7}$$

The system will be unstable for all the wave numbers provided by expression (7). We find that condition of Jeans instability and expression of critical Jeans wave number both are modified due to the presence of Alfven velocity, viscosity, shear viscosity and relaxation time. It is obvious that as the shear viscosity, viscosity or magnetic field in the strongly coupled plasma increases the corresponding critical jeans wave number decreases hence all these parameters have stabilizing influence on the growth rate of Jeans instability.

The effect of shear viscosity and sound speed is studied on the growth rate of Jeans instability of magnetized strongly coupled plasma. We now write dispersion relation (6) in the dimensionless form

$$\sigma^{*2}+\left(1+c_s^{*2}+\xi^*\right)k^{*2}-1=0, \tag{8}$$

where $\sigma^*=\sigma/(4\pi G\rho_0)^{1/2}, k^*=kV_A/(4\pi G\rho_0)^{1/2}$, $c_s^*=c_s/V_A$ and $\xi^*=(1/\tau\rho_0)(\xi+4\vartheta/3)/V_A^2$, are the dimensionless parameters.

In figure 1 and figure 2 we have plotted the dimensionless growth rate of Jeans instability against the dimensionless wave number for various values of shear viscosity and sound speed. It is obvious from figure 1 that shear viscosity has damping influence on the growth rate of Jeans instability similar to the classical fluid viscosity as in case of weakly coupled viscous plasma. We find that growth rate of Jeans instability decreases as the shear viscosity of the medium increases. Hence it has stabilizing influence on the growth rate of Jeans instability.

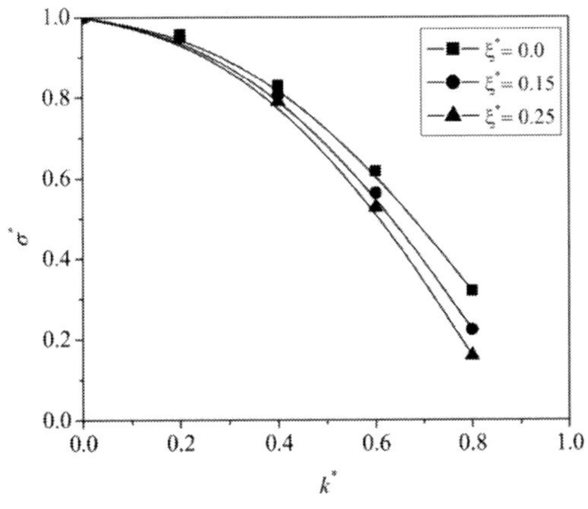

Figure 1. Effect of shear viscosity on the growth rate of Jeans instability keeping $c_s^*=0.25$.

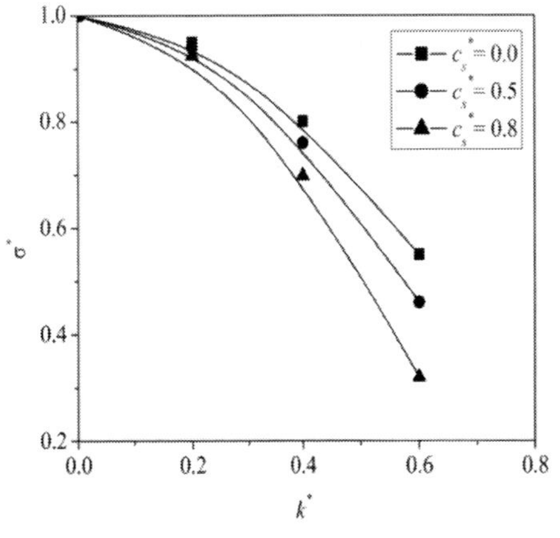

Figure 2. Effect of sound speed on the growth rate of Jeans instability keeping $\xi^*=0.25$.

2.2 Longitudinal mode of propagation $\mathbf{k} \parallel \mathbf{B}$:

In the case when $k_z = k,\ k_x = 0$ we get the following dispersion relation

$$\left[(1+\sigma\tau)(\sigma^2 + V_A^2 k^2) + \frac{k^2 \vartheta \sigma}{\rho_0}\right]^2 \times \left[(1+\sigma\tau)(\sigma^2 + k^2 c_s^2 - 4\pi G \rho_0) + \frac{\sigma k^2}{\rho_0}\left(\xi + \frac{4\vartheta}{3}\right)\right] = 0. \quad (9)$$

This dispersion relation has two factors. The first factor gives when equated to zero a non-gravitating damped viscous Alfven mode. In the classical limit $(\sigma\tau << 1)$ the above equation shows the dispersion relation for classical viscous magnetized plasma. In kinetic limit $(\sigma\tau >> 1)$ we obtain the following dispersion relation

$$\sigma^2 + V_A^2 k^2 + \frac{k^2 \vartheta}{\tau \rho_0} = 0. \quad (10)$$

The second factor of dispersion relation (9) gives

$$(1+\sigma\tau)(\sigma^2 + k^2 c_s^2 - 4\pi G \rho_0) + \frac{\sigma k^2}{\rho_0}\left(\xi + \frac{4\vartheta}{3}\right) = 0. \quad (11)$$

The above equation shows a gravitating shear viscous mode which is unaffected due to the presence of magnetic field. In the kinetic limit $(\sigma\tau >> 1)$ we obtain the following condition of Jeans instability $k < k_{J1} = [4\pi G \rho_0 /(c_s^2 + \frac{1}{\tau \rho_0}(\xi + 4/3\vartheta))]^{1/2}$. It is clear from the above expression that condition of Jeans instability and expression of critical Jeans wave number are both modified due to shear viscosity but they are unaffected due to the presence of magnetic field. Thus magnetic field play role in the condition of Jeans instability only in transverse mode of propagation and it has no influence in the longitudinal mode of propagation.

3. Conclusion

In the present paper we have investigated the Jeans instability of self-gravitating strongly coupled plasma. The dispersion relation of the problem is obtained which is discussed for both transverse and longitudinal mode of propagations. In the kinetic limits $(\sigma\tau >> 1)$ we obtain propagation of both longitudinal and transverse viscoelastic modes. In transverse mode we obtain a self-gravitating shear induced viscous mode modified due to Alfven velocity. The condition of Jeans instability is modified due to the presence of shear viscosity and Alfven velocity and both parameters have stabilizing influence on the growth rate of Jeans instability. In longitudinal mode of propagation a self-gravitating unmagnetized shear induced viscous mode. In this mode the condition of Jeans instability depends upon shear viscosity and it is unaffected due to the presence of magnetic field.

4. References

[1] Ikeji H 1996 *Phys. Fluids* **29** 1764
[2] Killian T C 2007 *Science* **316** 705
[3] Thomas H M and Morfill G E 1996 *Nature* **379** 806
[4] Robbins M O Krener K and Grest G S 1988 *J. Chem. Phys.* **88** 3286
[5] Kaw P K and Sen A 1998 *Phys. Plasmas* **5** 3552
[6] Rosenberg M and Shukla P K 2011 *Phys. Scr.* **83** 015503
[7] Baruah S and Das N 2010 *Phys. Plasmas* **17** 073702
[8] Ohta H and Hamaguchi S 2000 *Phys. Rev. Lett.* **84** 6026
[9] Piper J B and Goree J 1996 *Phys. Rev. Lett.* **77** 3137
[10] Yaroshenko V V Voitenko Y and Goossens M 2002 *Phys. Plasmas* **9** 1520
[11] Janaki M S Chakrabarti N and Benerjee D 2011 *Phys. Plasmas* **18** 012901

Effect of magnetic field on the conducting fluids streaming in porous medium

P K Sharma
BUIT, Barkatullah University, Bhopal (M.P.)-462026 India

E-mail : pks_buit30@yahoo.com

Abstract. The Kelvin–Helmholtz instability of two superposed fluids of plasma streaming in opposite directions through porous medium has been investigated in a uniform two dimensional horizontal magnetic field. The medium is assumed to be conducting and incompressible. By applying the normal mode technique, the dispersion relation has been derived. From the dispersion relation the Kelvin Helmholtz instability condition is obtained. It is observed that the instability condition is modified due to the simultaneous presence of two dimensional horizontal magnetic field, porosity and suspended particles. It is found that the conditions for the Kelvin Helmholtz instability depends on the magnetic field, on the suspended dust particles and on the relaxation frequency of the particles. Numerical analysis is performed to show the effect of various parameters on the growth rate of Kelvin Helmholtz instability. It is also found that magnetic field, suspended dust particles and medium porosity has a stabilizing effect on the growth rate of unstable K-H mode

1.Introduction

The Kelvin- Helmholtz (K-H) instability is a well-known instability that refers to the instability of plane interface between two superposed fluids flowing one over the other with relative velocity parallel to the interface. The hydromagnetic instabilities have wide applications in space plasma, astrophysical plasma and laboratory plasma. The flow problems of different fluids through porous medium have significant applications in petroleum production engineering, the chemical industry with regard to the recovery of crude oil from the pores of the reservoir rocks. Chandrasekhar [1] studied the classical K-H problems making use of single hydromagnetic equations.

The effect of suspended particle on the stability of superposed fluids might be of importance to industrial and chemical engineering. Saffman [2] had studied in detail, a dust gas in magnetohydrodynamics. Scanlon and Segel [3] have made a thorough study of the implication of suspended particles in hydromagnetics in the context of the Benard convection problem. Palaniswamy and Purushotum [4] have found the effect of fine dust to increase the region of instability in the shear flow of stratified fluids. Michael [5] has investigated the K-H instability of a dusty gas using Strokes drag force formula.

In the recent years, the wide applications of porosity in industrial, geophysical situation and chemical engineering has created considerable interest, particularly among geophysical fluid dynamics. Sharma and Spanos [6] have discussed the K-H and R-T instability of streaming fluids in a porous medium. Porosity is of use for the study of physical properties of comets, meteorites and interplanetary dust in astrophysical context McDonnel [7]. Prajapati et al. [8] have studied Kelvin Helmholtz and Rayleigh Taylor instability of two superposed magnetized fluids with suspended dust particles. Prajapati and Chhajlani [9] have studied Kelvin Helmholtz and Rayleigh Taylor instability of streaming fluids with suspended dust particles flowing through porous media.

Keeping in mind the importance of suspended particles and porosity in modern technology and industries we have examined the problem of the Kelvin-Helmholtz instability of the plane interface separating two incompressible superposed conducting fluids in porous medium in the presence of two dimensional uniform magnetic field with suspended particles.

2. Equations of the problem

Consider two superposed perfectly conducting semi infinite homogeneous incompressible fluids flowing through porous medium of porosity ε and medium permeability k1. Suppose that a plane interface of discontinuity exist at z=0. Let the mixture of hydromagnetic fluid and suspended dust particles streaming together with unperturbed velocity U(U,0,0) in the porous medium and two dimensional uniform horizontal magnetic field H(Hx,Hy,0). Let ρ, μ, p and u(u,v,w) denotes respectively the density, the viscosity, the pressure and velocity of the medium.

It is supposed that the suspended dust particles experience force KN (V-U) in the fluid, where K is a constant given by K = 6 π a μ (Stokes drag formula), 'a' being the particle radius. The linearized perturbation equations governing the fluids are written as follows:

$$\frac{\rho}{\varepsilon}\left[\frac{\partial \boldsymbol{u}}{\partial t}+\frac{1}{\varepsilon}(\boldsymbol{U}.\nabla)\boldsymbol{u}\right]=-\nabla\delta p+\frac{KN}{\varepsilon}(\boldsymbol{v}-\boldsymbol{u})+\frac{\mu}{k_1}\boldsymbol{u}+\frac{\mu_e}{4\pi}\left[(\nabla\times\mathbf{h})\times\mathbf{H}+(\nabla\times\mathbf{H})\times\mathbf{h}\right], \tag{1}$$

$$\varepsilon\frac{\partial}{\partial t}\delta\rho+(\boldsymbol{U}.\nabla)\delta\rho+(\boldsymbol{u}.\nabla)\rho=0, \tag{2}$$

$$\varepsilon\frac{\partial \mathbf{h}}{\partial t}=(\mathbf{H}.\nabla)\mathbf{u}-(\mathbf{u}.\nabla)\mathbf{H} \tag{3}$$

$$\nabla.\boldsymbol{u}=0, \tag{4}$$

$$\nabla.\boldsymbol{h}=0 \tag{5}$$

$$\left\{\tau\left[\frac{\partial}{\partial t}+\frac{1}{\varepsilon}(\boldsymbol{U}.\nabla)\right]+1\right\}\boldsymbol{v}=\boldsymbol{u}, \tag{6}$$

Where δp and δρ are the perturbations in fluid pressure and density of fluid respectively and τ = m/6πaμ denotes the relaxation time for the suspended dust particles.

Assuming the perturbation to be of the form f(z) exp(ik_xx+ik_yy+int) , where k_x and k_y are the horizontal wave numbers ($k^2=k_x^2+k_y^2$) and n is the growth rate of the harmonic perturbations.

3. Dispersion relation

The following dispersion relation, governing the vertical component of velocity w for the assumed configuration can be obtained by solving above equations and written as

$$\left[\frac{i}{\varepsilon}\left(n+\frac{k_xU}{\varepsilon}\right)+\nu/k_1+\left(\frac{\alpha_0}{\varepsilon}\right)\frac{(in+ik_xU/\varepsilon)}{\tau(in+ik_xU/\varepsilon)+1}\right]\left\{D(\rho Dw)-k^2\rho w\right\}+ \left[\frac{(H_xk_x+k_yH_y)^2}{4\pi(\varepsilon in+ik_xU)}\right](D^2-k^2)w=0 \tag{7}$$

Where α_0 =mN/ρ denotes the mass concentration of the particles and D=d/dz. Here we have assumed the same density of suspended particles in both the regions z<0 and z>0.

Consider the configuration of two superposed fluids of densities ρ_1 (lower fluid) and ρ_2 (upper fluid) slipping past each other at horizontal interface z=0. In case of two regions of constant density, Eq.(7) reduces to

$$(D^2 - k^2)w = 0. \tag{8}$$

After using the solutions of Eq.(8) for the upper and lower fluids and boundary conditions at the common interface of two fluids, we can obtain the following dispersion relation

$$n^2 + \left[\frac{2k_x}{\varepsilon}(\beta_1 U_1 + \beta_2 U_2) - \frac{i\varepsilon}{k_1}(\beta_1 \nu_1 + \beta_2 \nu_2)\right]n + \left\{\frac{k_x^2}{\varepsilon^2}(\beta_1 U_1^2 + \beta_2 U_2^2)\right.$$
$$-\frac{ik_x}{k_1}(\beta_1 \nu_1 U_1 + \beta_2 \nu_2 U_2) + \frac{\alpha_1 \beta_1 (n + k_x U_1/\varepsilon)^2}{[\tau(in + ik_x U_1/\varepsilon) + 1]} + \frac{\alpha_2 \beta_2 (n + k_x U_2/\varepsilon)^2}{[\tau(in + ik_x U_2/\varepsilon) + 1]} \tag{9}$$
$$\left. - 2(k_x V_A + k_y V_B)^2\right\} = 0,$$

Here $\alpha_1 = mN/\rho_1$, $\alpha_2 = mN/\rho_2$, $\beta_1 = \rho_1/(\rho_1 + \rho_2)$, and $\beta_2 = \rho_2/(\rho_1 + \rho_2)$.

For simplicity, we considered $V_{A,B}^2 = \dfrac{\mu_e H_{x,y}^2}{4\pi(\rho_1 + \rho_2)}$

The above dispersion relation represents the influence of the suspended particles, dynamic viscosity and two directional horizontal magnetic field on Kelvin Helmholtz instability of two superposed fluids flowing through porous medium.

4. Discussion

In this problem we consider two superposed fluids of same densities and permeability in presence of suspended dust particles and flow velocities U and –U respectively with horizontal magnetic field.

For the simplicity of the problem we put $\alpha_1 = \alpha_2 = \alpha_0, \beta_1 = \beta_2 = 1/2$, $U_1 = U, U_2 = -U$, $f_s = 1/\tau$, $\nu_1 = \nu_2 = \nu$ in dispersion relation (9) and get

$$\sigma^4 + \sigma^3\left[f_s(2+\alpha_0) + \frac{\varepsilon\nu}{k_1}\right] + \sigma^2\left[f_s^2(1+\alpha_0) + \frac{2f_s\varepsilon\nu}{k_1} + 2(k_x V_A + k_y V_B)^2\right]$$
$$+\sigma\left[\frac{f_s^2\varepsilon\nu}{k_1} + 4(k_x V_A + k_y V_B)^2 f_s + \frac{k_x^2 U^2 \nu}{k_1 \varepsilon} + \frac{f_s k_x^2 U^2}{\varepsilon^2}(\alpha_0 - 2)\right] - \tag{10}$$
$$\left[\frac{k_x^2 U^2}{\varepsilon^2}\left\{f_s^2(1+\alpha_0) + \frac{k_x^2 U^2}{\varepsilon^2}\right\} - 2(k_x V_A + k_y V_B)^2\left\{f_S^2 + \frac{k_x^2 U^2}{\varepsilon^2}\right\}\right] = 0,$$

The constant term in the dispersion relation (10) being negative means it will allow at least one real positive root leading to instability of the system.

The condition for the K-H instability of the system given by the constant term of Eq.(10) is

$$\frac{k_x^2 U^2}{\varepsilon^2}\left\{f_s^2(1+\alpha_0) + \frac{k_x^2 U^2}{\varepsilon^2}\right\} > 2(k_x V_A + k_y V_B)^2\left\{f_S^2 + \frac{k_x^2 U^2}{\varepsilon^2}\right\} \quad 11)$$

The instability criterion (11), is therefore independent of permeability of the medium as well as the viscosity of the fluid. So the stability does not depend on either the permeability of the medium or the fluid viscosity, but depends on density and relaxation frequency of suspended dust particles. It is also clear that the inequality (11) depends on the Alfven velocities and porosity of the medium. We also found that all these parameters have significant effect on the growth rate of unstable K-H mode.

In order to perform the numerical calculations on the growth rate of unstable K-H mode, we solve Eq.(10) numerically for positive roots in order to illustrate the influence of porosity, varying magnetic field and dust particle density. In fig.(1) and (2) we have plot the growth rate (σ*) of unstable K-H mode vs relaxation frequency of dust particles (fs*) for the

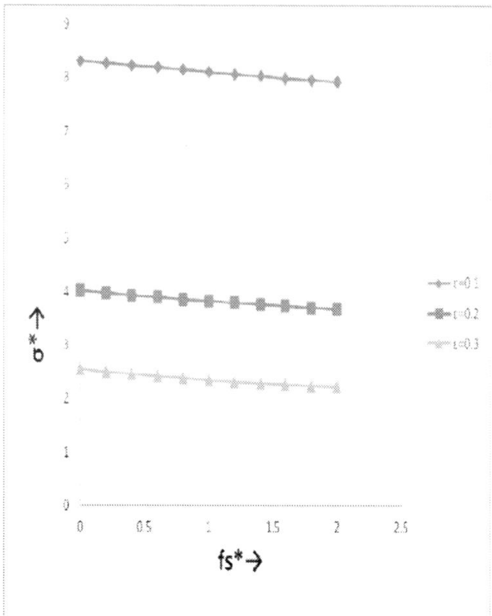

Figure 1. The growth rate σ^* (of unstable K-H mode) plotted against the relaxation frequency f_s^* of suspended dust particles for the medium porosity ε^*= 0.1, 0.2 and 0.3 with α=0.2, v*=0.2 , k_1=0.1, V_A=V_B=0.15, θ= 45^0

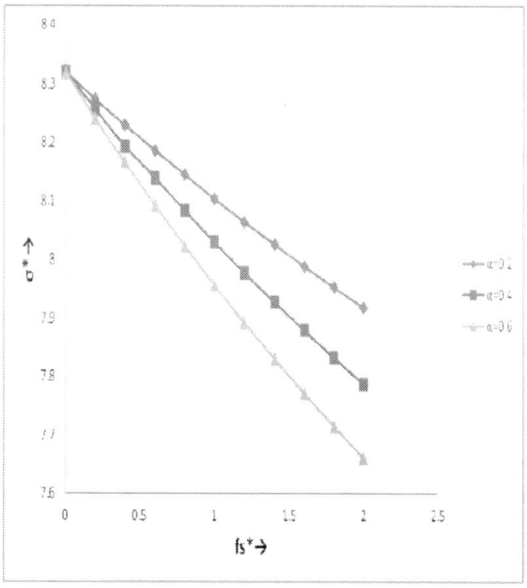

Figure 2. The growth rate σ^* (of unstable K-H mode) plotted against the relaxation frequency f_s^* of suspended dust particles for the density of suspended dust particles α^*=0.2,0.4 and 0.6 with ε=0.2, v*=0.2 , k_1=0.1, V_A=V_B=0.15, θ= 45^0

different values of medium porosity (ε) and dust particle density ($\alpha 0$). The numerical values of all the parameters are taken as arbitrary to study their effect on the growth rate. From the curves we find that growth rate of the system decreases as we increase the values of medium porosity and dust density, hence both the medium porosity and dust particle density have a stabilizing influence in presence of uniform two dimensional magnetic field.

5. Acknowledgement
The author is thankful to Prof. R.K Pandey, Director, BUIT and Prof. Nisha Dubey, Hon'ble V.C. Barkatullah University, Bhopal for their constant encouragement in this work. The author also expresses their sincere thanks to MPCST, Bhopal for providing Research Fellow and financial assistance in the research project.

6. References
[1] Chandrasekhar S 1991 *Hydrodynamic and Hydromagnetic Stabilit*, (Clarendon Press Oxfort)
[2] Saffmn P G 1962 *J. Fluids Mech.* **13** 120
[3] Scanlon J W and Segel L A 1973 *Phys. Fluids* **16** 1573
[4] Palaniswamy V I and Purushotham G M 1991 *Phys. Fluids* **24** 1224
[5] Michael D H 1965 *Proc. Comb. Phil. Sco.* **61** 569
[6] Sharma R C and Spanos T J T 1982 *Can. J. Phys.* **60** 1391
[7] McDonnel J A M 1978 *Cosmic Dust* (Toronto: John Wiley and Sons)
[8] Prajapati R P, Soni G D, Sanghvi R K and Chhajlani R K 2008 *Z. Naturforsch* **64a** 455
[9] Prajapati R P and Chhajlani R K 2010 *J. Porous Media* **13(9)** 765

Effect of magnetic field on Rayleigh-Taylor instability of two superposed fluids

P K Sharma[1], Anita Tiwari[2] and R K Chhajlani[3]
[1,2] BUIT, Barkatullah University Bhopal (M.P.)-462026 India
[3]School of Studies in Physics, Vikram University Ujjain (M.P.) India

E-mail : pks_buit30@yahoo.com

Abstract. The effect of two dimensional magnetic field on the Rayleigh-Taylor (R-T) instability in an incompressible plasma is investigated to include simultaneously the effects of suspended particles and the porosity of the medium. The relevant linearized perturbation equations have been solved. The explicit expression of the linear growth rate is obtained in the presence of fixed boundary conditions. A stability criterion for the medium is derived and discussed the Rayleigh Taylor instabilities in different configurations. It is found that the basic Rayleigh-Taylor instability condition is modified by the presence of magnetic field, suspended particles and porosity of the medium. In case of an unstable R-T configuration, the magnetic field has a stabilizing effect on the system. It is also found that the growth rate of an unstable R-T mode decreases with increasing relaxation frequency thereby showing a stabilizing influence on the R-T configuration.

1. Introduction

Currently, there is much interest in the evolution of the Rayleigh-Taylor (R-T) and hydrodynamic instabilities in both laboratory, inertial confinement, fusion and astrophysical plasmas. The Rayleigh Taylor instability occur when adverse density gradient exist, e.g., in the case of a heavy fluid supported by the lighter fluid in a gravitational field. Chandrasekhar [1] has discussed the R-T instability of two incompressible fluids taking into account different parameters, viz. magnetic field viscosity, surface tension and uniform rotation. Hans [2] has studied the effects of collisions with neutral atoms on the Rayleigh-Taylor in a composite medium. Many others e.g. Talwar [3], Bhatia [4], Sharma and Shrivastava [5], Sharma and Chhajlani [6] have extended his work using different parameters and assumptions.

The effect of suspended particle on the stability of superposed fluids might be of importance to industrial and chemical engineering. Saffman [7] has studied in detail a dust gas in magnetohydrodynamics. Sharma et al. Chhajlani [8] have investigated effect of rotation and suspended dust particles on the R-T instability of two superposed magnetizing conducing plasma.

In the recent years, the wide applications of porosity in industrial, geophysical situation and chemical engineering created considerable interest, particularly among geophysical fluid dynamics. Porosity is of use for the study of physical properties of comets, meteorites and interplanetary dust in astrophysical context McDonnel [9]. Wooding [10] has considered the Rayleigh instability of a thermal boundary layer in flow through a porous medium. Recently, Prajapati et al. [11,12] have studied the effect of porosity on R-T instability with suspended particles.

The aim of this paper is to study the effect of two dimensional horizontal magnetic field on the Rayleigh Taylor instability of two superposed incompressible conducting fluids in porous medium.

2. Lenearized equations of motion for the perturbation

The system under investigation is made of two semi infinite, non dissipative, incompressible, porous, magnetofluids under uniform horizontal magnetic field H(Hx,Hy,0) with suspended dust particles. Consider a Cartesian system of axes with the z direction as vertical. Suppose that a plane interface of discontinuity exist at z=0. The static interface between the two fluids is initially flat and the magnetofluid with density ρ2 in above the one with density ρ1 where ρ2 > ρ1. If the system is perturbed the variables representing the difference from the hydrostatic state are u(u,v,w), h(hx,hy,hz), δp, δρ the velocity, magnetic field, pressure and density.
It is supposed that suspended dust particles experience force KN (V-U) in the fluid, where K is a constant given by K = 6 π a μ (Stokes drag formula), 'a' being the particle radius.

$$\frac{\rho}{\varepsilon}\frac{\partial \boldsymbol{u}}{\partial t} = -\nabla\delta p + \frac{KN}{\varepsilon}(\boldsymbol{v}-\boldsymbol{u}) + g\delta\rho - \frac{\mu}{\boldsymbol{k}_I}\boldsymbol{u} + \frac{\mu_e}{4\pi}\left[(\nabla\times\mathbf{h})\times\mathbf{H} + (\nabla\times\mathbf{H})\times\mathbf{h}\right] \quad (1)$$

$$\varepsilon\frac{\partial}{\partial t}\delta\rho + (\boldsymbol{u}.\nabla)\rho = 0, \quad (2)$$

$$\varepsilon\frac{\partial \mathbf{h}}{\partial t} = (\mathbf{H}.\nabla)\mathbf{u} - (\mathbf{u}.\nabla)\mathbf{H} \quad (3)$$

$$\nabla.\boldsymbol{u} = 0, \quad (4)$$

$$\nabla.\boldsymbol{h} = 0 \quad (5)$$

$$\left[\tau\frac{\partial}{\partial t} + 1\right]\boldsymbol{v} = \boldsymbol{u}, \quad (6)$$

Where τ = m/6πaμ denotes the relaxation time for the suspended dust particles and ε denotes the medium porosity. As a result of this Darcy's law the usual viscous term in the equation of fluid motion is replaced by the resistance term $(-\mu/k_1)\mathbf{u}$ where μ is the viscosity of the fluid, k1 is the permeability of the medium.

A solution in normal modes is proposed to these equations :

$\exp(ik_x x + ik_y y + int)$ (7)

where k_x and k_y are the components of the wave numbers (k2=kx2 + ky2) and n is the growth rate of the harmonic perturbations.

A combination of the above equations and use of Eq.(7) leads to an equation of the z-component of the velocity w alone:

$$\left[\frac{i}{\varepsilon}n + v/k_1 + \frac{\alpha_0}{\varepsilon}\frac{(in)}{\tau(in)+1}\right]\left\{D(\rho Dw) - k^2\rho w\right\} + \left[\frac{(H_x k_x + k_y H_y)^2}{4\pi(\varepsilon in)}\right](D^2 - k^2)w + \frac{gk^2(D\rho)w}{(\varepsilon in)} = 0 \quad (8)$$

Where α_0 =mN/ρ denotes the mass concentration of the particles and D=d/dz. Equation (8) is the general dispersion relation incorporate the effect of the magnetic field and of the suspended dust particles in the porous medium. Here we have assumed the same density of suspended particles in both the regions z<0 and z>0.

3. Discussion

The solution of Eq.(8) by using the boundary conditions across the interface of two fluids can be written as

$$\sigma^3+\sigma^2\left[f_s\left(1+\frac{2mN}{\rho_1+\rho_2}\right)+\frac{\varepsilon}{k_1}(\beta_1 v_1+\beta_2 v_2)\right]+$$
$$\sigma\left\{\frac{\varepsilon f_s}{k_1}(\beta_1 v_1+\beta_2 v_2)+2(k_x V_A+k_y V_B)^2-gk(\beta_2-\beta_1)\right\}+ \tag{9}$$
$$f_s\left[2(k_x V_A+k_y V_B)^2-gk(\beta_2-\beta_1)\right]=0.$$

Here, $f_s = 1/\tau$, $\sigma = (in)$ and $\beta_1=\rho_1/(\rho_1+\rho_2)$, and $\beta_2=\rho_2/(\rho_1+\rho_2)$.
For simplicity, we considered the Alfven velocities of two plasmas are the same, so that

$$V_{A,B}^2=\frac{\mu_e H_{x,y}^2}{4\pi(\rho_1+\rho_2)} \tag{10}$$

Eq.(09) gives the dispersion relation of R-T configuration for two superposed fluids of different densities including permeability, porosity, suspended dust particles with two dimensional uniform horizontal magnetic field.

3.1 Stable configuration ($\beta_1>\beta_2$) or ($\rho_1>\rho_2$)
In this potentially stable arrangement when the lower fluid is heavier than the upper fluid, it is clearly represent the stable configuration. If we applied necessary condition of the Hurwitz criterion on equation (9) we find that all the coefficients are positive and real implying the stability of the considered system.

3.2 Unstable configuration ($\beta 2>\beta 1$) or ($\rho 2>\rho 1$)
The system is stable or unstable depends on the following conditions
3.2.1. If $2(k_x V_A+k_y V_B)^2>gk(\beta_2-\beta_1)$ the system is stable because there is no change in sign. So it has no positive root.
3.2.2. If $2(k_x V_A+k_y V_B)^2<gk(\beta_2-\beta_1)$ the system is unstable because constant term is negative therefore allows one change of sign and so has one positive root. The occurrence of a positive root implies that the system is unstable. Thus for the unstable case ($\beta 2>\beta 1$) the system is stable or unstable according as $2(V_A\sin\theta+V_B\cos\theta)^2$ is greater than or smaller than $gk(\beta_2-\beta_1)$. In the absence of magnetic field, equation (9) has one positive root, and so the system is unstable for ($\beta_2>\beta_1$). But the magnetic field has got a stabilizing effect and completely stabilizes the wave–number band $k>k_c$,

$$\text{Where } k_c=\frac{g(\beta_2-\beta_1)}{2[V_A\sin\theta+V_B\cos\theta]^2} \tag{11}$$

and θ is the inclination of the wave vector k to the direction of the magnetic field H i.e. $k_x=k\sin\theta$ and $k_y=k\cos\theta$. By this configuration we can say that the condition of R-T instability and cut-off wave-number do not depend upon medium porosity and suspended dust particles but depend on magnetic field only. We also find that the condition of R-T instability remains the same but there will be a change in the growth rate of R-T instability due to multiplication of relaxation frequency of suspended dust particles in the constant term of Eq. (9).

In order to perform the numerical interpretation on the growth rate of Rayleigh Taylor instability of two superposed magnetized plasma with suspended dust particles, medium viscosity and porosity we rewrite the dispersion relation (9) in dimensionless form as

$$\sigma^{*3}+\sigma^{*2}\left[f_S^*(1+2\alpha')+\frac{2\varepsilon v^*}{k_1}\right]+\sigma^*\left\{\frac{2\varepsilon f_S^* v^*}{k_1}+2(V_A\sin\theta+V_B\cos\theta)^2-(\beta_2-\beta_1)\right\}+ \tag{12}$$
$$f_S^*\left[2(V_A\sin\theta+V_B\cos\theta)^2-(\beta_2-\beta_1)\right]=0.$$

where $v' = \mu/(\rho_1+\rho_2)$ and $\alpha' = mN/(\rho_1+\rho_2)$.

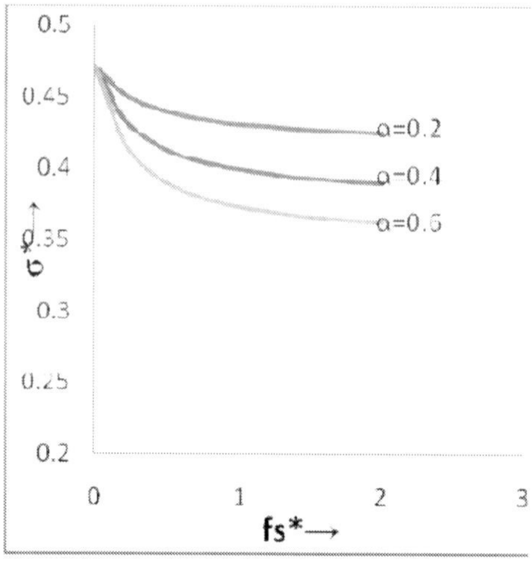

Figure1. The growth rate σ^* plotted against the relaxation frequency f_s^* of suspended dust particles for the density of suspended dust particles α^*=0.2,0.4 and 0.6 with ε=0.1, v*=0.2 , k_1=0.1, V_A=V_B=0.15, θ= 45⁰, (β_2-β_1)=0.5.

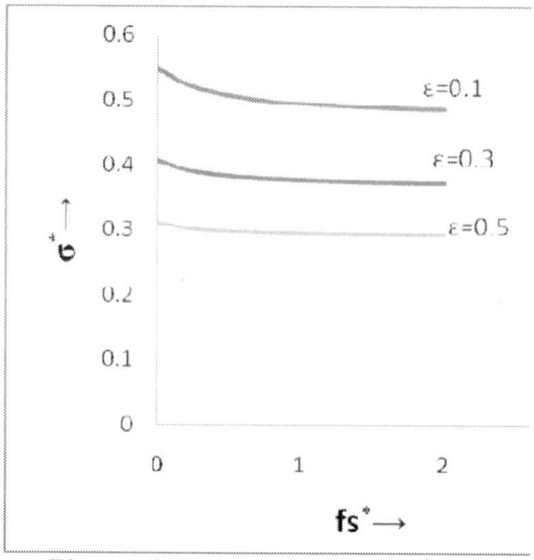

Figure 2. . The growth rate σ^* plotted against the relaxation frequency f_s^* of suspended dust particles for the medium porosity ε^*= 0.1, 03 and 0.5 with α=0.2, v*=0.2 , k_1=0.1, V_A=V_B=0.15, θ= 45⁰, (β_2-β_1)=0.5.

In Fig.(1-2), we have plot the growth rate (σ*) vs relaxation frequency of dust particles (fs*) for the different values of medium porosity (ε), and dust particle density (α0). The numerical values of all the parameters are taken as arbitrary to study their effect on the growth rate. From the curves we find that growth rate of the system decreases as we increase the values of medium porosity and dust density. According to discussion above, the Rayleigh-Taylor instability is suppressed by horizontal magnetic field, porosity and dust particle density

4. Acknowledgement

The authors (P.K.Sharma and Anita Tiwari) are thankful to Prof. R.K Pandey, Director, BUIT and Prof. Nisha Dubey, Hon'ble V.C. Barkatullah University, Bhopal for their constant encouragement in this work. The authors (P.K.S. and A.T.) are also expresses their sincere thanks to MPCST, Bhopal for providing Research Fellow and financial assistance in the research project.

5. References

[1] Chandrasekhar S 1991 *Hydrodynamic and Hydromagnetic Stabilit*, (Clarendon Press Oxfort).
[2] Hans H K 1968 *Nucl. Fusion* **8** 89
[3] Talwar S P *J.Geophysical Research* **69** 2707
[4] Bhatia P K 1970 *Cosmic Electrodynamics* **1** 269
[5] Sharma R C and. Shrivastava K M 1974 *Aust. J. Phys.***27** 53
[6] Sharma P K and .Chhajlani R K 1998 *Phys.Plasmas* **5** 2202
[7] Saffmn P G 1962 *J. Fluids Mech.* **13** 120
[8] Sharma P K, Prajapati R P and Chhajlani R K 2010 *Acta Physica Polonica A* **118** 576
[9] McDonnel J A M 1978 *Cosmic Dust* (Toronto: John Wiley and Sons)
[10] Wooding R A 1960 *J. Fluid Mech.* **9** 183
[11] Prajapati R P, Soni G D, Sanghvi R K and Chhajlani R K 2008 *Z. Naturforsch* **64a** 455
[12] Prajapati R P and Chhajlani R K 2010 *J. Porous Media* **13(9)** 765

Transient Brillouin gain in direct band gap semiconductors

Kirti Sontakke, Nischhal Yadav and S Ghosh
School of studies in Physics, Vikram University, Ujjain-456010 (M. P.), India

E-mail: kirtisontakke15@gmail.com

Abstract. The paper presents an analytical investigation of transient gain coefficient of the Brillouin scattered Stoke mode in direct gap semiconductors. The effect of density of state mass has been introduced through an empirical formula relating maximum possible plasma density with band structure and temperature. The transient Brillouin gain is determined through the effective Brillouin susceptibility derived with the help of coupled mode theory of plasmas. The numerical estimations are made for III-V (InSb and GaAs) and II-VI (CdS) compound semiconductors duly irradiated by a 10-nanosec pulsed 10.6μm CO_2 laser. It is found that the maximum plasma density increases with the amplitude of laser field for all three representative crystals which in turn increases the transient gain. For a fixed value of input intensity transient gain factor first increases with pulse duration and attains maximum value which remains invariant for a certain range of pulse duration. If pulse duration is further increased gain diminishes very rapidly and approaches to zero at different pulse durations for different crystals. This analysis establishes that the crystal having higher density of states mass is the automatic choice for the fabrication of Brillouin cells.

1. Introduction

Brillouin scattering exhibits a distinctive role in nonlinear optics. But to obtain appreciable steady state Brillouin gain a high power laser beam is a necessity. Hence, to rate out the possibility of optical damage of the medium short laser pulses are employed. However, if the pulse durations is comparable to, or less than the acoustic phonon life time, the transient phenomena become unavoidable and therefore we have motivated to study transient Brillouin gain characteristics in direct band gap semiconductors. To add novelty to this study we have included the effect of density of state mass on the study of transient Brillouin gain characteristics and found desired modifications.

2. Theoretical formulation

We employ the well known hydrodynamical model (for which $k_a l \langle\langle 1, (k_a$ and l being the acoustic wave vector and mean free path of an electron respectively). As the crystal is assumed to be centrosymmetric, the effect of any pseudopotential can be neglected for analytical simplicity. We have employed the coupled mode scheme to obtain the nonlinear polarization arising due to electrostrictive strain. Using phenomenological model for plasma thermodiffusion Forchel et al. [1] have given an empirical relation for maximum possible plasma density found to be in terms of state masses of electrons and holes of the medium as

$$n = 4.2\times10^{22}\left(\frac{m_{de}m_{dh}}{m_{de}+m_{dh}}\right)T^{3/2} \quad (1)$$

Following the method by developed by Ghosh et al. [2], one obtains the Brillouin susceptibility as

$$(\chi_B^{(3)})_{eff} = \frac{\gamma^2k^2\omega_p^2}{2\rho\varepsilon_0\left(\omega_a^2 - k^2v_a^2 + 2i\Gamma_a\omega_a\right)\omega_0\omega_1} + \frac{\gamma^2k^2}{2\rho\varepsilon_0\left(\omega_a^2 - k^2v_a^2 + 2i\Gamma_a\omega_a\right)} \quad (2)$$

all symbols have their usual meaning and are well explained in [2]. The threshold pump amplitude for the onset of SBS may be obtained by equating the value of induced polarisation is equal to zero as

$$\left|E_{0th}\right| = \frac{m}{ek}\left[\left(\delta_1^2 - i\omega_1 v\right)\left(\delta_2^2 + i\omega_a v\right)\right]^{1/2} \quad (3)$$

2.1. Transient gain

The transient gain coefficient g_{TB} can be obtained from Brillouin susceptibility using following relation [3].

$$g_{TB} = [2g_B x\Gamma\tau_p]^{\frac{1}{2}} - \Gamma\tau_p \quad (4)$$

where Γ being the optical phonon life time, x is the interaction length, τ_p is the pulse duration and g_B is the steady state Brillouin gain coefficient given by

$$g_B = \frac{-k}{2\varepsilon_1}(\chi_{eff})_{img}\left|E_0\right|^2 \quad (5)$$

This suggests that optimum pulse duration can be increased by increasing the intensity of the pump and the crystal having higher density of state mass achieves the large transient gain.

3. Results and discussion

This section aims to make a detailed numerical appreciation of the analytical investigation made in above section. The semiconductor bulk crystals used for this purpose are CdS, (Hexagonal) and GaAs, InSb (tetrahedral) respectively. We have considered the irradiation of these crystals by a pulsed 10.6 μm CO_2 laser. The other relevant parameters at room temperature used are; For InSb $\eta = 3.9$, $m_{de} = 4.2\times10^{22} m_0 m^{-3}$, $m_{dh} = 7.3\times10^{24} m_0 m^{-3}$, $\vartheta_a = 4.8\times10^3 ms^{-1}$, $v = 3\times10^{11} ms^{-1}$; For GaAs $m_{de} = 4.7\times10^{23} m_0 m^{-3}$, $m_{dh} = 9.8\times10^{24} m_0 m^{-3}$,; For CdS $m_{de} = 1.5\times10^{25} m_0 m^{-3}$, $m_{dh} = 1.2\times10^{25} m_0 m^{-3}$.

Figure 1, depicts the qualitative behaviour of the transient gain coefficient of signal wave as a function of pulse duration. The cell length is taken as $10^{-6} m$ and pulse duration within the range $10^{-10} \le \tau_p \le 10^{-2} s$. The density of states mass affects the g_{TB} implicitly and shifts the maximum transient gain point towards higher value of pulse duration. It is clear that behaviour of g_{TB} with respect to τ_p is identical in all the three semiconductor crystals. For a fixed value of I_{in} transient gain factor g_{TB} first increases with τ_p and attains maximum value which remains constant for certain range of τ_p. If τ_p is further increased g_{TB} diminishes very rapidly and approaches to

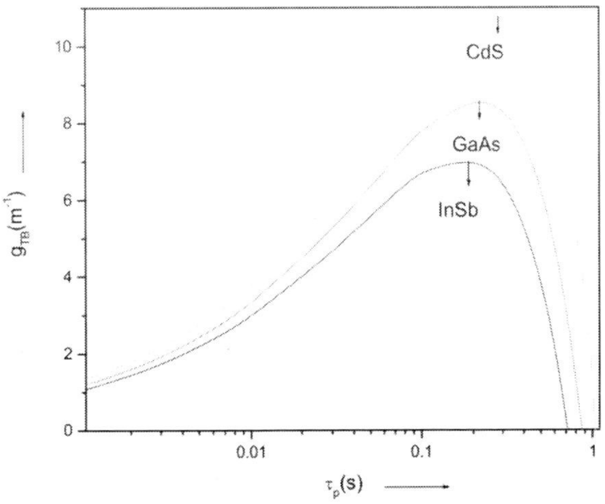

Figure 1. Variation of transient gain g_{TB} with pump pulse duration τ_p at $I_{in} = 9.558 \times 10^{15}\, Wm^{-1}$

zero at different pulse durations for the three (CdS, GaAs, InSb) direct gap semiconductor crystals. The pulse duration at which g_{TB} approaches zero is maximum for CdS and minimum for InSb.

Above study reveals that particularly in direct gap semiconductors while studying laser plasma interaction one must incorporate the density of states mass to achieve the better agreement between theoretically predicted and experimentally observed values. The above analysis concluded that CdS having higher density of states mass is the most prominent candidate for the fabrication of optoelectronic device.

References

[1] Forchel A, Schweizer H, Nather H and Romarhek K M, 1983 *J. Fischer Physica* **117B** and **118B** 336
[2] Neogi A and Ghosh S 1989 *Phys. Stat. Sol.* b **156** 705
[3] Wang C S, Rabin H and Tang C L *Quantum Electronics* 1975 (Academic Press, New York) 447

Optical phase conjugation reflectivity via stimulated Brillouin scattering in acousto-optic diffusive semiconductor plasma crystal

Nilesh Nimje[1] **and S. Ghosh**
School of Studies in Physics, Vikram University, Ujjain-456010, INDIA

[1]E-mail: nileshnimje.ssp@gmail.com

Abstract. Analytical investigation has been made for optical phase conjugation (OPC) reflectivity via stimulated Brillouin scattering (SBS) in acousto-optic diffusive semiconductor plasma crystal. Our analysis is based on hydrodynamic model and coupled mode approach of interacting waves. The numerical estimations are made for n-type InSb semiconductor plasma crystal duly irradiated by CO2 laser. The magnitude of the third-order nonlinear optical susceptibility for III-V semiconductor obtained from our analysis is found to agree well with the experimental values. Maximum OPC reflectivity is obtained when cyclotron frequency is in resonance with applied pump frequency.

1. Introduction

SBS has been extensively investigated due to its important property of optical phase conjugation (OPC), which can be exploited to recover the phase front of a beam [1-4]. Nimje et al. [5] investigated the effects of material parameters on interaction length to occur OPC-SBS in semiconductors and found that when cyclotron frequency is tuned with pump frequency interaction length becomes 10^4 time smaller than that obtained in absence of magnetic field. Sen et al. [6] studied phase conjugation in magnetised semiconductors via SBS and reported that when pump intensity is well above the threshold; reflectivity approaches its absolute limit $R=1$ at interaction length $800\mu m$. The OPC-SBS reflectivity being directly dependent on the third order Brillouin susceptibility, an enhancement in reflectivity is possible even at lower excitation intensity if one can achieve larger third-order susceptibility in the medium. Recently, Nimje et al. [7] have reported that magnetic field enhances the steady-state Brillouin gain through Brillouin susceptibility of acousto-optic semiconductors. In present paper, we aim to study the OPC reflectivity via SBS using hydrodynamic model and coupled mode theory of interacting waves under thermal equilibrium.

2. Brillouin Susceptibility

Here the electric field associated with pump wave is described by $\vec{E}_0 = \hat{x}E_0 \exp\{i(\omega_0 t - k_0 x)\}$. A magnetostatic field $(\vec{B}_0 = \hat{z}B_0)$ is applied across the direction of wave propagation $(\vec{k} = \hat{x}k)$.

We start with the calculation of the total current density for resonant Stokes component arises due to nonlinear interactions of the waves, followed by derivation of the effective Brillouin susceptibility through third-order nonlinear polarisation arising due to nonlinear induced current and acousto-optical strain. Nimje et al. [7] have already derived expression for Brillouin susceptibility in acousto-optic crystal for the above field geometry.

Thus, by following equation (19) of [7] in dispersionless acoustic wave regime, one may obtain the effective Brillouin susceptibility as

$$\chi_B = -\frac{i\varepsilon(\eta^2 - 1)^2 k^2 \omega_0^4}{4\rho\Gamma_a \omega_a \omega_1 (\omega_0^2 - \omega_c^2)^2}\left[1 + \frac{\omega_p^2}{\omega_0 \omega_1}|G|\right], \qquad (1)$$

where the symbols have their usual meaning and well defined in reference [7].

3. Optical phase conjugation reflectivity

In this section, an analytical expression for optical phase conjugation (OPC) reflectivity via stimulated Brillouin scattering (SBS) is derived. For this, we consider the irradiation of the semiconductor crystal by a slightly off-resonant laser of frequency $1.78\times10^{14}\,s^{-1}$ with excitation intensity well above the threshold value required for the onset of SBS in semiconductor crystal. The pump laser mode undergoes stimulated scattering processes in the medium. We have considered the transverse component of electric field that propagates along x-direction. The propagation of these modes through the transparent optical material can be represented by the general electromagnetic wave equation

$$\nabla^2 E - (\omega_L^2 / c^2)E = -\mu_0 \omega_L^2 P \qquad (2)$$

The field variation is given by the plane wave approximation $\exp\{i(\omega t - kx)\}$. P represents the total induced polarisation in semiconductor crystal comprising of linear as well as the nonlinear components and is given by

$$\vec{P}(x,t) = \varepsilon_0[\chi^{(1)} + \chi^{(3)}|E|^2 + \ldots\ldots\ldots\ldots\ldots]\vec{E} \qquad (3)$$

Here complex susceptibility $\chi^{(1)}$ corresponds to linear optical effects such as linear refraction and absorption of wave within the medium, while third-order nonlinear optical susceptibility $\chi^{(3)}$ is responsible for the nonlinear phenomena that gives rise to the phenomenon of OPC.

We assume that the pump wave $(\omega_0, \vec{k}_0)$ is propagating along the negative x direction while the Stokes mode of scattered electromagnetic wave is assumed to be propagating along positive x direction.

On applying slowly varying envelope approximation in equations (2) and (3), we can obtain the corresponding electromagnetic wave equation for $\vec{E}_0(\omega_0, \vec{k}_0)$ (pump wave) and $\vec{E}_1(\omega_0, \vec{k}_1)$ (Stoke mode) as

$$(\partial / \partial x)E_0 = \alpha_0 E_0 - i\alpha_{r0} E_0 - (\omega_0^2 / 2k_0^2)\chi_0^{(3)}|E_1|^2 E_0 \qquad (4a)$$

and

$$(\partial/\partial x)E_1 = \alpha_1 E_1 - i\alpha_{r1}E_1 - (\omega_1^2/2k_1^2)\chi_1^{(3)}|E_0|^2 E_1 \quad (4b)$$

where $\alpha_{0,1} = (\omega_{0,1}/2c)\chi_{0,1}^{(1)}$ and $\alpha_{r0,r1} = (\omega_{0,1}/2c)\chi_{r0,r1}^{(1)}$ stand for the absorption and dispersive properties of the crystal at frequency $\omega_{0,1}$. $\chi_{0,1}^{(3)}$ is the third-order optical susceptibility at frequency $\omega_{0,1}$. For phase conjugated wave, the phase matching conditions enable one to take $k_0 = -k_1$, such that $k_a = 2k_0$; also we take $|k_0| = |k_1| = k$ with $|k_a| = 2k$. We assume that $\omega_1 \approx \omega_0 = \omega$ and $\alpha_1 \approx \alpha_0 = \alpha$.

The phase-conjugate scattered radiations is a function of pump amplitude and is given by

$$E_1(x) = \beta(x)E_0^*(x), \quad (5)$$

where $\beta(x)$ is a measure of conjugacy with $|\beta(x)|^2$ being defined as the OPC reflectivity R via SBS. On simplifying equation (4a), one can obtained the pump electric field as

$$E_0(x) = [E_0(L)\exp\{-\alpha_E(L-x)\}]\exp\{i\alpha_r(L-x)\} \quad (6)$$

where $\alpha_E = \alpha_1 - (i\omega^2/2kc^2)\chi^{(3)}(\omega)|E_1|^2$ and $E_0(L)$ is the pump electric field at the entrance window $(L = x)$. Equation (6) reveals the x dependence of the pump amplitude as well as the nature of phase variation of the electric field. From equations (6) and (4b), we find the electric field associated with the Stokes mode of the backscattered electromagnetic wave as

$$E_1(x) = E_1(0)\exp[-\alpha_E x + (\kappa/2\alpha_E)\{1-\exp(2\alpha_E x)\}]\exp(i\alpha_e x) \quad (7)$$

where $\kappa = (\omega_0^2/2kc^2)\chi_B E_0^2(L)\exp(-2\alpha_E L)$ and $E_1(0)$ is the spontaneous noise of Stokes field amplitude at $x = 0$. The backscattered Brillouin mode $E_1(x)$ possesses a gain constant given by

$$-[\alpha_E x + (\kappa/2\alpha_E)\{1-\exp(2\alpha_E x)\}]$$

For finite gain, the condition should be satisfied

$$\alpha_E x + (\kappa/2\alpha_E)\{1-\exp(2\alpha_E x)\} < 0 \quad (8)$$

It is clear from equation (8) that the gain constant depends upon α_E and κ. For bulk semiconducting material of mm dimensions shined by a slightly off-resonant laser with phonon frequency less than the crystal band-gap energy, one may take $2\alpha_E x < 1$. Consequently, the threshold condition for onset of Brillouin gain is $\kappa \geq \alpha_E$.

Therefore, we can find the threshold value of the excitation intensity that is responsible for Brillouin gain as

$$I_{oth} \geq \eta\varepsilon_0 c^3 k\alpha_E/\omega^2|\chi_B| \quad (9)$$

For pump intensity well above the threshold, on can obtain finite OPC reflectivity R at the entrance window as

$$R = |\beta(x=L)|^2 = |E_1(0)/E_0(L)|^2 \exp[2(\kappa - \alpha_E)L] \quad (10)$$

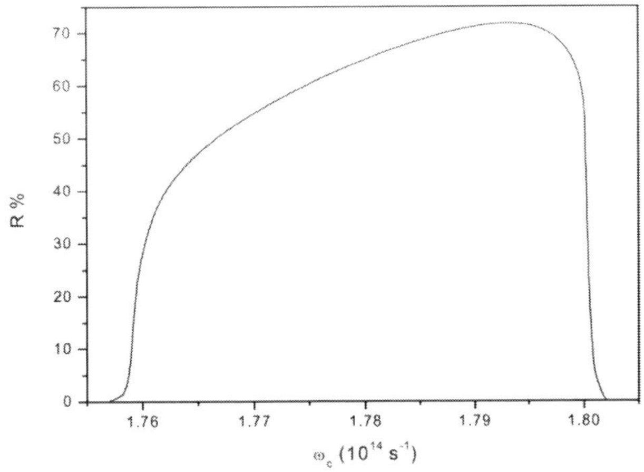

Figure. Dependence of OPC reflectivity R with applied magnetic field B_0 (in term of ω_c) at the interaction length $L = 10^{-7} m$.

where L is the length of semiconductor crystal, $|E_1(0)|^2$ being the noise intensity for the SBS process [8] and its magnitude is generally taken to be about 10^{-12} to 10^{-13} times of $|E_0(L)|^2$. Hence, significantly high OPC reflectivity ($R \approx 1$) can be achieved in the crystal only if $(\kappa - \alpha_E)L \approx 15$ i.e. when the gain is nearly equal to e^{30}.

4. Results and discussion:

The set of parameters used for numerical appreciation is taken from ref. [8]. The characteristic dependence of OPC reflectivity on magnetic field (in term of ω_c) is plotted in Figure. Figure shows the variation of OPC reflectivity R with cyclotron frequency ω_c. It may be inferred that the finite reflectivity is obtained only when magnetic field is in range from 13.99T to 14.35T. The corresponding cyclotron frequency range is from $1.75 \times 10^{14} s^{-1}$ to $1.81 \times 10^{14} s^{-1}$, respectively. Beyond this limit on both the ends the wave is completely absorbed by the semiconductor i.e. no OPC reflectivity is observed. This behaviour may be used for construction of magnetic switches. Thus 72% OPC reflectivity is obtained only when ω_c resonant with ω_0.

5. References

[1] Ostermeyer M, Kong H J, Kovalev V I et al. 2008 *Laser Part. Beams* **26** 1
[2] Bai J, Shi J, Ouyang M, Chen X, Gong W, Jing H, Liu J and Liu D 2008 *Opt. Lett.* **33** 1539
[3] Lu Z W, Hasi W L J, Gong H P, Li Q and He W M 2006 *Opt. Express* **14** 5497
[4] Wang Y L, Lu Z W, Li Y, Wu P, Zheng Z X and He W M 2010*Appl. Phys. B* **98** 391
[5] Nimje Nilesh, Yadav Nishchhal and Ghosh S 2012 *Phys. Lett. A* **376** 850
[6] Sen P K , Sen P and Vivek S 1996 *J. Phys. D: Appl. Phys.* **29** 1
[7] Nilesh Nimje, Dubey Swati and Ghosh S 2011 *Int J. Appl. Phys.* **2** 101
[8] Zel'dovich B Ya, Pilipetsky N F and Shkunov V V 1985 *Principles of Phase Conjugation*, (Springer-Verlag, Berlin)

Polaron induced parametric interactions in magnetized semiconductors

R Agrawal[1], S Dubey and S Ghosh
School of Studies in Physics, Vikram University, Ujjain-456010 (M.P.) India

E-mail: ratnagarwal@gmail.com

Abstract. In the framework of hydrodynamic model of semiconductor plasma, the parametric amplification due to polaron mode has been analytically investigated in a magnetized III–V semiconductor, viz. n-InSb. The origin of the phenomenon lies in the effective second-order optical susceptibility arising due to the induced nonlinear current density of the medium. Using the coupled-mode theory, the expressions for the parametric gain coefficient and the threshold amplitudes of the external electric field above which the parametric instability can occur are derived. The effect of the wave vector and the strength of the external magnetic field on the gain coefficient and the threshold amplitude of the pump electric field are analyzed. Reported results strongly suggest that at appropriate pump electric field, proper selection of the magnetic field not only lowers threshold pump field required for the onset of parametric excitation but also enhances gain effectively.

1. Introduction

The phenomena of parametric interaction exhibit a distinctive role in nonlinear optics. Parametric amplifiers, parametric oscillators, optical phase conjugation, pulse narrowing, squeezed state generation [1], etc. are some of the important devices and processes whose origins lie in the parametric interactions in a nonlinear medium. In optical parametric amplification, a weak signal is made to interact with a strong, higher frequency pump and both the generated difference frequency (known as the idler) and the original signal are amplified.

In a polar material, the electric field interacts with both optical phonons and electrons. The coupling of LO phonons and free carrier collective excitations by macroscopic longitudinal electric fields in polar semiconductors has been treated theoretically by a number of investigators [2, 3]. A polaron is a quasi-particle that arises due to the conduction electron (or hole) together with its self-induced polarization in an ionic crystal or in a polar semiconductor [4-6]. Physically the polaron is an electron in a localized state that induces a polarization of the medium. This polarization is local in character and is due to the displacements of ions from the equilibrium positions caused by the field produced by the electron density which gives rise to an electron density perturbation at the polaron frequency, which couples nonlinearly with the pump wave and drives the polaron wave at modulated frequency [7]. Thus in this paper using hydrodynamic model and coupled mode theory analytical expressions for the second order susceptibility arising from the nonlinear induced current density, gain coefficient of the amplified wave and threshold pump amplitude required to incite the parametric interaction in a semiconductor are derived. Numerical estimations have been made with a set of data appropriate for a polar semiconductor crystal (n-InSb) duly irradiated by a frequency doubled CO2 laser to establish the validity of the present work.

2. Theoretical Formulation

We consider a spatially uniform pump electric field $(|k_0| \approx 0)$ $\vec{E}_0 = \hat{x}\vec{E}_0 \exp(-i\omega_0 t)$ applied in homogeneous semiconductor plasma. It restricts our analysis to be valid only in the limit $kl << 1$ (k is the wave number and l is the carrier mean free path).

2.1 Effective second-order susceptibility. The basic equations employed in the present analysis are:

$$\frac{\partial^2 \vec{r}}{\partial t^2} + \left(\Omega_p^2 + \Gamma_e^2\right)\vec{r} + 2\Gamma_e \frac{\partial \vec{r}}{\partial t} = -\frac{e}{m_e}\left(\vec{E}_0 + \frac{\partial \vec{r}}{\partial t} \times \vec{B}\right), \tag{1a}$$

$$\frac{\partial^2 \vec{S}}{\partial t^2} + \left(\omega_L^2 + \Gamma_{ph}^2\right)\vec{S} + 2\Gamma_{ph} \frac{\partial \vec{S}}{\partial t} = \frac{q}{M}\left(\vec{E}_0 + \frac{\partial \vec{S}}{\partial t} \times \vec{B}\right) \tag{1b}$$

$$\frac{\partial n_1}{\partial t} + n_0 \frac{\partial \vartheta_0}{\partial x} + n_1 \frac{\partial \vartheta_0}{\partial x} + \vartheta_0 \frac{\partial n_1}{\partial x} = 0 \tag{2}$$

$$\frac{\partial E_{pl}}{\partial x} = -\frac{n_1 e}{\varepsilon_0} + \left(\frac{n_0 e}{\varepsilon_0} - \frac{Nq}{\varepsilon_0}\right)\frac{dR}{dx} \tag{3}$$

We assume that the electric fields associated with the electronic and lattice polarizations are parallel to each other and the polarizabilities of the electron and ion systems are additive. Thus, the simultaneous excitation of collective cyclotron vibrations and optical phonons results into coupling between them. The resulting coupled vibrations appear in the form of a new mode known as polaron mode. The equation of motion of a polaron mode is given by

$$\frac{\partial^2 \vec{R}}{\partial t^2} + \omega_{0,pl}^2 \vec{R} + 2\Gamma_{pl} \frac{\partial \vec{R}}{\partial t} = (NM)^{\frac{1}{2}}\left(-\frac{e}{m_e} + \frac{q}{M}\right)\left(\vec{E}_{pl} + \frac{\partial \vec{W}}{\partial t} \times \vec{B}\right) \tag{4}$$

Meanings of all the symbols are given in the reference [8]. We also assume that the energy transfer between the pump, polaron and side band waves satisfy phase matching conditions which are $\omega_0 = \omega_{pl} \pm \omega_s$ and $k_0 = k_{pl} \pm k_s$; under spatially uniform laser irradiations $|k_0| \approx 0$ so that $|k_{\pm}| \approx |k_0 \pm k_{pl}| \approx |k_{pl}| = k$ (say).

Following the procedure adopted by Dubey and Ghosh [8] and using equations (1-4), we may obtain second order susceptibility via nonlinear polarization at amplified frequencies considered as $P_{eff}^{(2)} = \varepsilon_0 \chi_{eff}^{(2)} E_0 E_{pl}$:

$$\chi_{eff}^{(2)} = \frac{-ike\,\omega_0 \Omega_p^2 A_1 A_3 X_1}{m_e \omega_s \left(\omega_0^2 - \omega_c^2\right)^2}\left[\delta'^2 + 2i\Gamma_{pl}\omega_{pl} - \frac{k^2 |A_2|^2 |\bar{E}|^2}{\delta^2 - 2i\Gamma_e \omega_s}\right]^{-1} = \left[\chi_{eff}^{(2)}\right]_r + i\left[\chi_{eff}^{2}\right]_i \tag{5}$$

Where

$$A_1 = \frac{\omega_{pl}^2}{\omega_{pl}^2 - \Omega_p^2 - \omega_c^2} \quad A_2 = \frac{\omega_0^2}{\omega_0^2 - \Omega_p^2 - \omega_c^2} \quad A_3 = 1 + \frac{\frac{n_1 eq}{M} + \frac{Neq}{m_e} - \frac{n_0 e^2}{m_e} - \frac{Nq^2}{M}}{\varepsilon_0 F_1} \quad X_1 = \frac{\omega_c^2}{4\Gamma_{pl}^2 - \omega_c^2}$$

$$\delta'^2 = A_1 \Omega_p^2 - \omega_{pl}^2 \quad \delta^2 = A_1 \Omega_p^2 - \omega_s^2 \quad F_1 = -\omega_{pl}^2 + \omega_{0,pl}^2 - 2i\Gamma_{pl}\omega_{pl} \quad \bar{E} = -\frac{e}{m_e} E_0 \frac{\omega_0^2}{\omega_0^2 - \omega_c^2}$$

Equation (5) characterizes the steady state optical response of the medium and reveals that the total crystal susceptibility is influenced by the equilibrium carrier concentration through $\Omega_p \neq 0$ and external magnetostatic field $\omega_c \neq 0$.

2.2 *Gain coefficient and Threshold.* It can be observed from equation (5) that there is an intensity dependent refractive index [via $\left(\chi_{eff}^{(2)}\right)_r$] leading to the possibility of a focusing or defocusing effect of the propagating beam and positive dispersive characteristics of the dissipative medium is possible at $\Omega_r >> \Gamma_c$. As $\left(\chi_{eff}^{(2)}\right)_r$ becomes more positive, one may expect more effective self-defocusing of the amplified polaron mode. In order to express the possibility of parametric amplification in semiconductor plasma, we employ the relation

$$\alpha_{eff} = \frac{k}{2\varepsilon_1}\left(\chi_{eff}^{(2)}\right)_i \tag{6}$$

Here, α_{eff} is the effective nonlinear absorption coefficient. The nonlinear growth of the amplified signal is possible only if α_{eff} obtainable from equation (6) is positive.

In order to determine the threshold value of the pump amplitude required for the onset of the parametric interaction, we set $P_{eff}^{(2)} = 0$ and obtain,

$$E_{th} = \left| \frac{m_c(\omega_s)^{\frac{3}{2}}(\omega_{pl})^{\frac{1}{2}}\left(\omega_0^2 - \omega_c^2\right)}{ekA_2\omega_0^2} \right| \tag{7}$$

It is observed from equation (7) that the parametric instability of the signal wave has a nonzero intensity threshold, even in the absence of damping. The threshold field E_{th} is found to have complex characteristics and is strongly dependent on the external magnetostatic field.

3. Results and Discussion

The above theory is now applied to study the threshold and amplification characteristics of polaron induced parametric interaction in magnetised semiconductor. The relevant parameters chosen for compound semiconductor InSb are listed in [8].

Figures 1 to 3 are plotted to have qualitative appreciation of threshold and amplification characteristics with respect to wave vector and cyclotron frequency. In figure 1 initially threshold pump amplitude shows a slow declination with increasing ω_c till ω_c approaches polaron wave frequency, at $\omega_{0pl} \approx \omega_0$, threshold field achieves a minimum value $\left(\approx 3\times10^3 Vm^{-1}\right)$. Further increasing ω_c results into rise in E_{th} but again at $\omega_c \approx \omega_0$, E_{th} decreases. However this dip corresponds to

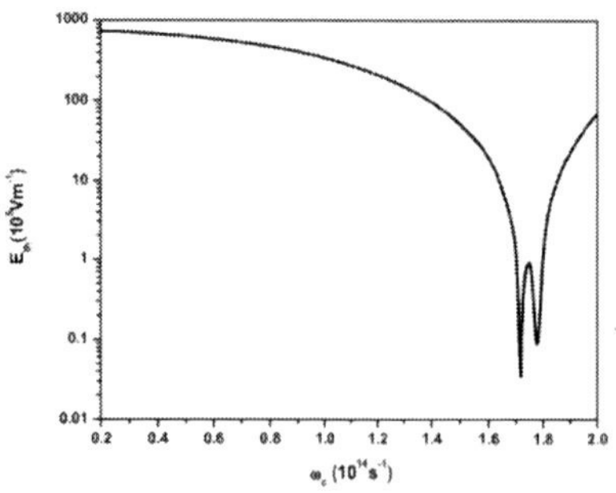

Figure 1. Variation of the threshold electric field E_{th} with cyclotron frequency ω_c

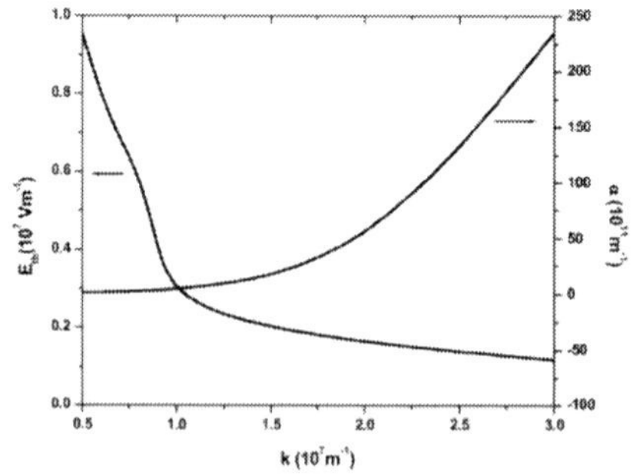

Figure 2. Variation of threshold and absorption coefficient with wave number k

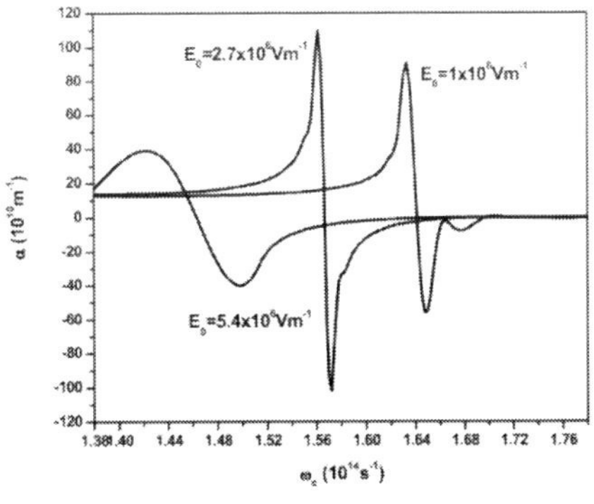

Figure 3. Variation of the absorption coefficient α with cyclotron frequency ω_c with the pump electric field E_0 as a parameter.

$\omega_c \approx \omega_0$ and we cannot increase the value of ω_c indefinitely otherwise cyclotron absorption has to be taken into consideration. Figure 2 displays the variation of pump amplitude and absorption coefficient α_{eff} Vs wave vector k and threshold pump amplitude E_{th} Vs cyclotron frequency ω_c. It is observed that threshold pump amplitude decreases whereas gain coefficient increases on increasing wave vector. Therefore lower threshold pump field amplitude and higher gain of the polaron mode is obtained at polaron wave number $k > 10^7 m^{-1}$.

The figure 3 displays the parametric absorption coefficient α as a function of cyclotron frequency ω_c with pump electric field E0 as a parameter. Here it is worth mentioning that absorption coefficient α can be both positive and negative under the anomalous regime. Resonance between polaron frequency and cyclotron frequency gives rise to amplification. A slight variation in ω_c beyond polaron wave frequency abruptly decreases the absorption coefficient resulting into absorption or loss of the polaron mode. Hence polaron frequency is a critical frequency beyond which amplification cannot be achieved at any of the pump field amplitudes. Application of higher pump field results into less gain at a lower critical magnetic field. Increment in the pump amplitude shifts the critical frequency towards the higher magnitudes of cyclotron frequency.

The magnetic field is found to augment the growth rate of the polaron mode. Considerable reduction in the threshold pump amplitude $(10^8 to 10^3 Vm^{-1})$ and gain $(10^{11} m^{-1})$ can be achieved by proper selection of pump amplitude and magnetic field. Hence, compound magnetized semiconductor namely InSb establishes its potential as a candidate material for the fabrication of optical parametric amplifiers.

4. References

[1] Tian L, Allman M S and Simmonds R W 2008 *New J. of Phys.* **10** 115001
[2] Kaplan R, Palik E D, Wallis R F, Iwasa S, Burstein E and Sawada Y 1967 *Phys. Rev.Lett.* **5** 159
[3] Hellwarth R W 1998 *Phys. Rev. B* **60** 299
[4] Landau L D 1933 *Phys. Z. Sowjetun.* **3** 664
[5] Devreese J T 1998 *Proceedings of the International School of Physics Enrico Fermi CXXXVI ed G Iodonisiet al Amsterdam IOSS Press*
[6] Dekorsy T, Kutt W, Pfeiffer T and Kurz H 1993 *Euro Phys. Lett* **3** 223
[7] Selyutin A Yu 2008 *Phys. of the Solid State* **1** 16
[8] Dubey S and Ghosh S 2009 *New J. of Phys.* **11** 093030

Quantum effect on parametric dispersion in presence of nonuniform size colloids in semiconductors

R Vanshpal[1], S Dubey, S Ghosh
[1] School of Studies in Physics, Vikram University, Ujjain-456010 (M.P.) India

E-mail: ravivanshpal@gmail.com

Abstract. Quantum effect on parametric dispersion characteristics in ion implanted semiconductors in presence of nonuniform size colloids is analytically investigated in the present report. Nonuniform size colloids are managed through polynomial distribution function in the analysis. Here the used quantum hydrodynamic model is described by a set of hydrodynamic equations (typically continuity and momentum transfer) that include quantum effects via Bohm potential. Bohm potential modified second order optical susceptibility is obtained through nonlinear induced current density in presence of electrons and negatively charged nonuniform size colloids. It is found that parametric dispersion characteristics are greatly influenced by the quantum modifications. The parametric dispersion of the generated signal mode reduces due to the presence of Bohm potential. The required pump intensity at which one achieves maximum dispersion shifts towards higher value in presence of quantum term. Moreover present study also establishes that quantum effect on colloids is inversely proportional to their size; smaller colloids induce more quantum modifications.

1. Introduction –

Ion implantation techniques are generally employed for the manufacturing of semiconductor components. The implanted ions in a host material can alter their properties through introduction of small quantities of dopant atoms. In this process implanted metal ions directed towards the target and being neutralized during slowing down process (electronic and nuclear stopping) and finally come at rest. The depth profile of dopant atoms may be explained by chemical binding effect associated with ion-ion and ion-target atom interaction at low energy. The energy ranges from several hundred to several million electron volts, resulting in ion distributions with average depth from < 10 nm to 10 μm. The colloid formation of metal ions (such as Ag+, Cu+, Fe+ etc) by ion implantation techniques in SiO_2 glasses has been carried out in a number of laboratory experiments.

The presence of metal colloids inside any semiconductor in addition to already present mobile charge carriers (electron and holes) converts the solid into multi-component semiconductor plasma medium. In recent past, the role of charged colloids in searching the novel modes of propagation and/or in modification of wave characteristics of existing modes in ion implanted semiconductor plasma (IISP) has been extensively explored [1-4]. In recent past, a number of workers [5, 6] have theoretically studied the long ranged order formation of colloid particles of small sizes. However, quantum effects were not taken into account in any of these studies.

The quantum effect is included in the present study by employing the quantum hydrodynamic model (QHD) for the carrier dynamics in the IISP medium. The QHD model is described by a set of hydrodynamic equations (typically continuity and momentum transfer) that include quantum effect via Bohm potential and Fermi pressure. The quantum effect on parametric amplification characteristics in piezoelectric semiconductor has been recently reported by the present authors [7].

Quantum effect on parametric dispersion characteristics in IISP in presence of nonuniform size colloids is analytically investigated in the present report. Nonuniform size colloids are managed through polynomial size distribution function in the analysis. In section 2 we first introduce the basic sets of QHD equations which are used to derive the second order susceptibility of IISP medium. Section 3 is devoted to the numerical analysis of the problem under study; the consequence of results are also discussed.

2. Theoretical Formulation

In this section, authors focus on the derivation of second-order optical susceptibility arising due to parametric interaction of a high frequency pump beam with internally generated acoustic mode in presence of nonuniform size colloids in IISP medium. The medium is subjected to a spatially uniform high frequency pump electric field $\vec{E}_0 \exp(-i\omega_0 t)$ (i.e. pump vector $|k_0| \approx 0$).

When plasma is cooled down to extremely low temperature the de-Broglie wavelength of the plasma particles may become comparable to the Debye length or other scale lengths of the plasma and quantum effects are expected to play a significant role in the behaviour of charged particles. Using the magnetohydrodynamic model of plasma, Haas, Manfredi and others [8] have developed the quantum hydrodynamic (QHD) model of plasma. The quantum statistics is included in the model through the equation of state which takes into account the Fermionic character of the electrons and colloids. The basic equations in such colloid laden semiconductor quantum plasma with different size colloids are as follows:-

$$\rho\frac{\partial^2 u}{\partial t^2} - 2\rho\gamma_s\frac{\partial u}{\partial t} + \beta\frac{\partial E_1}{\partial x} = C\frac{\partial^2 u}{\partial x^2} \qquad (1)$$

$$\frac{\partial \upsilon_{0l}}{\partial t} + \nu\upsilon_{0l} = \frac{-Z_l e}{m_l}E_0 \qquad (2)$$

$$\frac{\partial \upsilon_{1l}}{\partial t} + \nu\upsilon_{1l} + \left(\upsilon_{0l}\cdot\frac{\partial}{\partial x}\right)\upsilon_{1l} = \frac{-Z_l e}{m_l}E_1 - \frac{1}{m_l n_0}\frac{\partial P_l}{\partial x} + \frac{\hbar^2}{4m_l^{\,2} n_{0l}}\frac{\partial^3 n_{1l}}{\partial x^3} \qquad (3)$$

$$\upsilon_{0l}\frac{\partial n_{1l}}{\partial x} + n_{0l}\frac{\partial \upsilon_{1l}}{\partial x} = -\frac{\partial n_{1l}}{\partial t} \qquad (4)$$

$$\frac{\partial E_1}{\partial x} + \frac{\beta}{\varepsilon}\frac{\partial^2 u}{\partial x^2} = \frac{-z_l e n_{1l}}{\varepsilon} \qquad (5)$$

$$P_l = \frac{m_l V_{F_l}^2 n_{1l}^3}{3n_{0l}^2} \qquad (6)$$

where $l = e$ for electrons and d for colloids. Meaning of all the symbols is given in [7].

Following standard approach [9], with the aid of Eqs. (1) to (6), one gets second order optical susceptibility of the medium as

$$\chi_l^{(2)} = \frac{e\varepsilon_1 Ak}{2\gamma_s\omega_s\omega_0\omega_1}\left[\frac{Z_e\omega_{pe}^2}{m_e}R^{-1} + \frac{Z_d\omega_{pd}^2}{m_d}V^{-1}\right] = \frac{e\varepsilon_1 Ak}{2\gamma_s\omega_s\omega_0\omega_1}\left[\chi_e + \chi_d\right] \qquad (7)$$

In which $R = \left[(\omega_{pe}^2 + k^2 V'^2_{Fe}) - \omega_s^2 - i\omega_s v - \frac{k^2 |E_e|^2}{(\omega_{pe}^2 + k^2 V'^2_{Fe}) - \omega_1^2 + i\omega_1 v} \right]$, $V = \left[(\omega_{pd}^2 + k^2 V'^2_{Fd}) - \omega_s^2 - \frac{k^2 |E_d|^2}{(\omega_{pd}^2 + k^2 V'^2_{Fd}) - \omega_1^2} \right]$

and $\chi_{e,d}$ stand for susceptibility due to electrons and colloids respectively.

Rationalization of Eq. (7) gives the real and imaginary parts of the second-order optical susceptibility. It may be inferred from Eq. (7) that the susceptibility due to implanted charged colloid (χ_d) makes an important impact on the total optical susceptibility of the crystal.

2.1 Colloidal size distribution

The quantum effect on electron dynamics are identical due to constant size and mass of the electrons but quantum effect on colloids shall vary with slight change in their size and mass. The grains in colloidal plasma are found in a great variety of sizes, masses and charges. Quantum effect of the particles mainly depends upon their size and mass. In our work we consider a polynomial expressed distribution of colloid particles in quantum colloidal plasma medium consisting of different size colloids. The differential polynomial-expressed distribution function [10] is of the form.

$$N_{tot} = \int_{r_{min}}^{r_{max}} n(r)dr \ , \quad n(r)dr = [a_0 + a_1 r + a_2 r^2 + a_3 r^3 + \dots\dots\dots\dots]dr \qquad (8)$$

where N_{tot} is the total number density of colloid grains, r is the radius of colloids in a given range $[r_{min}, r_{max}]$, a_0, a_1, a_2, a_3.......... are all constants. Outside the limits $r < r_{min}$ and $r > r_{max}$, we considered $n(r) = 0$. The charge and the mass of the j-th colloid grain can be expressed as- $Z_{d_j} = k_z r_j, m_{d_j} = k_m r_j^3$ where $k_z = 4\pi\varepsilon_0 V_0 / e$, $k_m = 4\pi\rho_d/3$, V_0 is the electric surface potential at equilibrium, ρ_d is the mass density of the colloid grains (assumed to be constant and equal for all grains). We assume that the colloid size distribution is given by Eq. (8) and hence we can obtain susceptibility due to colloids as,

$$\chi_{d_j} = \int_{r_{min}}^{r_{max}} \frac{-Ae^3 k}{2\varepsilon_0 \gamma_s \omega_s \omega_0 \omega_1} \frac{k_z^3}{k_m^2} \frac{n(r)}{r^3} \left[\left(\frac{e^2}{\varepsilon} \frac{k_z^2}{k_m} \frac{n(r)}{r} + \frac{2k^2 K_B T_{Fd}}{k_m r^3} + \frac{k^4 \hbar^2}{4k_m^2 r^6} \right) - \omega_s^2 - \frac{k^2 |E_d|^2}{\left(\frac{e^2}{\varepsilon} \frac{k_z^2}{k_m} \frac{n(r)}{r} + \frac{2k^2 K_B T_{Fd_j}}{k_m r^3} + \frac{k^4 \hbar^2}{4k_m^2 r^6} \right) - \omega_1^2} \right]^{-1} dr$$

Thus the modified real part of total susceptibility becomes

$$\mathrm{Re}(\chi_l) = \frac{e\varepsilon 1 Ak}{2\gamma_s \omega_s \omega_0 \omega_1} \left[\mathrm{Re}(\chi_e) + \int_{r_{min}}^{r_{max}} \chi_{d_j} dr \right] \qquad (9)$$

It may be inferred from Eq. (9) that the total optical susceptibility is influenced and modified by the presence of implanted metal colloids and their size distribution; hence colloids and their size distribution is responsible for the modifications in dispersion characteristics of the scattered wave in a parametric process.

3. Results and Discussion

To have some numerical appreciation of the above formulations and to study the quantum effect on parametric dispersion in an IISP medium with non uniform size colloids, we have assumed this typical medium duly irradiated by a nano second pulsed 10.6 μm CO_2 laser. All the relevant parameters are given in [7].

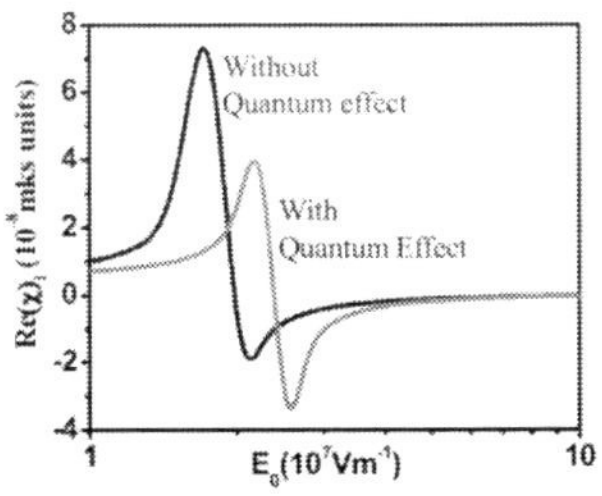

Figure 1. Variation of $\text{Re}(\chi_l)$ with E_0

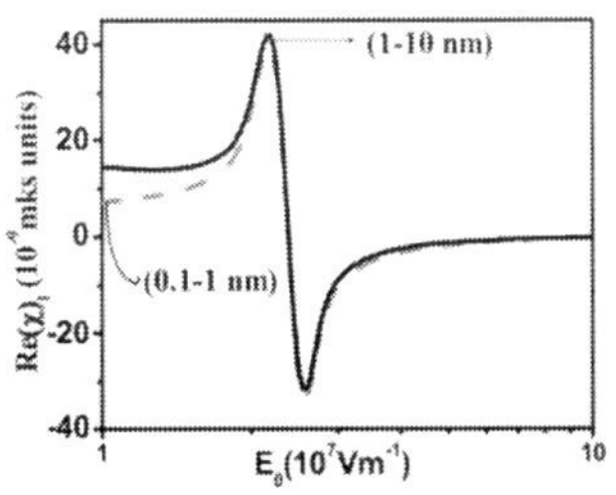

Figure 2. Variation of $\text{Re}(\chi_l)$ with E_0 in two radius ranges.

Figure 1 show the variation of real part of total crystal susceptibility $\text{Re}(\chi_l)$ with pump field amplitude E_0 in the presence and absence of quantum correction term. Here we have assumed range of colloid radius from 0.1 to 1 nm. The quantum correction shows the effective modification in total optical susceptibility in the crystal. The magnitude of $\text{Re}(\chi_l)$ first increases and reaches to a maximum value. The value of E_0 at which one gets maximum magnitude of $\text{Re}(\chi_l)$ shifts towards higher side whereas maximum possible magnitude of $\text{Re}(\chi_l)$ reduces due to the presence of quantum correction. A little departure from this peak point sharply diminishes the value of $\text{Re}(\chi_l)$ to zero. Then it crosses over to negative quadrature and approaches to a minimum. Again the magnitude of E_0 at which one gets minimum $\text{Re}(\chi_l)$ shifts towards higher point due to the presence of quantum correction term. Further a forward tuning of E_0 after the dip increases the $\text{Re}(\chi_l)$ which saturates afterwards. This saturation regime is achieved at E_0 where quantum effects become negligible.

Figure 2 shows the variation of real part of total optical susceptibility with E_0 in two different colloid radius ranges in IISP medium. The polynomial size distribution function gives the flexibility to choose the desired range of colloid radius. Larger colloid radius range increases the magnitude of total optical susceptibility $\text{Re}(\chi_l)$ as illustrated in the figure.

4. Conclusions

Therefore we can conclude that quantum correction term including Bohm potential effectively modifies the parametric dispersion characteristics of IISP medium having nonuniform size colloids. Modified electron plasma frequency and dust plasma frequency is responsible for these improved dispersion characteristics. Present study reports that polynomial size distribution function appears very useful in determining effect of nonuniform size distribution. The favourable range of radius of colloids can be obtained to have a better IISP medium for the fabrication of semiconductor components.

5. Acknowledgements

The financial assistance from the Madhya Pradesh Council of Science and Technology, Bhopal, India, under a research project is gratefully acknowledged.

6. References

[1] Ghosh S and Khare P 2005 *European Phys. J.* **D 35** 521
[2] Ghosh S and Thakur P 2006 *European Phys. J.* **D 37** 417
[3] Ghosh S and Thakur P 2006 *Ind. J. Pure & Appl. Phys.* **44** 188
[4] Ghosh S and Khare P 2006 *Acta Physica Polonica A* **109** 187
[5] Salimullah M, Shukla P K, Ghosh S K, Nitta H and Hayashi Y 2003 *J. Phys. D: Appl. Phys.* **36** 958
[6] Salimullah M, Ehsan Z, Zubia K, Shah H A and Murtaza G 2007 *J. Appl. Phys.* **102** 053301
[7] Ghosh S, Dubey S and Vanshpal R 2010 *Phys. Lett. A* **375** 43
[8] Haas F, Garcia L G, Geodert J and Manfredi G 2003 *Phys. Plasmas* **10** 3858
[9] Guha S, Sen P K, Ghosh S1979 *Phys. Stat. Sol.* **(a) 52** 407.
[10] He G J, Duan W S and Tian D X 2008 *Phys. Plasmas* **15** 043702.

International Conference on Recent Trends in Physics (ICRTP 2012) IOP Publishing
Journal of Physics: Conference Series **365** (2012) 012047 doi:10.1088/1742-6596/365/1/012047

Study of Radial Position of pole Piece Aperture in Focusing of Multibeam Electron Gun

A K Nehra[1,2*], R K Gupta[1], S M Sharma[1], P C Panda[1], Y Choyal[2] and R K Sharma[1]
[1]Microwave Tubes Division, CSIR-Central Electronics Engineering Research Institute, Pilani-333031, Rajasthan, India
[2]School of Physics, Devi Ahilya University, Indore-452001, M.P, India

*E-mail: ashokkumarnehra@gmail.com

Abstract. A four beam electron gun has been designed for multi-beam klystron using commercially available software OPERA 3D and the optimized beam focusing has been again examined through CST particle studio. This paper represents the effect of radial position of aperture in a pole piece in focusing of multi-beam electron gun. The operating voltage of electron gun is 4 kV with total beam current of 256 (64x4) mA and beam perveance of 1.01 μP which is equally divided among all the beamlets.

1. Introduction

Today, the microwave tubes designer faces a number of challenges like high operating voltage, narrow bandwidth and more weight. The basic idea of multi-beam electron gun is an attempt to increase the beam current without increasing the cathode loading. The multi-beam klystrons (MBKs) are the only microwave amplifiers that fulfil the most of present demand of users, such as high power, high efficiency, large bandwidth, low noise and low-voltage operation, as compared to the conventional single-beam klystrons. In addition, the reduced operating voltage significantly decreases the cost of the power supply system and reduces the device length thereby making the device more compact. Due to these features MBK is used for ground-based, airborne and space-based radars, radio-communication and high energy particle accelerators [1]-[7].

In an MBK each beamlet propagates along its own individual path through the RF section consisting of two or more cavity resonators. The most important advantage of an MBK is that the current and perveance per beamlet are low while total current and perveance remain high in this device. For higher electronic efficiency and gain of device bunching should be more, which can be achieved by low perveance per beamlet. Further, the overall higher perveance allows a lower voltage operation yielding higher beam power and RF power as compared to conventional single beam klystron. In addition, the reduced operating voltage significantly decreases the cost of the power supply system and reduces the device length thereby making the device more compact [2], [8]. The schematic of multi-beam electron gun with pole piece position is shown in figure 1and the gun design parameters of multi-beam electron gun are shown in Table 1.

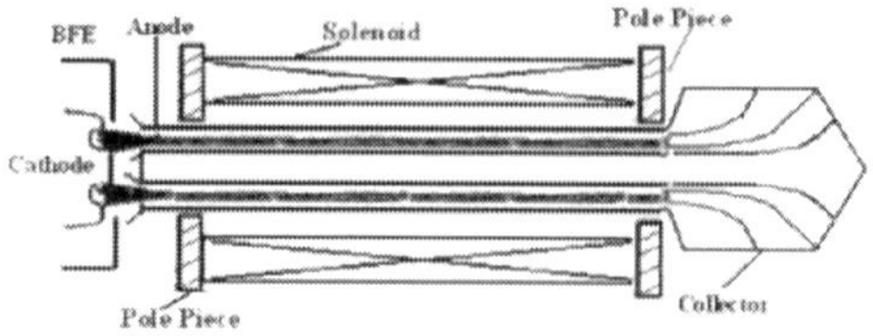

Figure 1. Schematic of multi-beam stick tube

Table 1. Gun design parameters

Parameter	Value
Cathode voltage	-4 kV
No. of beamlets	4
Total beam current	256 mA
Total beam perveance	1.01 μP
Individual beam perveance	0.252 μP
Cathode diameter	14.4 mm
Cathode loading	0.85 A/cm^2
Beam waist radius	0.45 mm
Beam throw	14 mm

2. Electron Gun Design and Effect of radial position of Pole Piece Aperture in Beam Focusing

Using OPERA 3D [9] code a confined flow multi-beam electron gun has been simulated with three different PCDs (pitch circle diameter) at 15 mm, 20 mm, and 25 mm at an angular distance of 90 degrees from each other to study the scalloping in off-axis beams relative to solenoid axis as shown in figure 2.

A study related to the effect of radial position of apertures in pole piece on beam focusing has been carried out using OPERA 3D and CST particle studio [10]. A four beam electron gun (Table I) has been simulated by keeping four beamlets at different PCDs and having a common solenoid focusing with optimum axial magnetic field Bz as 1200 Gauss. This field has been optimized for minimum beam scalloping after a number of iterations.

Detailed studies by keeping apertures in pole piece at three different PCDs (i.e. 15 mm, 20 mm and 25 mm) have revealed that apertures at smaller PCDs yield better beam focusing (less ripples) in comparison to the apertures at higher PCDs. This is corresponding to the fact that in case of apertures at smaller PCDs the electric field at cathodes is higher and transverse magnetic field at aperture is smaller. It has also been observed that for a common solenoid smaller the PCD better is the focusing. Beam scalloping is minimum at 15 mm PCD (Figure 3) in comparison to 20 mm PCD (Figure 4). The 25 mm PCD shows no beam focusing is possible using the same solenoid as the beam transmission through the tunnel is zero as shown in Fig. 5. This study shows that a maximum allowed PCD for a specific solenoid the clearance between solenoid inner diameter and the maximum PCD should be more than 40 % in order to achieve desired focusing with minimum scalloping. The same cases (Figures. 3, 4 and 5) have also been simulated using CST particle studio. The simulated results from both OPERA and CST codes are in good agreement and shown in Table 2.

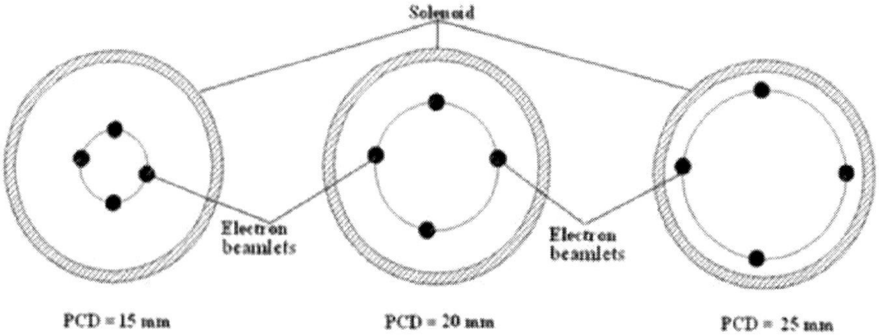

Figure 2. Schematic of a multiple-beam electron gun on different PCD

Figure 3. Electron trajectories at 15 mm PCD

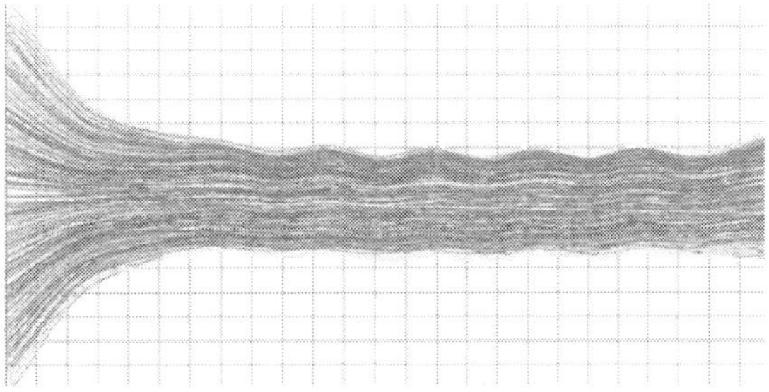

Figure 4. Electron trajectories at 20 mm PCD

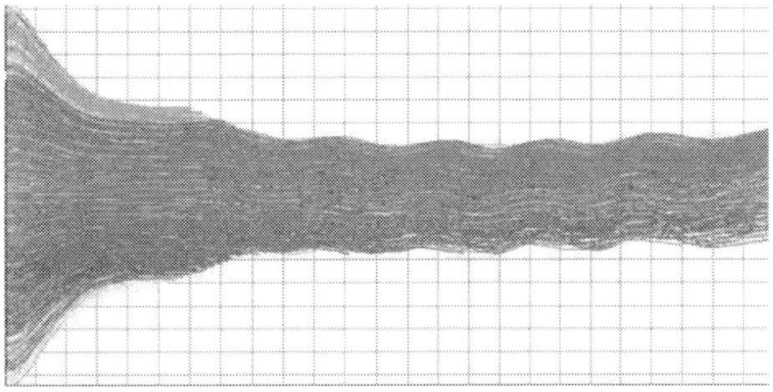

Figure 5. Electron trajectories at 25 mm PCD

Table 2. Comparison between OPERA-3D and CST-3D Simulated Beam Parameters

Parameters	OPERA-3D PCD (mm)			CST PCD (mm)		
	15	*20*	25	*15*	20	25
Beam Current (mA)	256 (64x4)	248 (62x4)	180 (45x4)	248 (62x4)	240 (60x4)	172 (43x4)
Beam Throw (mm)	13	12	------	13	12	------
Beam Waist Radius (mm)	0.35	0.33	------	0.36	0.34	-----

3. Conclusion

This paper represents that a maximum allowed PCD for a specific solenoid, the clearance between solenoid inner diameter and the maximum PCD should be more than 40 % in order to achieve desired focusing with minimum scalloping. The simulated results from both the codes OPERA-3D and CST particle studio are in good agreement as summarized in Table 2.

Acknowledgement

The authors are thankful to Dr Chandershaker, Director, CEERI, Pilani, for granting permission to publish this paper. They owe their gratitude to Dr SN Joshi, Dr V Shrivastava, and other team members for the constant encouragement and useful suggestions. Thanks are due to CSIR, New Delhi for funding the project.

References

[1] Korolyov A N, Gelvich E A, Zhary Y V, Zakurdayev A D, and Poognin V I, Jun. 2004 *IEEE Trans. Plasma Sci.,* **32, 3**, 1109.

[2] Nguyen K T, Abe D K, Pershing D E, Levush B, Wright E L, Bohlen H, Staprans A, Zitelli L, Smithe D, Pasour J A, Vlasov A N, Antonsen T M, Jr., Eppley K, and Petillo J J, Jun. 2004, *IEEE Trans. Plasma Sci.,* **32 3** 1119

[3] Abe D K, Pershing D E, Nguyen K T, Myers R E, Wood F N, and Levush B, May 2007 *IEEE Trans. Electron Devices*, 54 5 1253

[4] Gelvich E A, Borisov L M, Zhary Y V, Zakurdayev A D, Pobedonostsev A S, and Poognin V I, Jan. 1993, *IEEE Trans. Microwave Theory Tech.* **41** 15

[5] Korolyov A N, Gelvich E A, Zhary Y V, Zakurdayev A D, and Poognin V I., Dec. 2001, *IEEE Trans. Electron Dev.* **48** 2929

[6] Ding Y, Shen B, Shi S, Cao J, May 2005, *IEEE Trans. Electron Dev.* **52 5** 889

[7] Nguyen K T, Pershing D E, Abe D K, Levush B, Wood F N, Calame J P, Pasour J A, Petillo J J, Cusick M, Cattelino M J, and Wright E L, Jun. 2004, *IEEE Trans. Plasma Sci.* **32 3** 1212

[8] Gilmour A S, Jr, 2011 Boston, Artech House, Klystrons, Traveling Wave Tubes, Magnetrons, Crossed-Field Amplifiers, and Gyrotrons,

[9] Opera-3D, Cobham technical services, Vector fields software, 24 Bankside Kidlington Oxford OX5 1JE England.

[10] CST-Particle studio, User manual: 2010 version, Darmstadt, Germany.

Investigation of electron beam parameters inside the drift region of plasma cathode electron gun

Deepak K. Verma*, U. N. Pal, N. Kumar, J. Prajapati, M. Kumar, Ram Prakash and V. Srivastava
Microwave Tubes Division, CSIR-Central Electronics Engineering Research Institute (CSIR-CEERI), Pilani, Rajasthan-333031, India.

E-mail:deepakverma31@gmail.com

Abstract. This paper presents experimental studies for the production and propagation of an electron beam from a single gap pseudospark discharge based plasma cathode electron (PCE) gun. The generated electron beam has been successfully propagated for more than 25 cm in a gaseous environment without application of external guiding magnetic field at different operating conditions. The electron beam losses due to recombination with ions and collision with walls of drift space have been estimated. The electron beam profile has also been analyzed in the drift region of the gun.

1. Introduction

The pseudospark (PS) discharge is recognized as unique type of discharge [1] which is capable of producing electron beams with highest combined current density and brightness of any known type of electron source [2-7]. The PS discharge based plasma cathode electron (PCE) gun has potential applications in microwave generation, electron beam melting, welding, surface treatment, plasma chemistry, radiation technologies, laser pumping, where material cathode cannot be used [8,9]. This type of gun has longer life as compared to that of material cathode. The PS discharge operates in hollow cathode geometry on the left-hand side of the Paschen curve with axially symmetric parallel electrodes and central holes on the electrodes [10]. The discharge system for PCE-guns does not have hot filament or similar kind of thermionic solid cathode, and so it is ''filament less'' and known as ''cold cathode''. This is primary difference between a plasma cathode electron-gun and a thermionic electron-gun. Due to the absence of hot electrode, the plasma cathode system is basically more reliable, with a longer lifetime, and can generate electron beams at much higher background gas pressures, even in the fore-vacuum pressure ranges. In this paper experiments has been performed to investigate the electron beam propagation inside the drift region at different operating conditions. Variation in the electron beam current and discharge current at different location of drift region has been analyzed to understand the electron beam behavior in the drift space. This study is essentially required for the development of plasma assisted backward wave oscillator (BWO) [11].

* To whom any correspondence should be addressed.

2. Experimental Set up

The single-gap pseudospark based plasma cathode electron gun experiments have been performed at different operating conditions. The schematic view of the experimental setup and the fabricated PCE-gun are shown in Figure 1 (a, b).The hollow cathode has been connected to the dc power Supply (-25 kV, 1 mA) through a 5 MΩ current limiting resistor while the anode is grounded. There was no external applied guiding magnetic field for the focusing of electron beam. The hollow cathode and anode geometries have been used to design and fabricate the PCE-gun. The hollow cathode is cylindrical in shape having height 59.7 mm. The inner and outer diameters are 59.4 mm and 65.4 mm respectively. The thickness of the cylinder and the diameter of the aperture is 3 mm each. The anode dimensions are same as that of the cathode. The anode and cathode are assembled in a ceramic casing with 3 mm gap between them. The anode and cathode are assembled in a ceramic casing with 3 mm gap between them. A circular ring arrangement has been developed for beam current diagnostics in radial as well as in axial directions. The charging voltage has been measured using a capacitive voltage probe (Tektronix P6015A). The measurements of the beam current were realized using Rogowski belts, one located at the anode to measure the beam current while another current transformer was connected to the discharge circuit to measure the discharge current. These transformers are connected to the digital oscilloscope (Tektronix DPO 4054) which synchronously displays the voltage and current waveform. The discharge takes place in the region between hollow cathode and anode when the field reaches to the breakdown level of the gap. The beam current has been measured at different location of the drift space.

3.Result and discussion

Experiments have been performed for the analysis of beam current propagation inside the drift space of the PCE-gun. Initially the PCE-gun has been evacuated up to ~10^{-6} mbar. Then argon gas is filled inside the pseudospark chamber in a controlled manner.

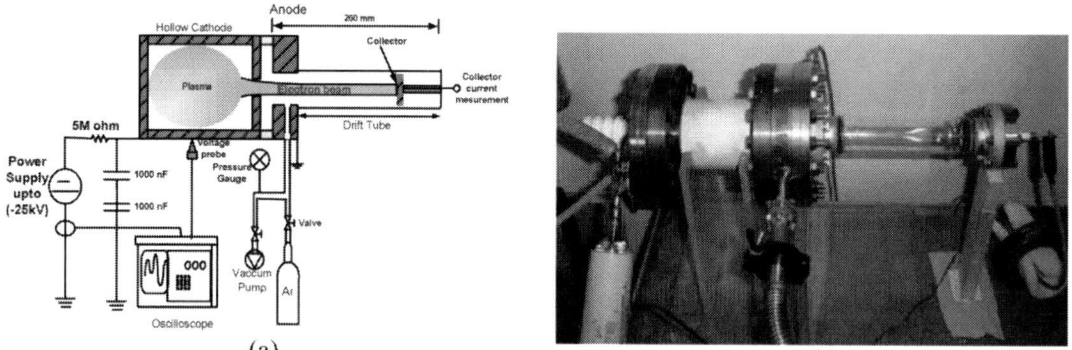

(a) (b)

Figure 1. (a) Schematic view of experimental set-up, (b) developed single gap PCE gun

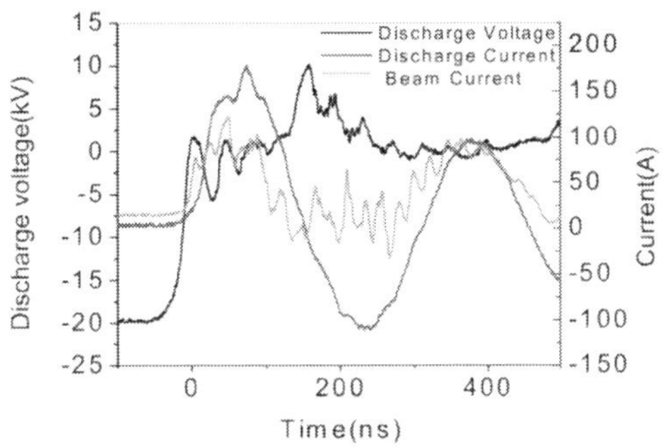

Figure 2. Experimental result showing hollow cathode voltage and discharge current and beam current.

The high voltage applied across the pseudospark chamber was then increased slowly until the breakdown occurred. It has been observed that the discharge took place around 80 Pa of working pressure and this has been repeated at different voltages. Figure 2 shows a typical trace of the experimental values of discharge voltage, discharge current and beam current which clearly demonstrate the pseudospark discharge characteristics at different times [10,12].When the electron beam is propagated in the pseudospark gaseous environment, it ionizes the gas and a plasma channel is formed in the beam path. The beam electrons would repeal the electrons of the plasma and an ion channel will be formed around the electron beam. These ions will attract the beam electrons and counteract the defocusing space charge forces of the electron beam. Therefore, there occurs a self-pinching of the pseudospark electron beam. This is due to the fact that the electron beam is propagating in the ion focusing regime [13]. The Figure 3 shows that at the fixed applied voltage, the percentage of the beam current with respect to the discharge current reduces from anode region to the collector region. It has also been seen that the maximum beam current is near anode region (z=0 mm) and the minimum beam current is near to the collector region (107 mm).

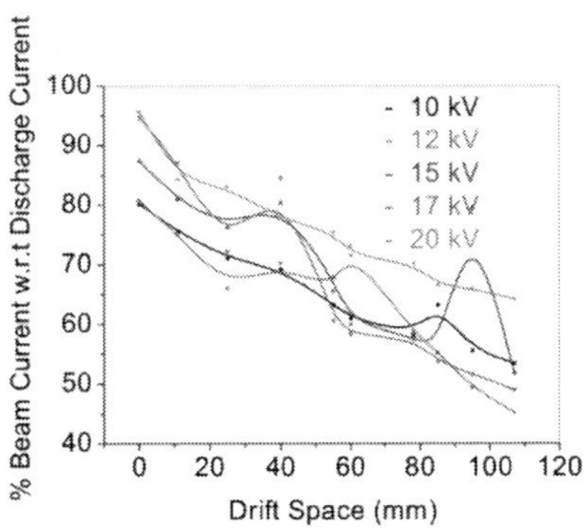

Figure 3. Experimental values of percentage of beam current measurements at different location of the drift space and at different anode voltages.

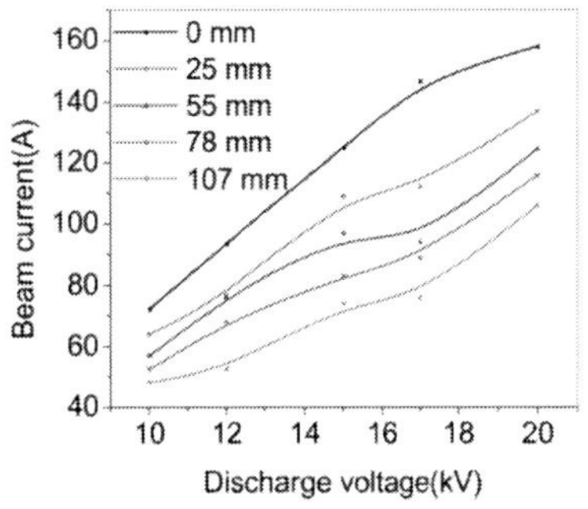

Figure 4. Experimental results of beam current variation with applied voltage at different location of the drift space.

The Figure 4 shows the beam current variation with applied voltage at different location of the drift space. It is observed that when applied voltage is 10 kV the obtained beam current is up to 70A. Furthermore, the beam current increases with applied voltage. The observed beam current reaches up to 150A at 20 kV applied voltage.

4. Conclusion

A pseudospark based PCE-gun has been designed and developed for the generation of electron beam useful for plasma assisted microwave sources and surface applications. The analysis of the electron beam profile inside the drift space has been carried out at different operating conditions. The generated electron beam has been successfully propagated for more than 25 cm in a gaseous environment without application of external guiding magnetic field. The beam current has been measured at different locations of the drift space which decreases with the distance. The electron beam losses due to the recombination with ions and collision with the walls of the drift space have been estimated.

5. Acknowledgement

The work has been carried out under CSIR Network project (NWP0024). The authors are thankful to Mr. B. L. Meena and Mr. Dharmendra Pal for useful discussions and support in the experiments.

References

[1] Gunderson M A and Schaefer G 1990 *NATO ASI Series B.* **219** 331
[2] Kirkman G F and Gunderson M A 1986 *Appl. Phys. Lett.* **49**, 494
[3] Riege H and Boggasch E 1989 *IEEE Trans. Plasma Sci.* **17** 775
[4] Boggasch E and Rhee M J 1990 *Appl. Phys. Lett.* **56**, 1746
[5] Choi P, Aliaga R and Blottiere B 1993 *Appl. Phys. Lett.* **63** 2750
[6] Alberta M P, Derouard J, Pitchford L C, Ouadoudi N and Boeuf J P *Phys. Rev. E* **50** 2239
[7] Anders A, Anders S and Gunderson M A 1993 *Phys. Rev. Lett.* **71** 364
[8] Cross A W, Yin H, He W, Ronald K, Phelps A D R and Pitchford L C 2007 *J.Phys.D:Appl.Phys.***40** 1953
[9] Goebel D M and Watkins R M 2000 *Rev. Sci. Instrum.***71** 388
[10] Frank K and Christensen J 1989 *IEEE Trans. Plasma Sci.***17** 748
[11] Baruch L, Thomas N A, Alan B, Lou W-R and Carmel Y 1992., *IEEE Trans. Plasma Sci,* **20** 3
[12] Pal U N, Dubey V P, Barik M K , Lamba V, Verma D K, Kumar N, Kumar M, Meena B L, Tyagi MS and Sharma A K 2011 *XII IEEE IVEC* Bangalore, India.
[13] Kumar N, Pal U N, Verma D K, Prajapati J, Kumar M, Meena B L, Tyagi M S and Srivastava V 2011 *J Infrared Milli Terahz Waves* **32** 1415

Dispersion characteristics of non-radiative coupled dielectric waveguide at millimeter wave frequencies

Alok Kumar Rastogi (FIETE) [a], Ayush Hajela[b], Manisha Nigam (MIETE)[b]
[a] Institute for Excellence in Higher Education, Bhopal, (M.P.), India
[b] Govt. Motilal Vigyan Mahavidhyalaya, Bhopal (M.P.), India

E-mail: akrastogi_bpl@yahoo.com

Abstract. The solutions of electromagnetic field equations for non-radiative coupled dielectric guides are derived using mode matching technique. Dispersion characteristics and wave impedance for the even and odd mode of non-radiative coupled dielectric guides are computed. Dispersion curves are plotted for various dielectric materials and dimensional parameters as a function of frequency. Normalized electric field for dominant electric component E^y_{11} of TM even and odd modes for various half spacing between two dielectric slabs of non-radiative coupled dielectric guide have been plotted. The non-radiative phenomenon in the coupled dielectric guide has been discussed in detail. The hybrid directional couplers in the above configuration have been designed by computing the normalized propagation constants. The superiority of such directional couplers is discussed.

1. Introduction

The dielectric image guide technology is an emergent area in the field of millimeter wave applications. The dielectric image guides are easy to fabricate, low cost, pose ease of fabrication and dimensional tolerance are not stringent. However the basic limitations of dielectric guide is that it radiates at bends. The losses due to radiation at curved section, junctions and discontinuity become inherent in dielectric image guides. This problem can be sorted out some how by using materials of high permittivity, but there are limitations of operating frequencies, size and so on. Radiations in dielectric image guide are directed outwards. These radiations are due to change in phase velocities at the corner of the dielectric image guide and the outer surface. Due to these radiations at bends the propagation characteristics also change. Radiation at bends, this phenomenon also occurs at couplers. To avoid these radiations at bends, trapped coupler can be fabricated, so that energy radiating at bends will be reflected from the metal trough and will be coupled back to dielectric slab [1], [2]. Non radiative dielectric guide was proposed by Yoneyama and Nishida [3], and Insulated non-radiative dielectric guide by Yoneyama, Fujita and Nishida [4]. These guides are reported to have negligible loss at curved sections.

2. Discussion

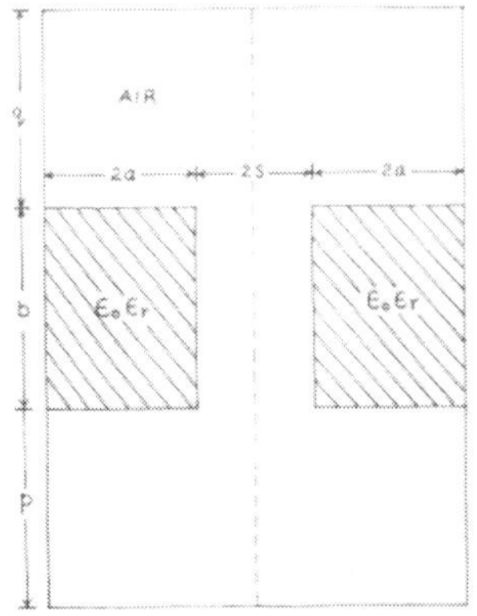

FIG. 1 CROSS-SECTIONAL VIEW OF NON-RADIATIVE COUPLED DIELECTRIC GUIDE

The non-radiative phenomenon in the coupled dielectric guide occurs when propagation frequency is well above the cutoff frequencies of the TM-mode parallel plate system, where the cutoff frequency is given by

$$f_c = \frac{C_0}{2(4a + 2d + 2s)} \quad (1)$$

Where C0 is the velocity of light, 'a' is half width of the dielectric slab and 's' is the half separation between two dielectric slabs and 'd' is the distance of dielectric slab from side metallic walls. This structure is known as insulated non-radiative coupled dielectric guide. If the side wall distance 'd' is kept equal to zero the structure reduces to non-radiative coupled dielectric guide as shown in Fig. 1.

Fig. 2 shows the variation of normalized propagation constant β / β_0 as a function of frequency for even and odd modes with various half spacing 's' between the two dielectric slabs. It is observed from the figure that in case of even coupled mode when half spacing s=0.1mm and s=0.5mm, β / β_0 increases with the increase in frequency. When s=3mm, which is sufficiently large, the β / β_0 increases with the increase in frequency and shows the sharp decrease in the dielectric image guide mode (β / β_0>1). Thus for even coupled mode the dispersion curves for various half spacing 's' cross each other at β / β_0=1. In the case of odd mode, dispersion curves do not cross-each other at β / β_0=1. Further for image guide mode (β / β_0>1), the even and odd mode propagation constant approaches each other with increasing half spacing value. It is also observed from the figure that the coupling space between the two dielectric strips decreases with an increase in half spacing and for sufficiently large values of half spacing, the even mode propagation constant becomes equal to the odd mode propagation.

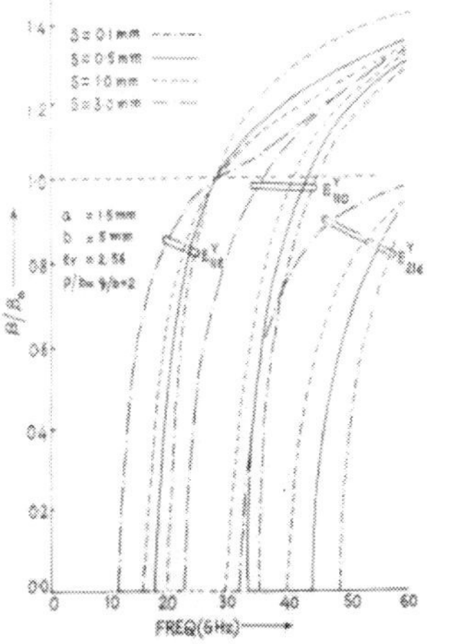

FIG. 2 VARIATION OF β/β_0 WITH FREQUENCY OF NONRADIATIVE COUPLED DIELECTRIC GUIDE $b/2a = 5/3$

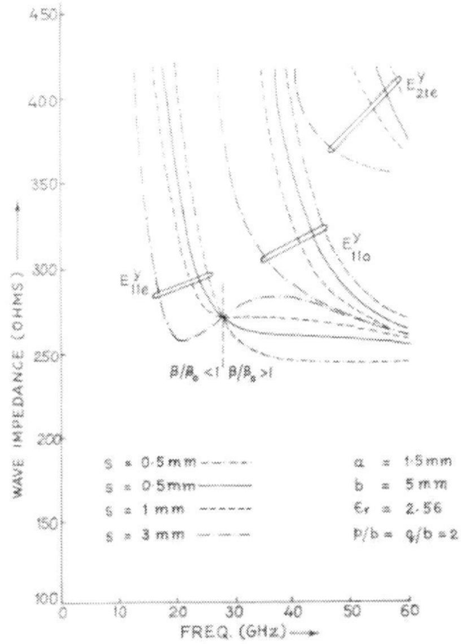

FIG. 3 VARIATION OF WAVE IMPEDANCE WITH FREQUENCY OF NONRADIATIVE COUPLED DIELECTRIC GUIDE $b/2a = 5/3$

Fig. 3 shows the variation of wave impedance with frequency for various half spacing 's'. It is observed from the figure that for the even modes, wave impedance curves crosses each other at $\beta/\beta_0=1$. For E^y_{11e} mode this crossing occurs at 27.8 GHz. For values s=0.1mm and s=0.5mm the wave impedance decreases with increase in frequency initially at a rapid rate and then tends to reach a constant value. For s=3mm the wave impedance goes through a minimum in the region $\beta/\beta_0<1$, increases with frequency in the intermediate range around $\beta/\beta_0=1$ and then decreases again. In the case of odd mode the wave impedance decreases with an increase in frequency for all values of 's'. It is also observed from the figure that the wave impedance of higher order modes is higher than those of the lower order modes.

Fig. 4 shows the normalized electric field curves for dominant electric field component of TM even and odd modes that is E^y_{11e} and E^y_{11o}, for various half spacing 's' between the two dielectric slabs. It is observed from the figure that for even mode, with the increase in value of 's' the field at the centre of the guide (in the air region) decreases and for sufficiently large value of s=3mm, the two dielectric slabs behaves as decoupled. Further as the metallic plane is just adjacent to one edge of the dielectric slab, the electric field is not symmetrical with respect to the dielectric slab.

Fig. 5 shows the variation of the effective coupling length with frequency of a non-radiative coupled dielectric guide coupler for 0dB and 3dB couplings for half spacing s=0.5mm and s=1.0mm. In the same

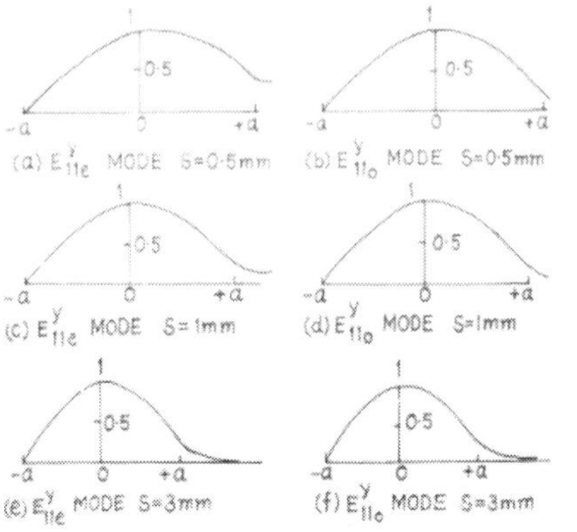

FIG 4 E^y_{11} EVEN AND ODD MODES Ey PLOTS FOR VARIOUS HALF SPACINGS OF NONRADIATIVE COUPLED DIELECTRIC IMAGE GUIDE (d=1mm, ϵ_r=2·56, p=q=6mm, b=3mm, a=2·5mm, FREQ=60GHz, y=p+0·8b)

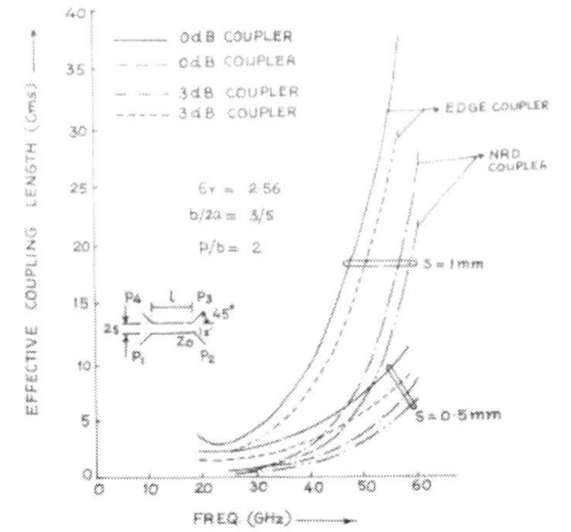

FIG 5 VARIATION OF EFFECTIVE COUPLING LENGTH WITH FREQUENCY FOR NONRADIATIVE COUPLED DIELECTRIC GUIDE COUPLER AND EDGE COUPLED COUPLER FOR 0dB AND 3dB

figure 0dB and 3dB coupling length curve of an edge coupled dielectric image guide coupler with the same parameters has been plotted for comparison. Viewing the curves, it is clear that for both half spacing s=0.5mm and s=1.0 mm and coupling ratios 0dB & 3dB, the effective coupling length of non-radiative guide coupler is less than edge coupled dielectric image guide coupler. In case of non-radiative guide coupler as the principal electric field is parallel to the metallic side walls, the conductor loss will be substantially less than in edge coupled dielectric image guide coupler. Further for a given frequency the coupling length of a coupler in non-radiative dielectric guide with edge coupled coupler are shown.

3. Conclusion

Non-radiative coupled dielectric waveguide has been proposed as a radiation free dielectric waveguide. Since the electric field is parallel to the metal plates, radiation if any suppressed due to the cutoff property of the side walls. Characteristics of curved non-radiative coupled dielectric guide directional couplers are analyzed by taking the effect of field deformation at bend into account.

The dispersion characteristics of non-radiative coupled dielectric guide are presented. The coupler characteristics of these guides have been presented with appropriate dimensional parameters and high dielectric constant for the dielectric slabs, the broadside coupled and insulated broadside coupled structures behave as non-radiative & insulated non-radiative dielectric guides respectively.

4. References

[1] Itoh T and Adelseck B,1980 *IEEE Trans., MTT-**28*** 1433
[2] Rastogi A K and Tiwari A K, 1993 *International Journal of Infrared and Millimeter waves,* **14** 1445
[3] Yoneyama T and Nishida S,1981 *IEEE Trans., MTT-***29** 1181
[4] Yoneyama T, Fujita S and Nishida S, 1983 *IEEE Trans., Symposium*

Dynamic Stark effects in a semiconductor quantum dot exhibiting a two-hole species valence band

Swati Goyal[1] and P. K. Sen
Department of Applied Physics, Shri G. S. Institute of Technology and Science, Indore 452003, India
[1] Department of Physics, Govt. Auotonomus Holkar Science College, Indore 452001, India

E-mail: goyalguptaswati@rediffmail.com

Abstract. We have analytically explored the nature of Stark shift and self energy correction in the presence of two discrete excitonic states i.e, the heavy-hole and light-hole valence bands, below the fundamental absorption edge of semiconductor quantum dots. The time dependent perturbation technique for a three-level system, under the near band gap resonant excitation regime, has been employed; with the incorporation of the relaxation and dephasing mechanisms phenomenologically. The detuning spectra of Stark shift and self energy correction and their dependence on a moderately strong electric field, have been analyzed. The numerical and analytical estimations in a weak noncentrosymmetric CdTe quantum dot show a decrease in Stark shift and an enhancement in self energy correction for both heavy-hole and light-hole excitons.

1. Introduction

During the past three decades, the physics of low-dimensional semiconductors have become a vital part of present-day research. Semiconductor quantum dots (QDs) represent ideal quantum systems because their electrons are three dimensionally confined, resulting in a discrete density of states. The combination of high nonlinear properties and fast response time make applications in all-optical switching devices possible with semiconductor QDs.

Electric field has become an interesting probe for studying the physical properties of low-dimensional structures, from both the theoretical and technical points of view. The ac Stark shift also known as dynamical Stark shift arises in an active medium when a rapidly oscillating field of an electromagnetic wave is applied to shift the atomic, molecular or electron energy levels. The measurement of Stark shift is utilized to calculate the transition dipole moments and relaxation parameters of the crystal [1]. The self energy correction indicates that the damping mechanisms are modulated in presence of the electric field amplitude of the pump radiation.

The observation of the Stark effect in semiconductors is normally hinded by the fast dephasing rates in the vicinity of the excitonic transitions. With the recent advances in ultrashort pulse spectroscopy, it is now possible to observe such phenomena induced by intense coherent laser beams in bulk as well as in low-dimensional semiconductor structures. In 1985, Frohlich and coworkers first demonstrated experimentally, the optical Stark effect in Cu2O and Miller et al explained theoretically the quantum-confined Stark effect (QCSE) in QWs [2]. In 2004, Unold et al [3] reported the first experimental study of the optical Stark effect in a single semiconductor QD.

As CdTe is a weak noncentrosymmetric crystal. The occurrence of noncentrosymmetry (NCS) has been classically explained by the local fields existing in the crystal and due importance should be given to its contribution while studying the stark effects in a specific semiconducting crystal. To the author's knowledge, so far no systematic attempt has been made to investigate the dependence of nonlinear transient's phenomena on applied electric field in NCS QDs.

In the present analytical study of optical nonlinearities of QDs, we have selected the QD sizes to be much smaller than the bulk exciton Bohr radii. Such QD sizes are classified as belonging to the strongly confinement regime (SCR). In a semiconductor QD, the confinement induced exciton binding energy increases abruptly when its geometrical radius becomes smaller than the bulk exciton Bohr radius a_B. Under the SCR, apart from exciton, biexcitons also play an important role in determining the optical properties of semiconductor QDs. It has been shown by Hu et al. [4] that the ground state exciton energy $\hbar\omega_{ab}$ scaling as R-2 dominates distinctly over the electron-hole (e-h) Coulomb interaction and remains practically independent of the e-h effective mass ratio for $R \le 0.6\ a_B$.

In most semiconductors, the conduction band can be approximated by a parabolic band model but the valence band represents a complicated structure with heavy-hole (hh) and light-hole (lh) bands and split-off bands. The split-off bands are far away from the band edge. The hh and lh bands are degenerate at k = 0 and strongly coupled at k ≠ 0. This coupling leads to a nonparabolic subband structure. The inclusion of valence-band mixing in the analysis of the optical properties of the active material makes the calculation complicated. So the intrasubband mixing in the valence bands as well as split-off energy band have been neglected. Hence the photo induced band-to-band electronic transitions are considered from both the heavy-hole (hh) and light hole (lh) valence bands to the lowest (1s) exciton state below the fundamental absorption edge. The special feature of the present analysis is the in-depth study of the transient optical phenomena viz. dynamical Stark shift and self energy correction in QDs under a moderately strong electric field.

2. Theoretical Formulations

The band structure of direct gap semiconductor is taken to be isotropic, parabolic and nondegenerate. The three energy levels considered are the crystal ground state $|a\rangle$, the lowest (n =1,1s) single electron-hole pair (exciton) state $|b\rangle$ is described in terms of both heavy-hole and light-hole subbands, and the biexcitonic state $|c\rangle$ arising due to the formation of bound exciton pairs. The energy of the lowest (1s) excitons in the presence of finite electron-hole coulomb interaction in a QD of radius R is expressed as $\hbar\omega_{ab} = \hbar\omega_g + (\hbar^2\pi^2/2m_\lambda R^2) - \{(3.527a_B/R)+0.248\}E_R$. Here, $\hbar\omega_g$ and $E_R\ (=\hbar^2/2m_\lambda a_B^{\ 2})$ are the bulk band gap energy and exciton Rydberg energy respectively. a_B is the Bohr radius and m_λ is the effective reduced mass of the exciton with λ = lh or hh. The light-hole and heavy-hole effective masses can be represented in terms of Luttinger parameters γ_1 and γ_2 as $m_{hh} = m_0/(\gamma_1 - 2\gamma_2)$ and $m_{lh} = m_0/(\gamma_1 + 2\gamma_2)$ with m_0 being the free electron mass. Also, the exciton-biexciton transition energy with biexciton binding energy ΔE_{xx} defined as $\hbar\omega_{bc} = \hbar\omega_{ab} - \Delta E_{xx}$. We have incorporated the relaxation and dephasing mechanisms phenomenologically through the dephasing parameters Γ_b and Γ_c.

Using time dependent perturbation treatment and rotating wave approximation, the equations of motion of the probability amplitudes in these three states of an NCS system are given by

$$\dot{a}(t) = i\Omega_{aa}e^{i\omega t}a(t) + i\sum_k \Omega_{ab}e^{-i(\omega_{ab}-\omega)t}b(\vec{k},t), \tag{1}$$

$$\dot{b}(\vec{k},t) = i\Omega_{bb}\, e^{-i\omega t} b(\vec{k},t) + i\Omega_{ab} e^{i(\omega_{ab}-\omega)t} a(t) + i\sum_k \Omega_{bc} e^{-i(\omega_{bc}-\omega)t} c(\vec{k},t) - \Gamma_b b(\vec{k},t), \quad (2)$$

$$\dot{c}(\vec{k},t) = i\Omega_{cc}\, e^{i\omega t} c(\vec{k},t) + i\sum_k \Omega_{bc}\, e^{i(\omega_{bc}-\omega)t}\, b(\vec{k},t) - \Gamma_c\, c(\vec{k},t). \quad (3)$$

In each of the above three equations, the first term with finite permanent dipole moments in all the 3 states represents the contribution of NCS effect through Ω_{aa}, Ω_{bb} and Ω_{cc} in the equations of motion. In an NCS system, the diagonal dipole moment operators $(\hat{\mu}_{ii})$ are finite [5] and we define $\hat{\mu}_{ij}(k) = \begin{bmatrix} \mu_{aa} & \mu_{ab} & \mu_{ac} \\ \mu_{ba} & \mu_{bb} & \mu_{bc} \\ \mu_{ca} & \mu_{cb} & \mu_{cc} \end{bmatrix}$. Here, $\mu_{ab,bc} = \dfrac{e|p_{ab,bc}(0)|}{m_o \omega_{ab,bc}}$; μ_{ii} and μ_{ij} $(i, j = a, b, c)$ are the permanent and transition dipole moments, respectively. Ω_{ab} and Ω_{bc} are the field dependent Rabi flopping frequencies corresponding to the level-to-level transitions $|a\rangle \leftrightarrow |b\rangle$ and $|b\rangle \leftrightarrow |c\rangle$ respectively, defined by $\Omega_{ab,bc} = \mu_{ab,bc} E_0 / 2\hbar$. Assuming the transition dipole moment operators μ_{ab} and μ_{bc} to act parallel to the applied electric field $\vec{E}(t) = (1/2)E_0 \exp(i\omega t) + \text{c.c.}$. Using the standard approach, [6] a cubic equation in Ω is obtained (Ω is introduced in the solution for $a(t)$ given by $a(t) = A\exp[\Omega_{aa}(\omega t)/\omega - i\Omega t])$ as

$$\Omega^3 - \Omega^2(\omega_{NB} + \omega_{NC}) + \Omega(\omega_{NB}\omega_{NC} + \omega\omega_{Nab} - \Omega_{ab}^2 - \Omega_{bc}^2) + \omega_{NC}\Omega_{ab}^2 = 0, \quad (4)$$

where $\omega_{NC} = \omega_{bc} + \omega_{ab} - 2\omega + (\Omega_{aa} - \Omega_{cc})\cos\omega t - i\{(\Omega_{aa} - \Omega_{cc})\sin\omega t - \Gamma_c\}$; $\omega_{NB} = \omega_{ab} - \omega + (\Omega_{aa} - \Omega_{cc})\cos\omega t + i\{(\Omega_{aa} - \Omega_{cc})\sin\omega t - \Gamma_b\}$; $\omega_{Nab} = (\Omega_{aa} + \Omega_{bb})\cos\omega t + i(\Omega_{aa} - \Omega_{bb})\sin\omega t$.

The solution of the above equation yields three complex roots. Among these three roots, the magnitude of two are comparable to the dephasing frequency Γ_b and third one being numerically at least three orders of magnitude lower than the other two becomes almost trivial. Thus to solve Eq. (4) analytically, we have taken $\Omega^3 > \omega_{NC}\Omega_{ab}^2$ such that the cubic equation reduces to the following quadratic equation,

$$\Omega^2 - \Omega(\omega_{NB} + \omega_{NC}) + (\omega_{NB}\omega_{NC} + \omega\omega_{Nab} - \Omega_{ab}^2 - \Omega_{bc}^2) = 0. \quad (5)$$

Equation (8) yields two complex solutions as

$$\Omega_{1,2} = \frac{\omega_{NB} + \omega_{NC}}{2} \pm \sqrt{\frac{\omega_{NB}^2}{4} + \frac{\omega_{NC}^2}{4} - \omega\omega_{Nab} + \Omega_{bc}^2 + \Omega_{ab}^2}. \quad (6)$$

Following the approach of Zimmermann [7], the negative solution i.e., Ω_2 may be utilized to calculate the Stark broadening in a semiconductors. The real and imaginary part of the Ω_2, yield the Stark shift and self energy correction respectively.

3. Results and Discussions

In order to explore the occurrence of Stark shift and self energy correction in QDs, we have made numerical estimates for a weak NCS material viz., CdTe. The material parameters of CdTe are: $\hbar\omega_g = 1.607eV$, $a_B = 2.9nm$ and $m_e = 0.096m_o$ and Luttinger parameters are taken $\gamma_1 = 5.29$ and $\gamma_2 = 1.89$. We have assumed the average size of the QDs to be R = $0.6a_B$ = 5nm with pump frequency $\omega\ (= \omega_{ab} - 10\Gamma_b) = 2.73\times10^{15}\ \text{sec}^{-1}$, $T_2 = 150$ fs. The contribution of NCS effect is introduced through $\Omega_{aa,bb,cc}$ assuming $\Omega_{aa} = \gamma\,\Omega_{ab}$ and $\Omega_{bb} = \Omega_{cc} = 0.8\Omega_{aa}$; the NCS factor $\gamma = 0.01$. The system becomes centrosymmetric for $\gamma = 0$.

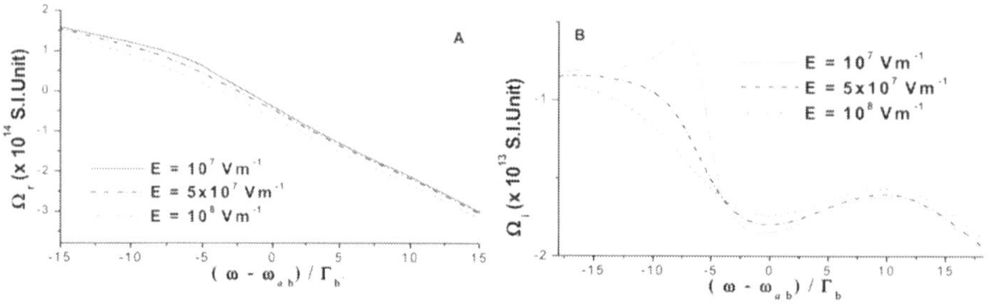

Figures 1. Detuning spectra of Stark shift Ω_r (A) and self energy correction Ω_i (B), at t = 2fs for different strength of electric field.

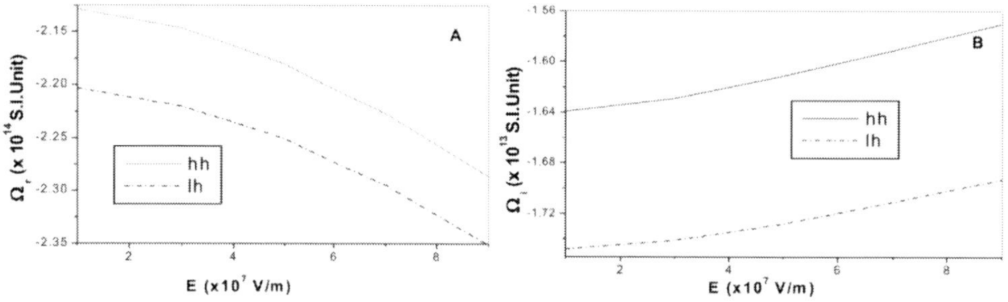

Figures 2. Electric field dependence of Stark shift Ω_r (A) and self energy correction Ω_i (B) for the transitions from hh and lh valence subband.

Figures 1 show that both the nonlinear optical processes are diminished through the increase in electric field intensity. The behavior of Stark shift is symmetric with respect to the central frequency $(\omega = \omega_{ab})$ while the self energy correction deviates from such symmetric responses and its magnitude is large under off-resonant transitions below the band edge $(\omega < \omega_{ab})$ than the corresponding magnitude above the band edge $(\omega > \omega_{ab})$. Figures 2 depict a decrease in Stark shift and an enhancement in self energy correction for both hh and lh excitons. It is clear that hh contribution dominates over lh contribution significantly.

4. Conclusions

To conclude, we have analyzed theoretically the role of excitons and biexcitons on the Stark shift and self energy correction in a weak NCS CdTe QD under strong confinement regime. The two topmost valence subbands have been assumed to participate in the photoinduced band-to-band transitions in the direct gap semiconductor QDs. Increase in electric field strength reduces both Stark shift and self energy correction. The self energy correction show distinct asymmetry. The asymmetric signal has its origin in the strong biexcitonic and excitonic resonances. It is worthy to note that hh contribution dominates over lh contribution significantly. A more realistic theoretical treatment incorporating intrasubband mixing in the valence band will be the subject matter of a future publication.

5. References

[1] Joshi R R, Puri A and Lawande S V 1988 *Mod. Phys. Lett. B* **2** 457
[2] Fröhlich D *et al* 1985 *Phys. Rev. Lett.* **55** 1335; Miller D A B *et al* 1985 *Phys. Rev. B* **32** 1043
[3] Unold T *et al* 2004 *Phys. Rev. Lett.* **92** 157401
[4] Hu Y Z *et al* 1990 Phys. Rev. B **42** 1713
[5] Banerjee S, Sen P and Sen P K, 1999 *J. Opt. B: Quantum Semiclass. Opt.* **1** 505
[6] Goyal S and Sen P K 2011 *I. J. Nanosci.* **10** 675
[7] Zimmermann R 1990 *Festkorperprobleme/Adv. in Solid Stat.Phys.*(Vieweg, Braunschweig) 295

Comparative simulation studies of plasma cathode electron (PCE) gun

Jitendra Prajapati*, U. N. Pal, Niraj Kumar, D. K. Verma, Ram Prakash and V. Srivastava
Microwave Tube Division, CSIR-Central Electronics Engineering Research Institute (CSIR-CEERI), Pilani, Rajasthan-333031

E-mail: jitendra1189@gmail.com

Abstract. Pseudospark discharge based plasma cathode has capability to provide high current density electron beam during discharge process. In this paper an effort has been made to simulate the breakdown processes in the pseudospark discharge based plasma cathode electron gun. The two-dimensional plasma simulation codes VORPAL and OOPIC-Pro have been used and results are compared. The peak discharge current in the plasma cathode electron gun is found to be dependent on aperture size, hollow cathode dimensions, anode voltage and seed electrons energy. The effect of these design parameters on the peak anode current has been analysed by both the codes and results matches well within 10% variation. For the seed electron generation an electron beam trigger source is used to control the discharge process in the hollow cathode cavity. The time span of trigger source has been varied from 1-100 ns to analyze the effect on the peak anode current.

1.Introduction

The pseudospark (PS) discharge is recognized as unique type of discharge [1] which is capable of producing electron beams with highest combined current density and brightness of any known type of electron source [2-6]. The PS discharge based plasma cathode electron (PCE) gun has potential applications in microwave generation, electron beam melting, welding, surface treatment, plasma chemistry, radiation technologies, laser pumping, and where material cathode cannot be used. This type of gun has longer life as compared to that of material cathode. PS discharge is a specific type of gas discharge, which operates in hollow cathode geometry on the left-hand side of the Paschen curve with axially symmetric parallel electrodes and central holes on the electrodes [7].

There are two mechanisms to generate pseudospark discharge; one is self-breakdown discharge that follows the Paschen's curve for breakdown of gases. Another mechanism is triggering discharge in which some initial electrons are provided to start and stimulate the discharge process. In triggered discharge process breakdown of gases can be achieved below the required voltage as calculated by Paschen's curve. In our case typically 1x1011 seed electrons are injected at the centre of the cavity of the hollow cathode with appropriate energies to initiate the discharge. It is fixed for all simulation data in the present analysis. The initial electron which comes from trigger source collides with the gas molecule and generates low density discharge and accelerates towards anode leaving a cluster of positive ions which further acts as virtual anode. Due to this virtual anode formation electric field start penetrating inside the hollow cathode structure and stimulate further

* To whom any correspondence should be addressed.

Published under licence by IOP Publishing Ltd

discharge process. This penetrating electric field helps to create dense plasma with high current density. By this method current can be achieved from few amperes to several thousand amperes using suitable dimensions of PCE-gun [8].

In this work an effort has been made to simulate breakdown processes in the PS discharge based PCE-gun. The two-dimensional plasma simulation codes VORPAL and OOPIC-Pro are used and results are compared. It has been found that the peak current arriving at the anode depends on the hollow cathode dimensions, potential applied at the anode and average velocity of the seed electrons.

In this simulation study secondary electrons generation due to wall impact by electrons is also considered [9].

2. Model Description

The simulation study has been carried out through the two-dimensional electrostatic kinetic particle-in-cell (PIC) simulation codes VORPAL and OOPIC-Pro developed by Tech-X Corp. Both the codes use elastic and ionizing collision using Monte Carlo Collision (MCC). A cross-section model is taken from the Tech-X Corp. in-built library for the electron ionization process. The input parameters like trigger position, seed electron velocity, gas pressure, time constant, etc. are kept similar in both the codes. The schematic view of the model used in the simulation is given in Figure 1. This model has fixed hollow cathode height, Ymax=50 mm and d= 5 mm separation between anode and cathode. The hollow cathode length is also fixed, D+W = 37mm where W is the hollow cathode thickness and D is depth of the hollow cathode. The hollow cathode dimensions have been varied to check the dependency of the peak current on these parameters. The position of the trigger source is kept constant for all the operating parameters. The trigger source is used to supply the seed electrons for the initiation of the discharge process in the hollow cathode. The injection period of the seed electrons is kept 1ns for high energetic electrons and 10 ns for low energy electrons. The seed electron velocity is varied in two cases corresponding to 1keV kinetic and 2.5eV thermal energy and 100eV kinetic and 2.5eV thermal energy respectively. Argon gas has been used in the PCE-gun for all the operating conditions. The neutral gas is kept at room temperature and operating pressure is kept 60 Pa. The applied potential at the anode is varied from 0 to 10 keV to analyse its effect on the peak anode current.

3. Simulation Results

The variation of the applied potential and anode current at different times is shown in figure 2. At the starting of the simulation, the anode current is very small but when the voltage starts penetrating inside the hollow cathode cavity, the potential difference between anode and cathode starts decreasing and the current starts increasing [5]. It has been observed that the peak amplitude of anode current changes with change in the hollow cathode dimensions.

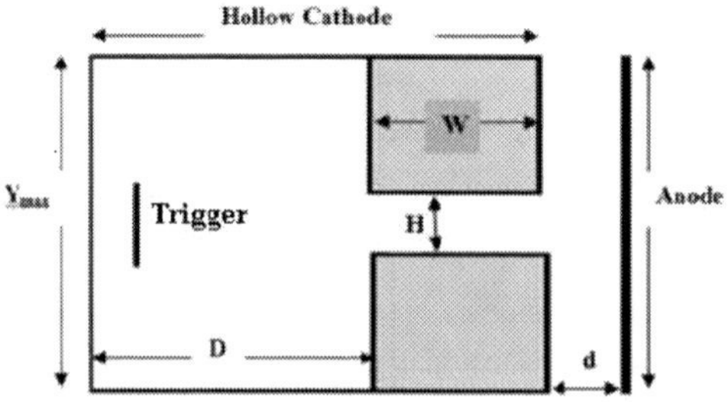

Figure 1. Simulation model for PCE Gun

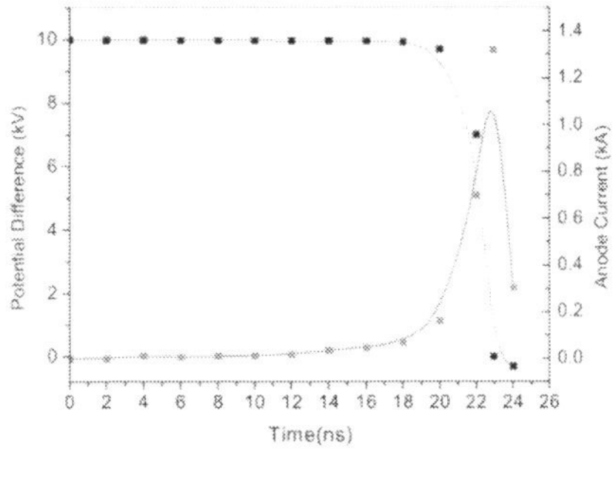

Figure 2. Variation of potential and anode current using VORPAL

Figure 3 shows voltage penetration inside the hollow cathode cavity. The applied electric field accelerates seed electrons which generate ions and also extract the electrons out of the hollow cathode, leaving behind a net positive charge region. This positive charge region grows inside the hollow cathode cavity and acts as a virtual anode. The electric field penetration inside the hollow cathode vicinity is due to the virtual anode formation.

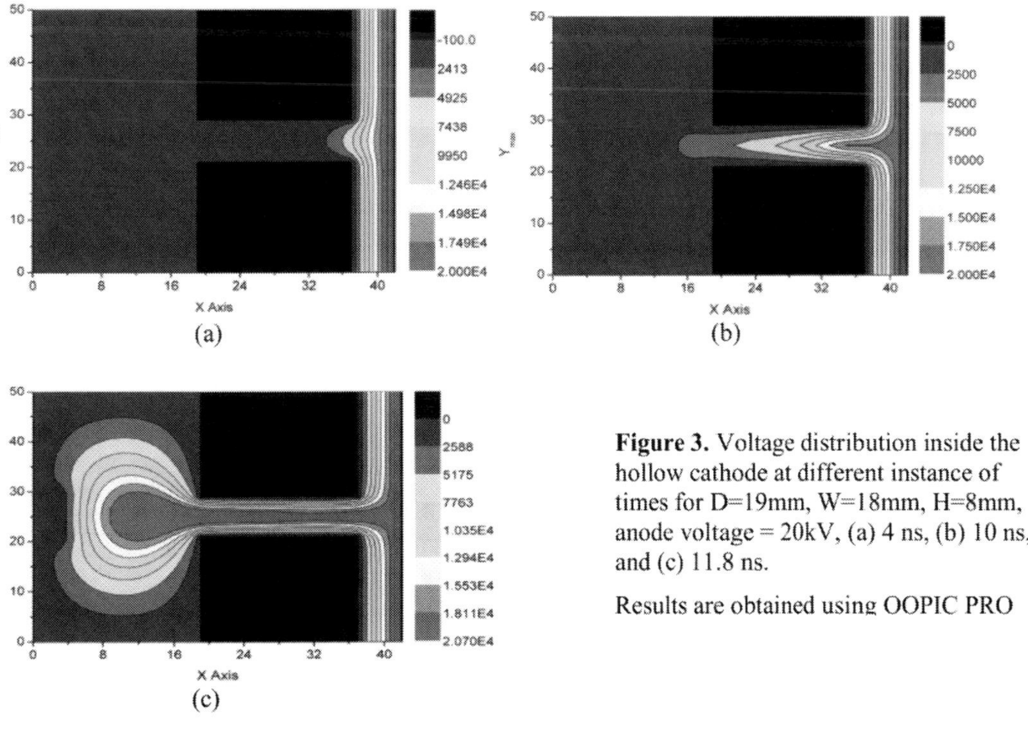

Figure 3. Voltage distribution inside the hollow cathode at different instance of times for D=19mm, W=18mm, H=8mm, anode voltage = 20kV, (a) 4 ns, (b) 10 ns, and (c) 11.8 ns.

Results are obtained using OOPIC PRO

The anode current variation with the velocity of the seed electron change is shown in the figure 4. It is found the more energetic seed electron energy leads more anode current. In fact, increase in the seed electrons energy generates denser plasma due to the more collisions between neutral atoms and electrons.

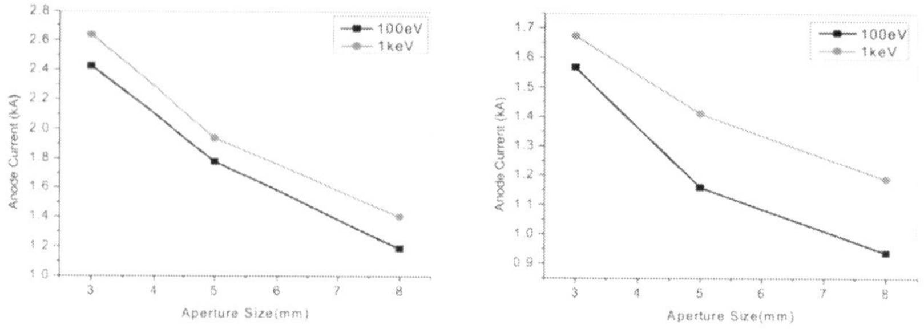

(a) (b)

Figure 4. Anode current at different energy of seed electron for D=32mm, W=5mm, (a) anode voltage=10 kV, (b) anode voltage= 20 kV

The variation of the peak anode current with aperture size of the hollow cathode at different dimensions of the PCE-gun is shown in figure 5. In this figure the results obtained by two codes are also compared which matches well within 10% variation. In all the cases the anode current amplitude decreases with increase in aperture size due to less discharge time and density.

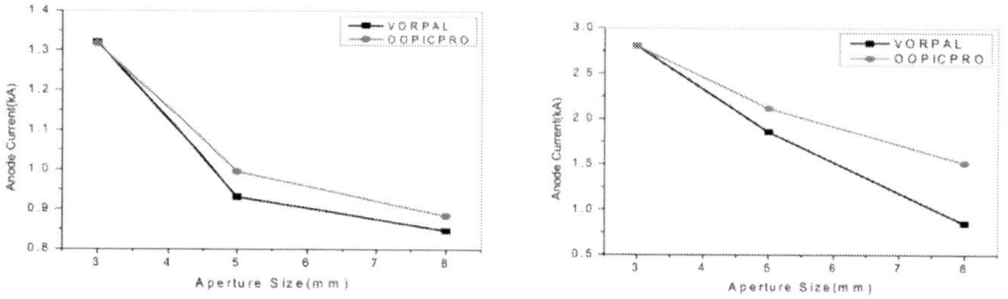

(a) (b)

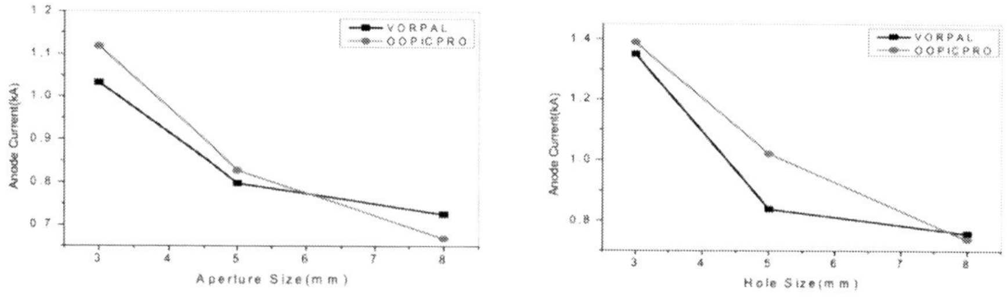

(c) (d)

Figure 5. Comparison of peak anode current for different aperture sizes by both the softwares OOPIC Pro and VORPAL (a) D=19 mm W=18 mm, anode voltage =10 kV, (b) D=19 mm W=18 mm, anode voltage =20 kV, (c) D=32mm W=5 mm, anode voltage =10 kV, and (d) D=32 mm W=5 mm, anode voltage =20 kV

4.Conclusion

The change in peak anode current by variation in the input parameters –like, seed electron velocity, hollow cathode depth, hollow cathode width, aperture size and anode voltage has been analysed by 2-D PIC simulation codes VORPAL and OOPIC-Pro. Both the codes show good agreement in the variation of the peak anode current. The virtual anode formation in the hollow cathode basically leads to penetrate the electric field inside the hollow cathode vicinity. It is found that the increase in seed electrons velocity generates denser plasma.

5.References

1. Gundersen M A and Schaefer G 1990 *NATO ASI Series B.* **219** 331
2. Cross A W, Yin H, He W, Ronald K, Phelps A D R and Pitchford L C 2007 *J.Phys.D: Appl. Phys.* **40** 1953
3. Yin H, Cross AW, He W, Phelps ADR and Ronald K 2004 *IEEE Trans on Plasma Science,* **32** 233
4. Gu X, Meng L, Sun Y and Yu X 2008 *Int J Infrared Milli Waves*, **29** 1032
5. Kumar N, Pal U N, Verma D K, Prajapati J and Srivastava V 2011 *International Journal of Engineering* **5** 4.
6. Goebel DM and Watkins R M 2000 *Rev. Sci. Instrum.* **71** 388
7. Frank K and Christiansen J 1989 *IEEE Trans. Plasma Sci.* **17** 748
8. Pal U N, Dubey V P, Barik M K, Lamba V, Verma D K, Kumar N, Kumar M, Meena B L, Tyagi MS and Sharma AK 2011 *XII IEEE International Vacuum Electronic Conference.*
9. Cetiner S O, Stoltz P and Messmer P 2008 *J. Applied Physics* **103** 023304

An overview of the undulator field measurement studies at insertion device lab, DAVV

[1]Mona Gehlot*, [1]G Mishra, [1]G Sharma, [2]J Hussain.
[1] School of physics, DAVV, Indore, India.
[2] Department of Applied Physics, RGPV, Bhopal, India.

E-mail: mona_gehlot@yahoo.co.in

Abstract. This paper describes undulator field measurement procedures by hall probe and the pulsed wire method. The Hall probe method gives the direct field mapping and calculates the field integrals through double numerical integration. A pulsed wire magnetic measurement method is an optical method which has been independently developed by DAVV. It gives directly the field integrals and calculates the field mapping via a double numerical differentiation. The measurement methods are illustrated with field measurement of a six period PPM undulator and a third harmonic undulator.

1. Introduction

Wigglers/undulators are the most used insertion devices in applied physics research in synchrotron radiation and free electron laser science. The fundamental issue in the design of a planar undulator/wiggler is to produce a periodic magnetic field with a high peak field in a given period and given gap. The performance of a synchrotron radiation source or a free electron laser depends on undulator properties which depend on several parameters such as the gap mismatch along the undulator length, uniform periods, precise length and precise undulator parameter, mechanical arrangement, transverse misalignments, construction errors, radiation damage and its end field configuration. The fabrication of a permanent undulator/wiggler is therefore a complicated process that involves selecting individual magnets, assembling the magnets and testing & correcting the assembled undulator. It is therefore required to obtain precise values of the undulator field parameters by precise and accurate measurements of the undulator to assess their performances.

There are several magnet measurement systems to characterize the undulator magnet. The simplest procedure with a hall probe is to sample the magnetic field at many points along the undulator and numerically calculating the quantities of interest.There is a pulsed wire system which gives directly the the first and second field integrals of the undulator field. In this paper we review the working principles of the Hall probe [1] and pulsed wire method [2] and discuss the novel features of both the measurement procedures and relevant physical consequences of the measurements.

2. Measurement procedure of Hall probe Method

The measurement of field integrals is straightforward in Hall probe method. The method makes use of the Lorentz force equation to express the field integrals as,

$$x'(z) = -\frac{e}{\gamma mc}\int_0^z B_u(z')dz' \tag{1}$$

$$x''(z) = -\frac{e}{\gamma mc}\int_0^z dz''\int_0^z B_u(z')dz' \tag{2}$$

Using a 1 D Hall Probe mounted on a three axis bench one builds up the magnetic field profile. In the standard practice, the Hall probe is moved continuously in steps for taking longitudinal measurements. The two field integrals are obtained from this data by a simple FORTRAN code using trapezoidal rule.

In terms of the practical parameters, $e/mc = 0.058\times10^4\ 1/T-m$.The calculation of the field integrals when multiplied by this factor gives the electron trajectory. The second field integral data in Eq.(2) in terms of the practical parameter can be converted to electron trajectory,

$$x_{tr}(z) = \frac{0.3}{E(Gev)}\int_0^z dz''\int_0^z B_u(z')dz' \tag{3}$$

3. Measurement procedure of Pulsed wire Method

The pulsed-wire technique is an alternate magnetic field measurement method for small gap undulators. In the pulsed wire method, a thin wire is stretched along the undulator axis. Under the combined action of the current and magnetic field, mechanical vibrations are generated which propagates along the wireThe transverse motion equation of the traveling acoustic wave in the wire is given by

$$\frac{\partial^2 x}{\partial z^2} - \frac{1}{v^2}\frac{\partial^2 x}{\partial t^2} = -\frac{B_u(z)i(t)}{T} \tag{4}$$

Where x is the displacement, $i(t)$ is the current in the wire at time t, $B_u(z)$ is the undulator magnetic field at z, T is the wire tension, the wave velocity v is given by

$$v = \sqrt{\frac{T}{\mu}} \tag{5}$$

μ is the mass per unit length of the wire. For a short current pulse, the wire displacement at sensor location is given by

$$x'(t) = -\frac{i_0\Delta t}{2\mu v}\int_0^z B_u(z')dz' \tag{6}$$

i_0 is the peak pulse current, Δt is the current pulse width. For a longer pulse, the wire displacement is given by

$$x''(t) = -\frac{i_0}{2\mu v^2}\int_0^z dz''\int_0^z B_u(z')dz' \quad . \tag{7}$$

Eqs. (7) and (8) gives the first integral and second integral of the undulator field strength respectively and can be determined at appropriate pulse widths.

An oscilloscope connected to the sensor reads the wire displacements in volts versus time. The output voltage can be translated into wire displacement by a calibration factor. To construct the calibration curve, measure the output voltage without applying current to the wire while the wire is moved over the laser output path typically in steps of $10\mu m$. From the output voltage versus position, identify a region where the output voltage is linear with the wire position and the slope of the calibration curve gives the sensitivity in volts versus wire displacement. This provides a suitable factor to express the wire displacement from the oscilloscope output in voltage. By determining the calibration factor helps to determine the second field integral as,

$$\int_0^z dz''\int_0^z B_u(z')dz' = x''(z)\frac{2\mu v^2}{i_0} \tag{9}$$

And with a similar procedure one calculates the first field integral. The second field integral can be differentiated to find the magnetic field profile.

4. A pulsed wire Measurement system at DAVV : The entire magnetic field measurement system [3] for the pulsed wire method. The optical sensor is the key component of the pulsed wire measurement system. In this set up each laser-photodiode pair is mounted on translation stages .The laser ,the wire and the photodiode are aligned to obtain approximately one half of the nominal voltage from the laser. A suitable sensitivity factor for a given wire diameter is chosen for best optimum results. The sensitivity of the optical sensor is defined by the voltage recorded in the oscilloscope with the wire displacement across the laser output.

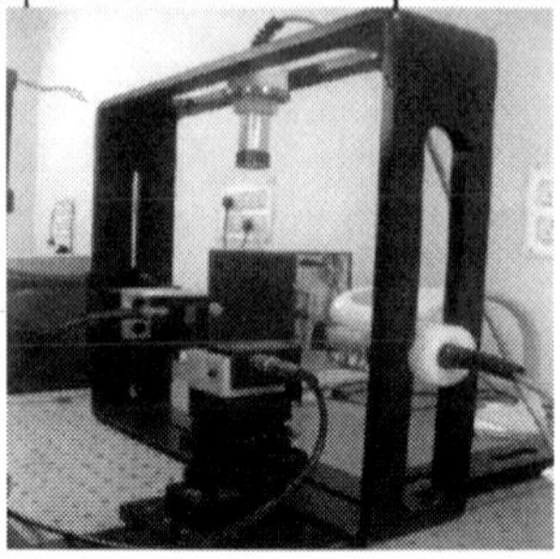

Figure 1 An optical sensor at DAVV

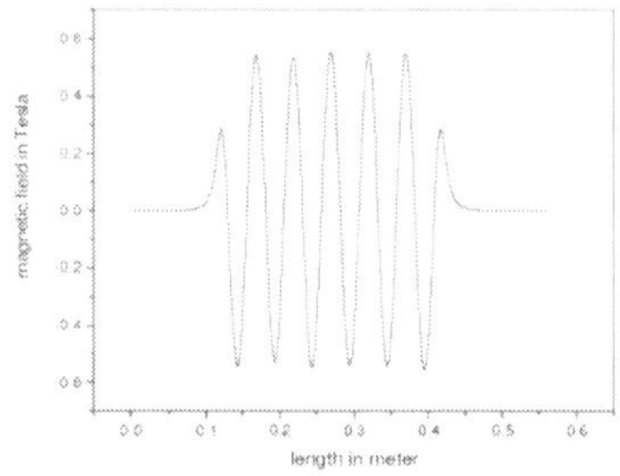

Figure2 Hall probe data for magnetic field profile

The sensor of the pulsed wire set up at DAVV as shown in Figure1 is an orthogonal pair of laser and photo diode movable on a rail .It consists of coherent's ULN series diode laser model No. 31-0144-000 modules for the laser which gives 5 mW power at $635nm$. The detector for the system is Newport's silicon biased detector model no. 818-BB-21. It has both rise time and fall time <300 ps and responsivity at $830nm$ is $0.4\frac{mA}{W}$.

5. Example

To illustrate the integral measurement procedures consider a typical magnetic profile of a six period undulator measured with the help of a Hall probe and represented in Figure 2.

One period of the undulator is 5 cm. The undulator is of PPM type with six periods. The first and second field integrals are obtained when integrated the data of the Figure 2 and represented in Figure3a),b) respectively. The field integrals in the pulsed wire method are determined by the use of appropriate pulse widths.

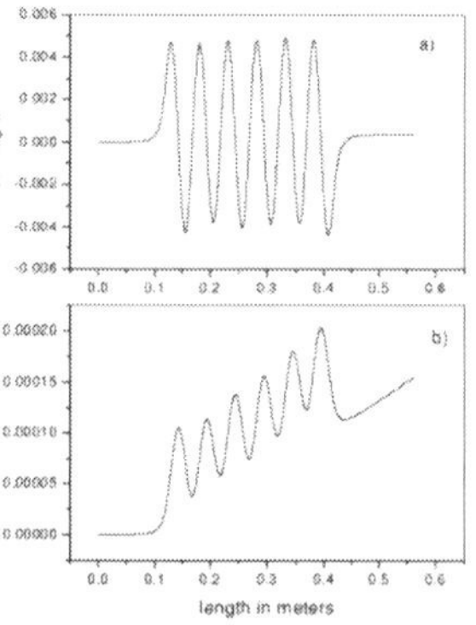

Figure3. first field (a) and second field integral (b) data from magnetic field profile.

A typical output of a oscilloscope connected with the PWM gives the data for the second field integral with a pulse width of 4.32 ms in Figure4. When this data is differentiated twice with above algorithm, the magnetic field profile of the undulator is reproduced in Fig 5.

6. Third harmonic undulator:We use CRGO shims for third harmonic undulatotor [4] of $1.5mm\times9.0mm\times28mm$ in thickness, height and length respectively. We placed shims at each joints of magnets in every period as shown in figure (6). The distance between the centres of the two shim groups is $12.5mm$.

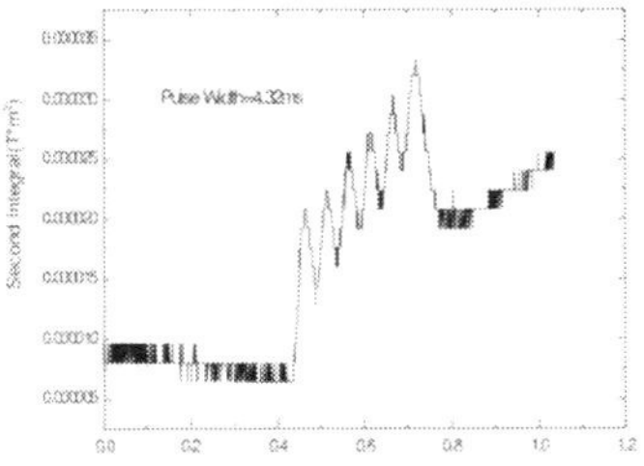

Figure 4 second field integral from Pulsed wire method

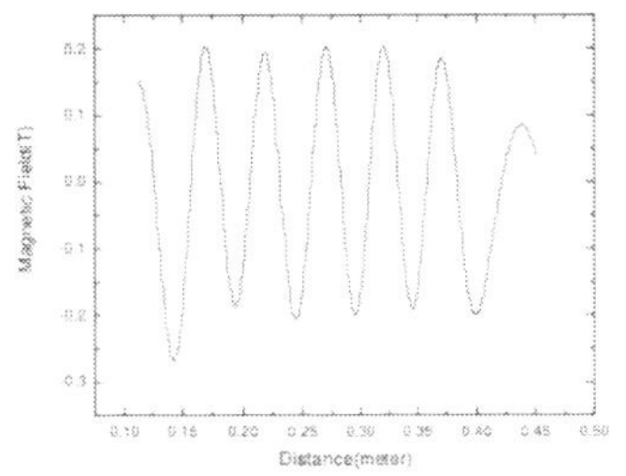

Figure 5 magnetic field profile from the Pulsed wire data

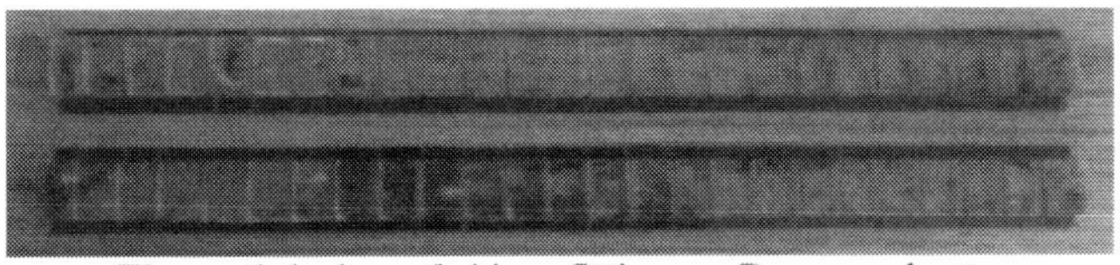

Figure 6 design of shims fitting on Perspex sheets.

The substrate thickness is $2.0\,mm$. Figure (7) gives the hall probe measurement of the magnetic field of the modified undulator .The meaurements are done at shims gaps of $7.0\,mm$. By shim gap we mean the spacing between the shims that are placed opposite to each other. Figure (8) shows the modification ratio of computed by Fourier transforming the measured field profile. The third harmonic component with a shim gap of $7.0\,mm$ is $\approx 18\%$ of the fundamental where as the fifth harmonic is $\approx 9\%$ of the fundamental.

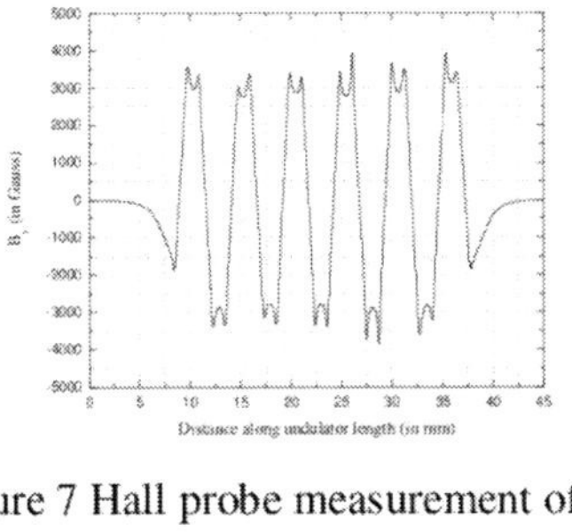

Figure 7 Hall probe measurement of third harmonic undulator

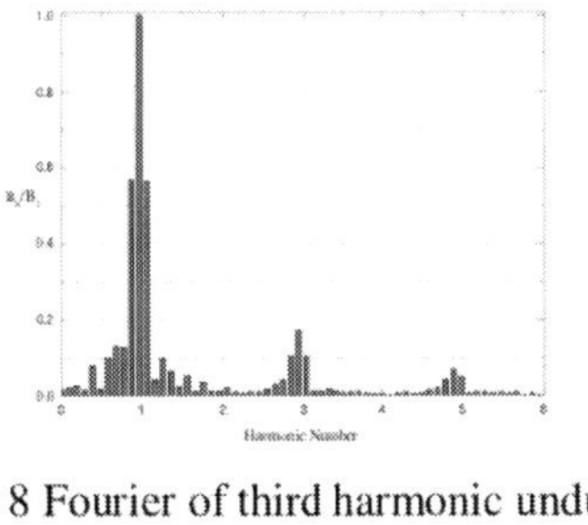

Figure 8 Fourier of third harmonic undulator

5. Conclusions

We have discussed undulator field measurement procedures, numerical algorithms for the pulsed wire set up at DAVV. The results of the measurement of a six period undulator are presented in this set up. Lists of error sources are identified. The results show that the pulsed wire parameters are required to be optimized for matching the results with the hall probe data. The third harmonic undulator is very well characterised.

References

1. R. S. Popovich, 9th International Magnetic measurement workshop, CEA/Saclay, June, 1995.
2. R.W. Warren, Nuclear Instruments and Methods in Physics Research A 272(1988), p-257.
3. Sumit Tripathi, Mona Gehlot, Jeeva Khan Hussain, G.Mishra, Sanjay Chouksey, Vinit Kumar, Umesh Kale, Pravin Nerpagar, IL Nuovo Cimento, B, Vol.124, No.3, 2009.
4. Nakao Naoya et.al,Jpn. J. App. Phy.vol 37(1998) .

Journal of Physics: Conference Series **365** (2012) 012053 doi:10.1088/1742-6596/365/1/012053

Gravitational instability of radiative plasma with finite larmor radius corrections

Sachin Kaothekar[1] and R K Chhajlani
School of Studies in Physics, Vikram University, Ujjain -456010, M.P., India.
[1]Mahakal Institute of Technology and Management, Ujjain-M.P., India.

E-mail: sac_kaothekar@rediffmail.com

Abstract. The effect of radiative heat-loss function and finite ion Larmor radius (FLR) corrections on the gravitational instability of infinite homogeneous viscous plasma has been investigated incorporating the effects of thermal conductivity and finite electrical resistivity. We find that radiative heat-loss function and thermal conductivity modify the Jeans criterion into radiative instability criterion. Numerical calculations have been performed to show the effect of various parameters on the growth of the gravitational instability.

1. Introduction

Plasma instability is one of the most fascinating fields of physics. It is a well established fact that gravitation plays a key role in understanding the process of fragmentation of interstellar medium, formation of stars, planets, asteroids, comets and other astrophysical objects. Shaikh et al. [1] have investigated the gravitational instability of thermally conducting plasma in a variable magnetic field. Recently Dhiman and Dadwal [2] have discussed the gravitational instability of a non-uniform rotating heat-conducting medium. Janaki et al. [3] have investigated the Jeans instability in a viscoelastic gravitating fluid. Along with this, the recent observations from solar corona and interstellar medium have shown that the radiative heat-loss mechanism plays a significant role in the process of molecular clouds condensation and in star formation, in connection with thermal instability. Bora and Talwar [4] have investigated magneto-thermal instability with generalized ohm's law. More recently Prajapati et al. [5] have investigated the self-gravitational instability of rotating plasma with arbitrary radiative heat-loss function. In addition to this, in many plasma situations such as in magnetic reconnection, solar corona and interstellar plasmas the FLR plays a key role in stability investigations. Vaghela and Chhajlani [6] have investigated the stabilizing effect of FLR on magneto-gravitational stability of resistive plasma with thermal conduction. Recently Devlen and Pekunlu [7] have investigated the effects of FLR on weakly magnetized, dilute plasmas. Thus in the present work, we have examined the effect of radiative heat-loss function, FLR corrections, thermal conductivity and finite resistivity on gravitational instability of plasma. The above work is applicable to structure formation in interstellar medium.

2. Linearized perturbation equations of the problem

Let us consider an infinite homogeneous, gravitating, radiating, thermally conducting, viscous plasma of finite electrical resistivity in the presence of magnetic field H (0, 0, H). The linearized perturbation equations of the problem with these effects are written as –

$$\frac{\partial \mathbf{u}}{\partial t} = -\frac{1}{\rho}\nabla \delta p - \frac{\nabla \cdot \mathbf{P}}{\rho} + \nabla \delta U + \frac{1}{4\pi\rho}(\nabla \times \mathbf{h}) \times \mathbf{H} + \upsilon\left[\nabla^2 \mathbf{u} - \frac{\mathbf{u}}{K_1}\right], \tag{1}$$

$$\frac{\partial \rho}{\partial t} = -\rho \nabla . \mathbf{u}, \tag{2}$$

$$\frac{1}{\gamma - 1}\frac{\partial \delta p}{\partial t} - \frac{\gamma}{\gamma - 1}\frac{p}{\rho}\frac{\partial \delta \rho}{\partial t} + \rho(L_\rho \delta\rho + L_T \delta T) - \lambda \nabla^2 \delta T = 0, \tag{3}$$

$$\frac{\delta p}{p} = \frac{\delta T}{T} + \frac{\delta \rho}{\rho}, \tag{4}$$

$$\frac{\partial \mathbf{h}}{\partial \mathrm{t}} = \nabla \times (\mathbf{u} \times \mathbf{H}) + \eta \nabla^2 \mathbf{h}, \tag{5}$$

$$\nabla . \mathbf{h} = 0, \tag{6}$$

$$\nabla^2 \delta U = -4\pi G \delta \rho. \tag{7}$$

where $p, \rho, \upsilon, L_T, L_\rho, \lambda, T, \eta, c, G$ and γ are the fluid pressure, density, kinematic viscosity, temperature dependent heat-loss function, density dependent heat-loss function, thermal conductivity, temperature, electrical resistivity, sound velocity, universal gas constant and ratio of two specific heats. We seek plain wave solution of the form

$$e^{i(k_x x + k_z z + \sigma t)}, \tag{8}$$

where σ is the frequency of perturbation and k_x, k_z are the components of the wave vector $\boldsymbol{k}$, in x, z directions so that $k^2 = k_x^2 + k_z^2$.

3. Dispersion relation and discussion

The substitution of equation (8) into equations (1) - (7) gives the general dispersion relation.

$$\begin{aligned}
&[\omega^2 + \omega\Omega_\upsilon + \Omega_T^2]\left[\omega + \Omega_\upsilon + \frac{\mathrm{V}^2 k^2}{d}\right]\Big\{[\omega + \Omega_\upsilon]\left[\omega + \Omega_\upsilon + \frac{\mathrm{V}^2 k_z^2}{d}\right] + 4\upsilon_0^2 k_x^2 k_z^2\Big\} - \frac{2\upsilon_0^2 k_x^2 k_z^2}{k^2}\Omega_T^2\Big[\omega + \Omega_\upsilon \\
&+ \frac{\mathrm{V}^2 k^2}{d}\Big](k_x^2 + 4k_z^2) + [\omega + \Omega_\upsilon][\omega^2 + \omega\Omega_\upsilon + \Omega_T^2]\upsilon_0^2(k_x^2 + 2k_z^2)^2 + 2\upsilon_0^2 k_x^2 k_z^2 (k_x^2 + 2k_z^2)\frac{V^2}{d}\Omega_T^2 - \frac{\upsilon_0^2 k_x^2}{k^2}\Omega_T^2(k_x^2 + 2k_z^2) \\
&\times [\omega + \Omega_\upsilon](k_x^2 + 4k_z^2) - \frac{V^2 k_x^2}{d}\Omega_T^2[\omega + \Omega_\upsilon]\left[\omega + \Omega_\upsilon + \frac{\mathrm{V}^2 k_z^2}{d}\right] - 4\upsilon_0^2 k_x^4 k_z^2 \frac{V^2}{d}\Omega_T^2 = 0.
\end{aligned} \tag{9}$$

In order to examine the effects of various parameters on the condition of gravitational instability we consider two special cases of propagation:

3.1. Longitudinal mode of Propagation (k||H)

In this case the perturbations are taken parallel to the direction of the magnetic field (*i.e.* $k_x = 0,\ k_z = k$) the dispersion relation becomes

$$[\omega + \Omega_\upsilon] \times \left\{\left[\omega + \Omega_\upsilon + \frac{\mathrm{V}^2 k^2}{d}\right]^2 + 4\upsilon_0^2 k^4\right\}\left[\omega^2 + \omega\Omega_\upsilon + \frac{\Omega_I^2 + \omega\Omega_J^2}{B + \omega}\right] = 0. \tag{10}$$

Equation (10) has three independent factors. The first factor $[\omega + \Omega_\upsilon] = 0$, represents a stable damped mode modified by the presence of viscosity and permeability of the medium. The second factor gives the fourth degree dispersion relation in ω, this is an Alfven mode which is modified by the presence of FLR corrections, viscosity, permeability and finite electrical resistivity of the medium.

The third factor gives the third degree dispersion relation in ω, which is modified by the presence of radiative heat-loss function, self-gravitation, thermal conductivity, viscosity and permeability. The condition of instability obtained from the constant term is given as

$$\left\{k^2(\gamma-1)\left(TL_T-\rho L_\rho+\frac{\lambda k^2 T}{\rho}\right)-4\pi G\rho(\gamma-1)\left(\frac{T\rho L_T}{p}+\frac{\lambda k^2 T}{p}\right)\right\}<0, \tag{11}$$

From equation (11) we see that the fundamental Jeans criterion of gravitational instability is modified into radiative instability criterion. Thus the dispersion relation and growth rate is modified due to FLR corrections, viscosity and permeability. The condition of instability (11) can be solved to yield the following critical wave number.

$$k_{J1}^2=\frac{1}{2}\left\{\left(\frac{4\pi G\rho}{c'^2}+\frac{\rho^2 L_\rho}{\lambda T}-\frac{\rho L_T}{\lambda}\right)\pm\left[\left(\frac{4\pi G\rho}{c'^2}+\frac{\rho^2 L_\rho}{\lambda T}-\frac{\rho L_T}{\lambda}\right)^2+\frac{16\pi G\rho^2}{\lambda c'^2}L_T\right]^{1/2}\right\}. \tag{12}$$

The medium is unstable for wave number $k<k_{J1}$. Here it may be noted that the modified critical wave number involves the derivatives of temperature dependent, density dependent heat-loss functions and thermal conductivity of the medium.

In absence of thermal conductivity and radiative heat-loss function $(\lambda=L_T=L_\rho=0)$ the condition of instability and critical Jeans wave number is given as

$$(c^2k^2-4\pi G\rho)<0, \tag{13}$$

$$k_J^2=4\pi G\rho/c^2. \tag{14}$$

The medium is unstable for $k<k_J$, where k_J is critical Jeans wavenumber. On comparing equations (11) and (13) we conclude that inclusion of radiative heat-loss functions and thermal conductivity changes the fundamental criterion of Jeans gravitational instability into radiative instability criterion.

3.2 Transverse mode of propagation (k||H)

In this case the perturbations are taken perpendicular to the direction of the magnetic field (*i.e.* $k_x=k,\ k_z=0$). The dispersion relation becomes

$$\{\omega+\Omega_\upsilon\}^2\left[\{\omega+\Omega_\upsilon\}\left\{\omega^2+\omega\,\Omega_\upsilon+\frac{\Omega_I^2+\omega\Omega_J^2}{B+\omega}+\frac{\omega V^2k^2}{d}\right\}+\omega\upsilon_0^2k^4\right]=0. \tag{15}$$

Equation (15) has two independent factors. The first factor is discussed earlier. The second factor gives the fifth degree dispersion relation in ω, which is modified by the presence of radiative heat-loss function, FLR corrections, self-gravitation, thermal conductivity, finite electrical resistivity, viscosity, permeability and magnetic field. The condition of instability obtained from the constant term of the dispersion relation is same as discussed in equation (11).

In absence of viscosity $(\upsilon=0)$ the condition of instability is given as

$$\left\{\upsilon_0^2k^4(\gamma-1)\left(\frac{T\rho L_T}{p}+\frac{\lambda k^2 T}{p}\right)+k^2(\gamma-1)\left(TL_T-\rho L_\rho+\frac{\lambda k^2 T}{\rho}\right)-4\pi G\rho(\gamma-1)\left(\frac{T\rho L_T}{p}+\frac{\lambda k^2 T}{p}\right)\right\}<0. \tag{16}$$

From equation (16) we conclude that FLR correction tries to stabilize the system. Also we see that inclusion of viscosity removes the effects of FLR correction from condition of instability.

In absence of viscosity and radiative heat-loss function $(\upsilon=L_T=L_\rho=0)$ the condition of instability is given as $\left(\upsilon_0^2k^4+c'^2k^2-4\pi G\rho\right)<0,$ (17)

from equation (17) we conclude that FLR have stabilizing effect on the instability of the system. On comparing equations (16) and (17) we conclude that condition of instability given by Vaghela and Chhajlani [6] is modified by inclusion of radiative effects.

In absence of viscosity and finite resistivity ($\upsilon = \eta = 0$) the condition of instability is given as

$$\left\{\left(\upsilon_0^2k^4+V^2k^2\right)\left(\frac{T\rho L_T}{p}+\frac{\lambda k^2 T}{p}\right)+k^2\left(TL_T-\rho L_\rho+\frac{\lambda k^2 T}{\rho}\right)-4\pi G\rho\left(\frac{T\rho L_T}{p}+\frac{\lambda k^2 T}{p}\right)\right\}<0. \qquad (18)$$

From equation (18) we conclude that FLR correction and magnetic field try to stabilize the system. Thus from equation (18) we see that condition of instability given by Bora and Talwar [4] excluding electron inertia in that case is modified by inclusion of finite Larmor radius corrections in our case. Thus the present results are the improvement of Bora and Talwar [4]. Numerical calculations were performed for equation (15) by making it dimensionless in terms of self-gravitation to determine the roots of ω^*, as a function of wave number k^* taking $\gamma = 5/3$. Out of the five modes only one mode is unstable for which the calculations are presented in Figures 1-4, where the growth rate ω^* (positive real value of ω) has been plotted against the wave number k^* to show the dependence of the growth rate on the different physical parameters.

It is clear from figure 1 and figure 3 that growth rate decreases with increasing temperature dependent heat-loss function and thermal conductivity. Thus the effect of temperature dependent heat-loss function and thermal conductivity is stabilizing, where as from figure 2 we conclude that growth rate increases with increasing density dependent heat-loss function. Thus the effect of density dependent heat-loss function is destabilizing. One can observe from figure 4 that the growth rate decreases with increase in FLR corrections. Thus the effect of FLR correction is stabilizing.

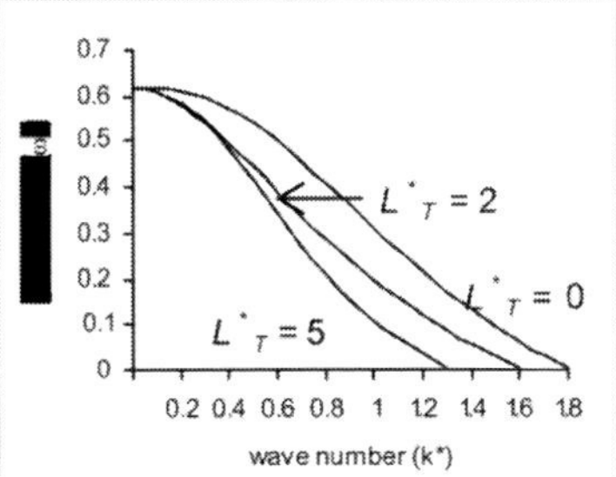

Figure 1. The normalized growth rate v/s normalized wavenumber with variation in temperature dependent heat-loss function L_T^*.

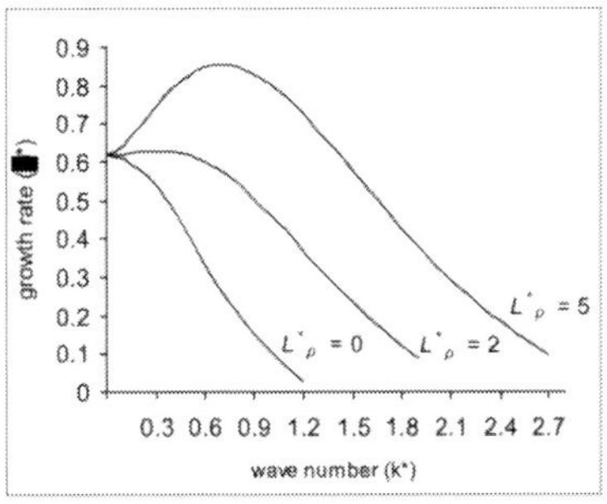

Figure 2. The normalized growth rate v/s normalized wavenumber with variation in density dependent heat-loss function L_ρ^*.

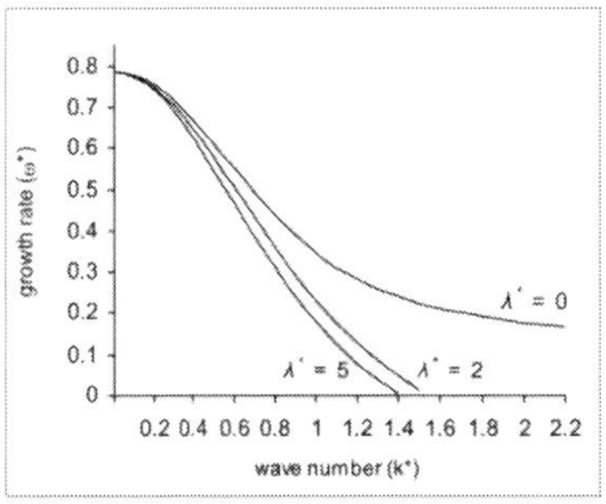

Figure 3. The normalized growth rate v/s normalized wavenumber with variation in thermal conductivity λ^*.

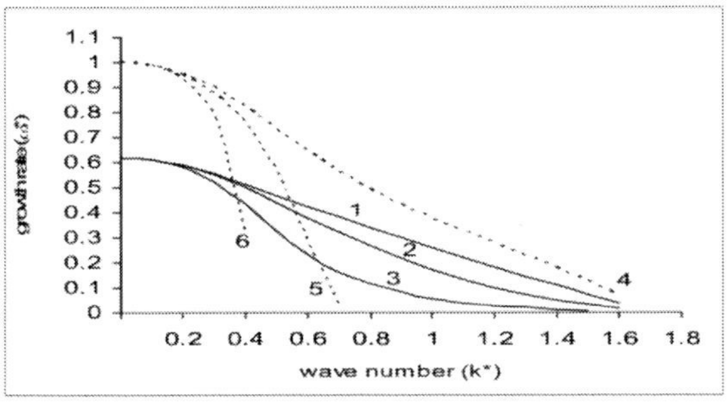

Figure 4. The normalized growth rate v/s normalized wavenumber with variation in FLR corrections υ_0^*

______ , $\upsilon^* = 1.0$: (1) $\upsilon_0^* = 0$, (2) $\upsilon_0^* = 2$ (3) $\upsilon_0^* = 5$.

-------- , $\upsilon^* = 0$: (4) $\upsilon_0^* = 0$, (5) $\upsilon_0^* = 2$ (6) $\upsilon_0^* = 5$.

4. Conclusion

Thus in the present paper, the effects of radiative heat-loss function, FLR corrections, thermal conductivity, finite electrical resistivity and viscosity on gravitational instability of plasma is investigated. We find that Jeans criterion remains valid and gets modified because of radiative heat-loss function, FLR corrections, thermal conductivity and magnetic field. The effect of viscosity is found to stabilize the system in both the longitudinal and transverse mode of propagation.

References

[1] Shaikh S, Khan A and Bhatia P K 2007 *Astrophys. Space Sci.* **312** 35
[2] Dhiman J S and Dadwal R 2010 *Astrophys. Space Sci.* **325** 195
[3] Janaki M S, Chakrabarti N and Banerjee D 2011 *Phys. Plasmas* **18** 012901
[4] Bora M P and Talwar S P 1993 *Phys. Fluids B* **5** 950
[5] Prajapati R P, Pensia R K, Kaothekar S and Chhajlani R K 2010 *Astrophys. Space Sci.* **327** 139
[6] Vaghela D S and Chhajlani R K 1989 *Contrib. Plasma Phys.* **29** 77
[7] Devlen E and Pekunlu E R 2010 *Mon. Not. R. Astron. Soc.* **404** 830

A comparison of paraxial and extended paraxial approach for laser beam propagation in plasma

Prerana Sharma[1], **Shweta Jain**[1], **R.P.Sharma**[2]
[1]Physics Department, Ujjain Engineering College, Ujjain, M. P. - 465010, India
[2]Centre for Energy Studies, IIT, New Delhi – 110016, India

E-mail :preranaiitd@rediffmail.com

Abstract. The propagation of laser beam in the plasma has been investigated for two cases, first with imposing the restriction of paraxial approximation on the beam profile, and second relaxing the restriction on the profile up to a certain extent. Besides the standard ponderomotive nonlinearity the work also includes the nonlinearity that arises from the relativistic electron velocities. The expression for the laser beam width parameter has been derived by expanding the eikonal, dielectric constant and other relevant quantities up to the fourth power of r, therefore, the contribution of both r2 and r4 terms is incorporated in the present analysis. The dynamical equation governing the beam width parameter has been numerically solved for typical laser plasma parameters. The result shows that the focusing of the laser beams has been affected and becomes fast due to the contribution of the off-axial rays in the extended-paraxial region. The intensity of the laser beam has also been examined in the light of these two cases and a notable change is found in the magnitude in the laser beam intensity with modified-paraxial case as compared to the paraxial case.

1.Introduction

To explore maximum energy from the laser beam in all laser-matter interaction applications it is vital to understand the dynamics of laser beams as they propagate [1] through the plasma. The high power laser beam is capable of introducing various types of nonlinearities (thermal, pondermotive and relativistic) during its propagation in the plasma [2]. but both relativistic and ponderomotive nonlinearities are operative when time scales ranges from $\tau_{pe} < \tau < \tau_{pi}$, , here τ is laser pulse duration, τ_{pi} - ion plasma period, τ_{pe} - electron plasma period [3]. In this case, the relativistic change in the electron mass alters the plasma frequency; this in turn alters the dielectric properties of a plasma medium through the modification of the index of refraction [4]. Therefore, in the presence of an ultra intense Gaussian laser beam, the plasma gets depleted from the high field region to the low field region, (on account of the relativistic and pondermotive force), and hence a transverse (with respect to the direction of propagation) density gradient is established which affects the laser plasma coupling efficiency..

In addition to this, when the ultra intense laser beam propagates in the plasma it can lead to break-up into intense filaments [5], as it has fluctuations of intensity over its profile. Conventionally, solutions for the laser beams can easily be obtained in the paraxial approximation of the wave equation and it is assumed that the paraxial wave equation gives an accurate description for wave beams near the axis throughout the propagation. But it is important to mention here that once the laser beam gets filamented the paraxial

approximation breaks down in filament where the focusing intensity is above a certain level. In this situation paraxial approximation is less accurate and some corrections are necessary. The theory given by Akhmanov et al [6] have demonstrated that in the limit when eikonal is expanded only up to second power of r, the shape of the radial intensity profile remains unchanged. This approximation is inadequate when the radial electron density varies sharply with r in the near axis region and slowly afterwards.
In the present paper, we examined the propagation of laser beam by modifying the paraxial profile of the beam up to a certain level [7], when both relativistic and pondermotive nonlineraties are operative simultaneously. The higher order teams in the expansion of dielectric constant and eikonal have been taken into account. The laser intensity profile and other relevant quantities of plasma were expended up to the 4th power of radial distance r. The paper is organized as follows. In section 2 we have derived the expression for the effective dielectric constant of the plasma when relativistic and ponderomotive nonlinearities are operative. The dynamical equation governing the laser beam propagation has been derived in section 3 when restriction on the beam has been relaxed up to some extent and a brief discussion of results follows in section 5.

2. Effective dielectric constant of the plasma

Consider the propagation of a Gaussian laser beam of frequency ω0 along the z- direction. The initial intensity distribution of the beam is given by $E_0.E_0^* = E_{00}^2 e^{-r^2/r_0^2}$

where r is the radial coordinate of the cylindrical coordinate system and ro is its initial beam width. The dielectric constant of the plasma is given by $\varepsilon_0 = 1 - \omega_{p0}^2/\omega_0^2$

where ω_{p0} is the plasma frequency given by $\omega_{p0}^2 = 4\pi n_0 e^2/m_0\gamma$ (with e the charge of an electron, mo its rest mass and n_o the density of plasma electrons in the absence of laser beam) and relativistic factor is given by

$$\gamma = \left[1 + \frac{e^2 E\ E^*}{c^2 m_0^2 \omega_0^2}\right]^{1/2} \tag{1}$$

The above expression is valid, when there is no change in the plasma density. The relativistic ponderomotive force is given by [3] $F_p = -m_0 c^2 \nabla(\gamma - 1)$. Using the electron continuity equation and current density equation for second order correction in the electron density equation (with the help of ponderomotive force) and total density is given by [3]

$$n = n_0 + n_2 = n_0\left[1 + \frac{c^2}{\omega_{p0}^2}\left(\nabla^2\gamma - \frac{(\nabla\gamma)^2}{\gamma}\right)\right] \tag{2}$$

The effective dielectric constant of the plasma at frequency ω_0 is $\varepsilon = \varepsilon_0 + \phi(E_0.E_0^*)$ where $\phi(E_0.E_0^*) = \omega_{p0}^2/\omega_0^2(1 - n/n_0\gamma)$. Expending ε_0 around r = 0 by Taylor expansion, we write $\varepsilon = \varepsilon_f + \gamma(\omega_0)r^2$ where

$$\varepsilon_f = \varepsilon_0 + \frac{\omega_{p0}^2}{\omega_0^2}\left[-\frac{1}{2}\frac{e^2E_{00}^2}{m_0^2\omega_0^2c^2f_0^2} + \frac{c^2}{\omega_{p0}^2}\frac{e^2E_{00}^2}{\gamma_{r=0}^2 m_0^2\omega_0^2c^2r_0^2f_0^4}(\alpha_{00} - 1)\right] \tag{3a}$$

$$\gamma(\omega_0) = -\frac{\omega_{p0}^2}{\omega_0^2}\frac{e^2E_{00}^2k_0^2r_0^2}{2\gamma_{r=0}^3 m_0^2\omega_0^2c^2f_0^2}\left[(\alpha_{00} - 1) + \frac{c^2}{\omega_{p0}^2r_0^2f_0^2}(4\alpha_{02} + \alpha_{00})\right] \tag{3b}$$

all the symbols used are defined in the next section.

3. Laser beam propagation

The wave equation governing the vector potential of the laser in plasma is $\nabla^2 A + (\omega_0^2/c^2)\varepsilon A = 0$ and to solve this equation we write [6]

$$A = A_0(r,z)\exp(i\omega_0 t - ik_0(z + s_0(r,z)))$$

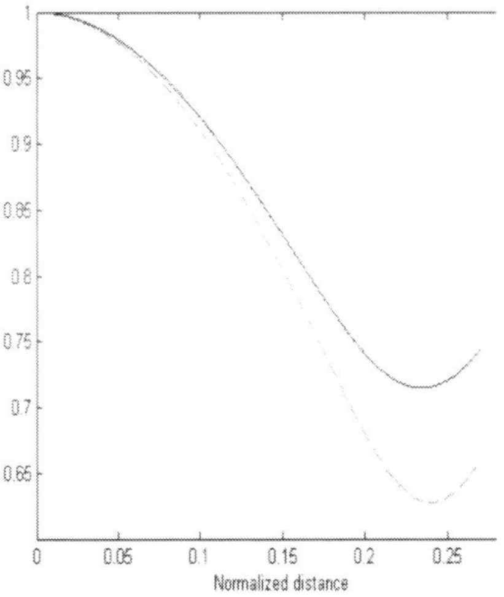

Figure 1. Variation in laser beam width parameter with normalized distance where red dashed line and solid black line is for on axis ($\alpha_{01}=\alpha_{02}=0$) and off-axis case ($\alpha_{01}\neq\alpha_{02}\neq 0$)

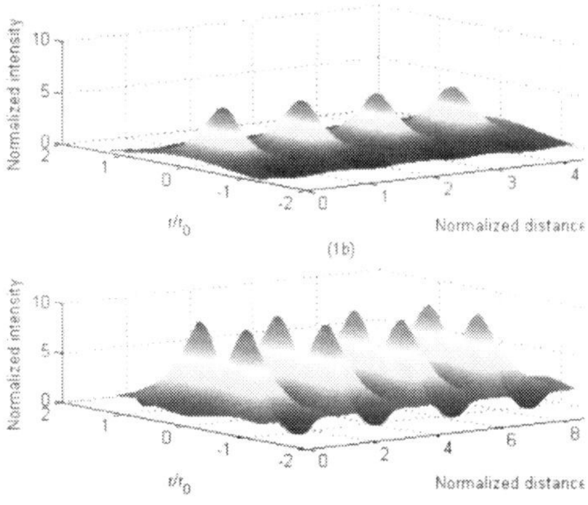

Figure 2. Variation of laser intensity with normalized distance and radial distance (a) in paraxial case ($\alpha_{00}=\alpha_{02}=0$) and (b) in modified-paraxial case ($\alpha_{00}\neq\alpha_{02}\neq 0$)..

Substituting for A in equation (5) and separating the real and imaginary parts, one obtains:

$$2\frac{\partial S_0}{\partial z}+\left(\frac{\partial S_0}{\partial r}\right)^2=\frac{\omega_0^2}{c^2k_0^2}\gamma_1(\omega_0)+\frac{1}{k_0^2A_0}\left(\frac{\partial^2A_0}{\partial r^2}+\frac{1}{r}\frac{\partial A_0}{\partial r}\right), \tag{4a}$$

$$\frac{\partial A_0^2}{\partial z}+A_0^2\left(\frac{\partial^2S_0}{\partial r^2}+\frac{1}{r}\frac{\partial S_0}{\partial r}\right)+\frac{\partial S_0}{\partial r}\frac{\partial A_0^2}{\partial r}=0, \tag{4b}$$

To solve the coupled equations, assuming the new solutions for A and eikonal S

$$A_0^2=\left(1+\frac{\alpha_{00}r^2}{r_0^2f_0^2}+\frac{\alpha_{02}r^4}{r_o^4f_0^4}\right)\frac{E_{00}^2}{f_0^2}e^{-r^2/r_0^2f_0^2}\ \text{and}\ S=\frac{S_{00}}{r_0^2}+\frac{S_{02}r^4}{r_0^4}\quad with\ S_{00}=\frac{r^2}{2f_0}\frac{df_0}{dz} \tag{5}$$

where E_{00} are the axial amplitude of the laser beam, α_{00},α_{02} are the coefficients of r2 and r4 respectively and S_{00}, S_{02} are the slowly varying function of r and z. Substituting the value of $\gamma(\omega_0)$ from equation (3b) and using equation (5) in equation (4a) and further equating the coefficients of r^2 and r^4 with the normalized distance $\xi=zc/\omega_0r_0^2$, the laser beam width parameter f_0 and equation for S_{02} comes out as

$$\frac{d^2f_0}{d\xi^2}=\frac{1}{f_0^3}\left(-3\alpha_{00}^2+g\alpha_{02}+1-2\alpha_{00}\right)-\frac{\omega_{p0}^2}{\omega_0^2}\frac{e^2E_{00}^2k_0^2r_0^2}{2\gamma^3m_0^2c^2f_0^2}\left[\left(\alpha_{00}-1\right)+\frac{c^2}{\omega_{p0}^2r_0^2f_0^2}\left(4\alpha_{02}+\alpha_{00}\right)\right] \tag{6}$$

$$\frac{\partial S_{02}}{\partial z}=\frac{-2\alpha_{02}}{k_0^2r_0^2f_0^6}-\frac{3\alpha_{02}\alpha_{00}}{2k_0^2r_0^2f_0^6}-\frac{3\alpha_{00}^2}{4k_0^2r_0^2f_0^6}+\frac{eE_{00}^2r_0^2}{m_0^2\omega_0^2c^2k_0^2\gamma^3f_0^6}\left(\frac{-\alpha_{00}+2\alpha_{02}}{r_0^4f_0^4}\right)+\frac{eE_{00}^2}{4m_0^2\omega_0^2k_0^2c^2\gamma^2}\left(\frac{16\alpha_{02}}{r_0^2f_0^4}-\frac{8\alpha_{00}}{r_0^2f_0^4}\right) \tag{7}$$

Using equation (5) into the imaginary part equation (4b) and equating the coefficient of r^2 and r^4 on both sides of the resulting equation, we obtain α_{00} and α_{02} as

$$\frac{\partial\alpha_{00}}{\partial z}=-\frac{16S_{02}f_0^2}{r_0^2}\quad and\quad \frac{\partial\alpha_{02}}{\partial z}=\frac{8S_{02}f_0^2}{r_0^2}-\frac{24\alpha_{00}S_{02}f_0^2}{r_0^2} \tag{8}$$

Equation (6) expresses the variation of laser beam width parameter f_0 with the distance of propagation. The nonlinearities considered here are relativistic and pondermotive. it is clear by equation (6) that f_0 depends on the coefficients (α_{00} and α_{02}) of r^2 and r^4 in extended-paraxial region. In order to have a numerical appreciation of contribution of off-axial rays in the beam width parameter (when both the relativistic and ponderomotive nonlinearity are taken into account), we have performed numerical computation of equation (6) with the help of equations. (7-8). The coupled equations have been solved for an initial plain wave front the initial conditions used here are $f_0 = 1$ and $df_0/dz = 0$ and $S_{00} = S_{02} = 0$ at z=0. The results are presented in the form of figure (1). In figure (1) beam width parameter has been plotted for both the cases: paraxial and modified paraxial. The results for paraxial case can be reproduced from equation (6) by substituting the coefficients $\alpha_{00} = \alpha_{02} = 0$. It is also observed, (by solving equation (6), results are not presented here) that in paraxial case, the intensity of laser beam is maximum at r=0 along the distance of propagation. It can clearly be observed from the figure (1) that focusing becomes faster in extended-paraxial case in comparision to paraxial case due to the participation of off-axis parts ($\alpha_{00} \neq \alpha_{02} \neq 0$).

The laser beam intensity is expressed by equation (5) and it is clear that intensity also depends on the coefficients α_{00} and α_{02}. To observe the effect of these coefficients on laser beam intensity, equation (5) has been solved numericaly with the help of equation (6-8). Figure (2) depicts the variation in laser beam intensity with normalized distance and radial distance.. It is clear from the figure (1) that as the laser beam propagates through the plasma it gets self focused / filamented, figure (1a) clearly expresses the generated filaments of the laser beam when the paraxial approximation is taken into consideration, (by substituting α_{00} and $\alpha_{02} = 0$ in equation (5) and equation (6)) . Figure (1b) shows the variation of the laser beam intensity including the contribution of off-axial rays. It can be clearly seen from figure (1b) that the filaments are splitted and the point at which the laser focuses is also altered. this result is very important to all laser plasma applications like laser induced fusion, lase based acceleraters as it affects the focusing locations of the laser beam. It is obvious from figure (1a) that in paraxial region the intensity of laser beams is maximum at r =0. While in extended paraxial region the laser intensity becomes minimum at r=0, and maximum at r = $\pm$0.45, as also verified analytically by equation (5). The following set of the laser beam parameters have been used in the numerical calculation: the vacuum wavelength of µthe laser beam (λ=1064nm), the initial radius of the laser beams (20µm), laser power flux (10^{18}W/cm^2) at plasma density $\omega_p = 0.3\omega_0$ and v_{th} = 0.1c.

Conclusion:

It is observed that in extended paraxial case the dynamics of the laser beam gets modified. it is observed that the filamentation process becomes faster in comparison to the case when the off-axial contribution is not taken into account. Due to the contribution of the off-axial part the filamentary structures of the laser beam gets splitted and it makes a significant difference in the coupling efficiency of the laser and plasma. The modified structures can also alter the intensity distribution, affecting the uniformity of energy deposition of laser beam in the plasma.This will be an important problem from theoretical point of view, particularly which involves laser beam propagation

Acknowledgment:

Authors are greatful to the organizing committee ICRTP and the work was partially supported by the MAPCOST, Bhopal India.

References

[1] Ying Wang and Zhongxiang Zhou 2011 *Phys. plasma* **18**, 043101
[2] Salih H A, Tripathi V K and Pandey B K *IEEE Trans plasma Sc.* 2003 **31** 304
[3] Brandi H S, Manus C, and Mainfray G 1993 *Phy.Rev. E*, **47** 3780
[4] Walia K and Singh A 2011 Contrib. Plasma Phys. **51** 375
[5] Sarri G Cecchetti C A, Jung R Hobbs P 2011 *Phys. Rev. Lett.* **106** 095001
[6] Akhmanov S A, Sukhorukov A P and Khokhlov R V 1968 *Sov. Phys. Usp.* **10** 609
[7] Faisal M and Sodha M S 2007 *phys plasmas* **14** 103103

Simulation study of resonant reflector for S-band BWO

Y Choyal[1], Nidhi Parmar[1], Ajay Kumar Saini[2], SK Chhotray[2], KS Bhat[2], Lalit Kumar[2]
[1]Devi Ahilya University, School of Physics, Khandwa Road, Indore 452001, India
[2]Microwave Tube Research and Development Center, BEL Complex, Jalahalli, Bangalore -560013, India

1ychoyal@yahoo.com

Abstract. This paper presents the result of simulation studies of resonant reflector used for reflection of backward wave in relativistic BWO. The resonant reflector is modelled and analyzed by CST MWS for TM_{01}. A TM_{01} mode is fed at the output end of the BWO and signal is observed at the cathode end. Results show that 90 percent of the backward TM_{01} wave is get reflected back by the locked TM_{02} mode in the resonant reflector.

1. Introduction

Relativistic backward wave oscillator (BWO) is a Cherenkov device used to generate high power microwave in centimeter and millimeter wave frequencies with power level ranging from 0.1-1GW [1-3]. It consists of sinusoidal corrugated periodic metallic structure which is fed by high current relativistic electron beam, confined by applying a strong axial magnetic field. In the device electron beam and electromagnetic wave travels in the opposite directions. As the electron beam passes through the periodic metallic structure it gives energy to one of the slow wave mode.

This backward wave suffers reflections at the beam entrance of the slow wave structure and becomes a forward wave. There are many techniques used by which this backward wave reflection occurs effectively. One of the techniques is to use a resonant reflector at the entrance of SWS where wave reflection from the reflector is similar to the reflection from open circuit [4]. G.G.Denison et al. have studied the problem of optimization of reflector dimensions using special software based on the scattering matrix method [5]. In this paper above problem is solved using CST MWS.

2. Theory and calculation

Resonant reflector is a piece of a cylindrical waveguide having a radius 'b' and length c (~ $\lambda/3$) such that a higher order mode which is having the same in azimuthal index is locked inside the waveguide section within the required frequency range. A waveguide with radius 'a' connected between reflector and SWS as shown in figure1.

At least one oscillation of TM_{02} mode exists over the range of wavelength $\chi 02/b < k < \chi 02/a$ where $k=2\pi/\lambda$ and $\chi_{02}=5.52$, is the second root of zeroth order Bessel function $J_0(x) = 0$, $x = r^*\chi_{02}/a$, $2\pi b/\lambda_r \sim 5.52$.

Under a zero diffraction loss approximation the resonance frequencies are derived from the equation $\cos(\varepsilon) = s^*\varepsilon$, where $s = 2/[(k_{\square 2}L(1-a^2/b^2)^{1/2}]$, $\quad = L(k^2 - k_{\square 2}{}^2)^{1/2}/2$, $k_{\square 2} = \chi_{02}/b$, L is the

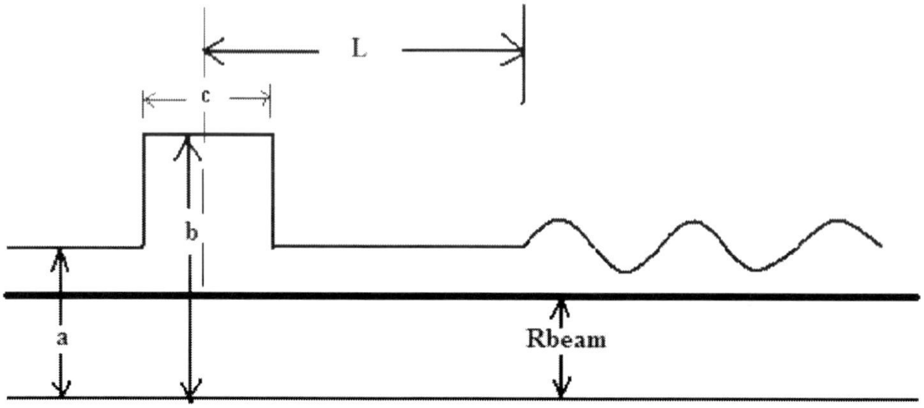

Figure 1 showing the Schematic diagram of the simulation geometry

distance between starting point of SWS and midpoint of the reflector (c/2) and ε is a dimensionless quantity which depends on L.For frequency range 2.8 GHz to 3.2 GHz, b ~94.1 mm and a~ 79.9 mm. These dimensions are further these optimized in CST MWS

3. CST simulation

CST is general purpose electromagnetic simulators based on finite integration technique (FIT) .we have performed the CST simulation study of the designed resonant reflector. Simulation configuration has been shown in the figure 2.A corrugated sinusoidal structure of 5 periods with resonant reflector at its entrance has been modelled in CST MWS.

Corrugated structure is described by the equation R(z)=Ro+ h*sin(kz/d) where Ro is the average radius of SWS, 'h' is the corrugation depth and 'd' is the period of corrugation.

A TM01 wave with frequency 3GHz fed at the port 1(right hand side of the structure) and transmitted signal is observed at port 2 (left side of the structure)a shown in figure 3(a) and 3(b). The lower the strength of transmitted signal indicates the higher effectiveness of the resonant reflector.

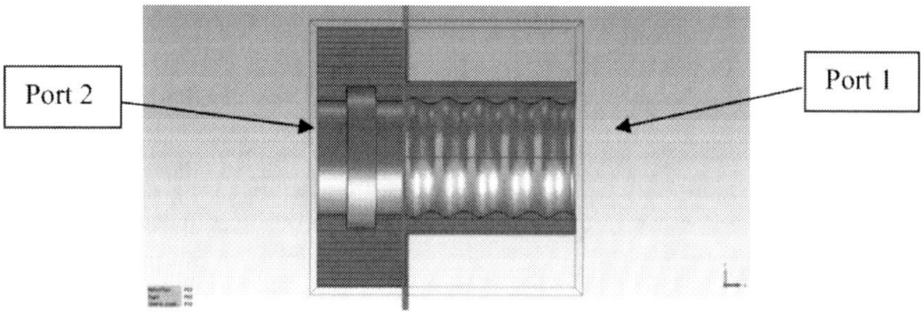

Figure 2. Simulation model of structure

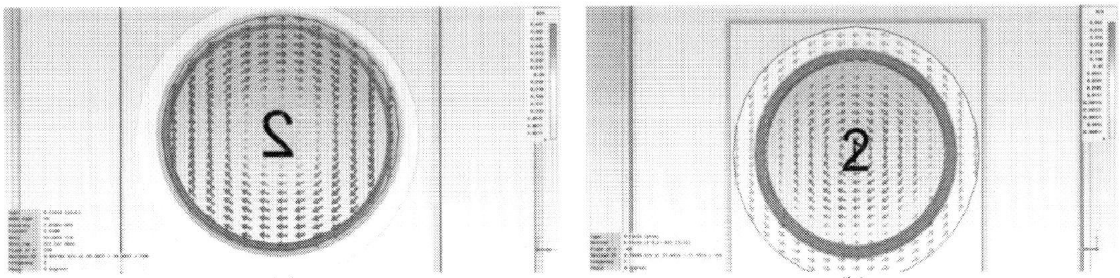

(a) (b)

Figure 3 (a) Field pattern of TM01 mode at port 1, (b) field pattern of TM_{02} mode inside reflector.

4. Simulation results

Optimization of the reflector dimensions has been done in two stages. In first stage outer radius of the reflector has been changed by keeping the inner radius constant. After many iterations maximum reflections has been found at b=93mm. Further keeping the reflector radius constant at b=93mm waveguide radius has been optimized.

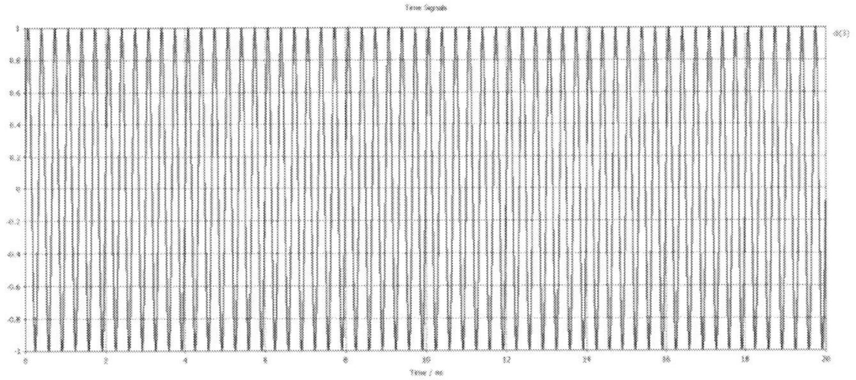

(a)

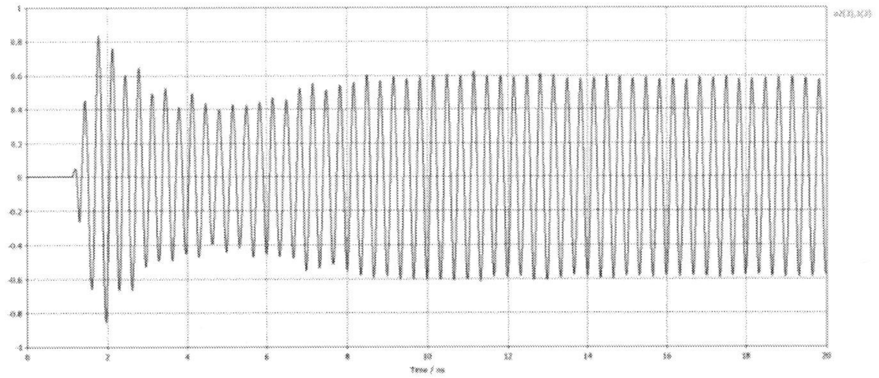

(b)

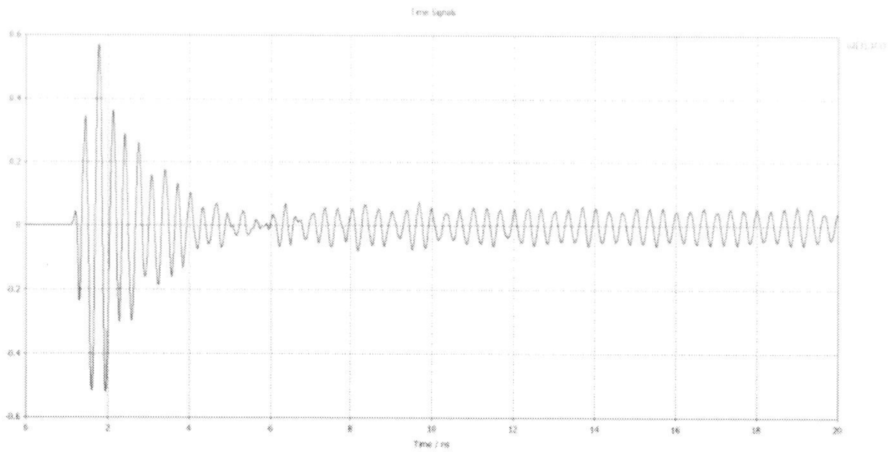

(c)

Figure 4 (a) Input signal of normalized amplitude 1 at port 1 (b) recieved signal at port 2 with b=93mm, a=82.4mm,L=60mm (c) recieved signal at port 2 with b=93mm, a=75.93mm,L=60mm

5. Conclusions

A resonant reflector was designed for an overmoded relativistic S-band BWO. It is optimized in CST MWS in two stages to get the maximum reflection of backward wave at the cathode end. In the first stage the radius of the resonant reflector has been optimized and maximum reflection at b=93mm is obtained. In the second stage radius of the waveguide has been optimized by keeping b constant. The result of the simulation demonstrate that maximum reflection occurs at b=93mm, a=75.93 and L=60mm for the present case.

References

[1] Gunin A V, Korovin S D, Kurkan I K, Pegel I V, Rostov V V and Totmeninov E M 1998 *proc.12th Int. conf. High-Power part. Beam,* **11** 849, 1998 *IEEE Trans. Plasma Sci.* **26,no-3** 326-331

[2] Gunin A V, Klimov A I, Korovin S D, Kurkan I K, Pegel I V, Polevin S D, Roitman A M, Rostov V V, Stepchenko A S and Totmeninov E M 1998 *IEEE Trans. Plasma Sci.,* **26,no-3**, 326

[3] Korovin S D, Rostov V V and Tot'meninov E M 2005 *Tech Phys. Lett.,* **31,no.5** 411

[4] Teng Y, Liu G, Shao H and Tang C 2009 *IEEE Trans. Plasma Sci.,* **37,no.6,**

[5] Denisov G.G, Lukovnikov D A and Samsonov S V 1995 *Inf. J. millimeter waves,* **16, no-4** 745

High Power Microwave Pulse Compression Using a Helical Slow-Wave Structure

Vasudha Rajput [1,2], N. Parmar[2], K.S.Bhat[3], K.P.Maheswari [4], Y.Choyal[2]
[1]Department of Physics, M.B. Khalsa College, Indore-452001, India
[2]School of Physics, Devi AhilyaVishwavidyalaya, Indore-452001, India
[3]Microwave Tube Research and Development Centre, Defence Research and Development Organization, Bangalore 560013, India
[4]Department of Physics, Vardhman Mahaveer Open University, Kota,-324006, India

Abstract. One interesting method of microwave pulse compression is passive–frequency-based microwave pulse compression where a dispersive medium is used to compress a frequency swept pulse. Oversized circular waveguide having helical corrugation supports eigenmodes that are suitable for pulse compression for frequency modulated input pulse is as shown by Brat et.al. [IEEE Trans. Plasma Sci., vol. 33, no.2, pp. 661–667, 2005.]. In this paper, theoretical studies on microwave pulse compression in helically corrugated waveguide is presented. The numerical results show a eleven fold peak power compression.

1.Introduction

Ultra high power short microwave pulses have been produced in recent years due to its wide range of applications [1]. These high power pulses can be produced by microwave pulse compression. One of the common methods to achieve necessary power levels is to use electron beam -wave interaction to amplify the microwave pulse to higher powers. To further increase the peak output power, an economically viable method is to compress a long duration lower power pulse into short duration higher power pulse. The principles and method of pulse compression vary depending on their application [2].

Compression of smoothly frequency modulated electromagnetic pulses in dispersive medium is well known and actively used in microwave electronics and laser physics. Hollow metal waveguides representing dispersive media are attractive for the microwave pulse compression because of their capabilities of handling very high power, relative compactness and simplicity. In a dispersive medium, the group velocity of any wave propagating through it is dependent on the frequency of the wave. Therefore, if a microwave pulse is produced in which the wave is swept from a frequency with a low-group velocity to a frequency with a high-group velocity, the tail of the pulse will move to overtake the front of the pulse, resulting in pulse shortening and a corresponding growth in amplitude if the losses are sufficiently small. If a cylindrical waveguide is used as the dispersive medium, the optimum region of operation is where the largest change in-group velocity with frequency occurs, which is near cut-off for a smooth waveguide.

One of the solutions is to use a helically corrugated waveguide, which was previously studied intensively for use as an interaction region for a gyro-TWA [3]. The helical wall perturbation can provide selective coupling between a higher and a lower circularly polarized mode avoiding the Bragg reflection zones, which would inevitably appear in the case of an axisymmetric corrugation because of

Published under licence by IOP Publishing Ltd

coupling between forward and backward propagating spatial harmonics. This coupling results in an eigenmode where the dispersion characteristics of one mode gradually transform into the other. If the parameters of the corrugation are chosen correctly, this can give the eigenwave a group velocity that decreases with frequency, and avoids regions with zero or negative group velocity, in its operating bandwidth, which is far from cut-off.

As the helically corrugated waveguide operates far from cutoff, the compressor provides much lower reflections at its input than the smooth circular waveguide. This allows the compressor to be used at the output of a high-power Cherenkov TWT. The second advantage is that the optimum frequency sweep in a helical compressor is from a high frequency to a lower frequency, and can be controlled by the corrugation parameters. This makes the helically corrugated waveguide compressor attractive for use at the output of a powerful relativistic BWO as this would require a beam voltage that decays with time for compression to take place [4].

This paper deals with the theoretical study to obtain a multigigawatt radiation power, which is based on passive compression of the microwave pulse compression generated in helically corrugated waveguide. Section 2 presents the general principles and relationships determining compression of frequency modulated pulses and considers the features of a dispersive medium in the form of a helically corrugated waveguide. Section 3 presents results and comparison to earlier experimental results.

2. Dispersion relation and pulse compression

In this section, the dispersion characteristics of the helically corrugated waveguides are presented. Considered a hollow metallic waveguide with inner wall radius $R(z,\theta) = R_0 + h\cos(k_0 z + q_0\theta)$. Here $k_0 = 2\pi/z_0$, z_0 being the periodicity in axial direction is while $\varphi_1 = 2\pi/q_0$ is the periodicity in azimuthal direction and q_0 is the order of folds of helical corrugation.

In a periodically corrugated waveguide, the electromagnetic field can be represented as a superposition of the spatial harmonics. Using the boundary conditions that the tangential component of electric field must vanish on the helical boundary, we get a dispersion relation, to be published elsewhere, whose solution is:

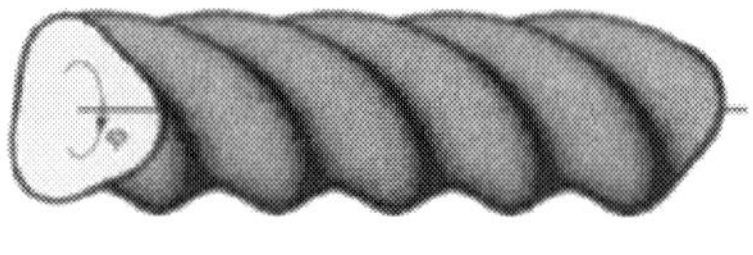

Figure. 1 Schematic view of a waveguide with a three-fold right-handed (q=3) helical corrugation.

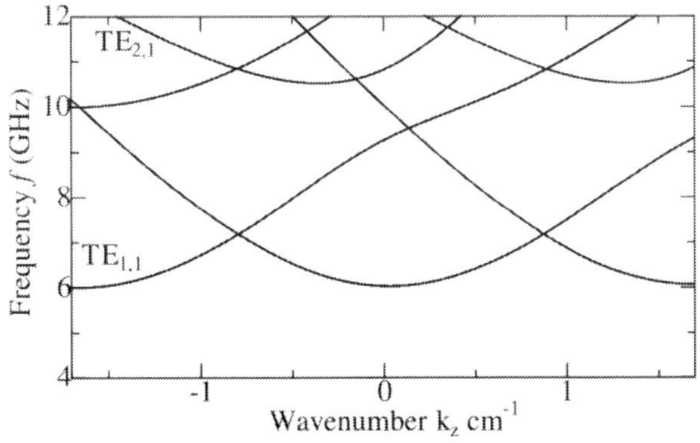

Figure. 2 Dispersion diagram for three-fold helically corrugated waveguide: Mean radius R_0=1.47cm, corrugation period z_0=2.89, and amplitude h=0.19cm.

We assume that a quasi-monochromotic pulse at the input of the waveguide compressor. The shape of the pulse envelope is close to a Gaussian, which is given as

$$u(z,t)=\frac{1}{2\sqrt{2\pi}}\int_{-\infty}^{\infty}A(\omega)\exp\left[i\left(k_z t-\omega t\right)\right]d\omega \tag{1}$$

It is easy to show that the amplitude $A(\omega)$ is given by

$$A(\omega)=\frac{1}{\sqrt{2\pi}}\int_{-\infty}^{\infty}e^{i\omega t}\left[u(0,t)-\frac{i}{k_z(\omega)}\frac{\partial u}{\partial z}(0,t)\right]dt \tag{2}$$

We assume

$$u(0,t)=S(t)\sin\left[\left(\omega_0-\omega_0\beta t\right)t\right] \tag{3}$$

as the initial shape of the pulse. Here shape function $S(t)$ has been taken as Gaussian as well as a step function. The parameter β is the measure of frequency variation with time. Substituting Eq. (3) in Eq. (2) we get $u(z,t)$. The numerical results are given in next section.

3. Results

In this section, we examine the operation of helically corrugated waveguide for given set of parameters: Mean radius R_0=1.47cm, corrugation period z_0=2.89, corrugation amplitude h=0.19cm, three fold periodicity q_0=3, and β=0.01 ns^{-1}.

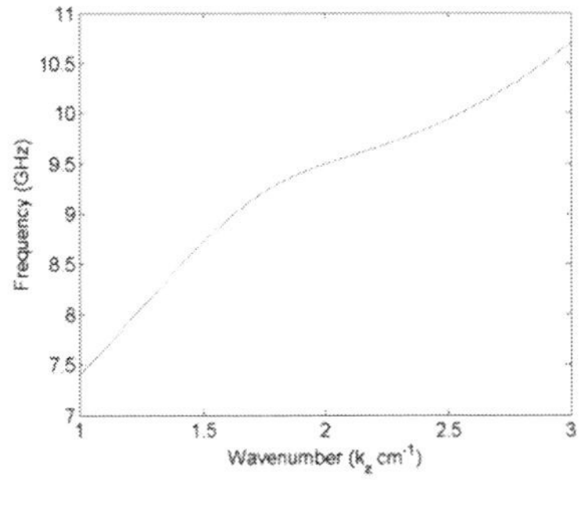

Figure. 3 Dispersion diagram of the helical waveguide. Central frequency 9.3 GHz is considered

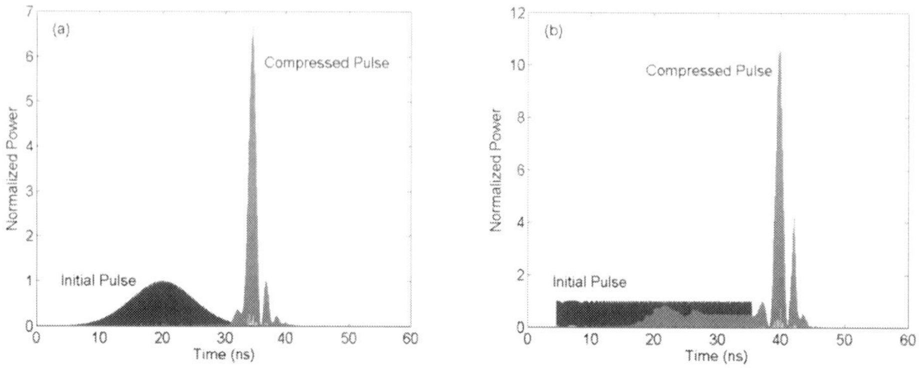

Figure. 4 Pulse compression. (a) Gaussian pulse and (b) Rectangular pulse

In Fig. 3, frequency as a function for TE_{11}-TE_{21} (coupled first and zeroth floquet harmonic) is given for the chosen parameters. In the neighbourhood of frequency 9.3 GHz, decrease in frequency increases the group velocity of the mode. In a pulse whose frequency decreases with time, the tail will move faster, resulting into the compression of pulse.

In Fig. 4, pulse compression for two different pulse profiles are given. In Fig. 4(a), shape function $S(t)=e^{-0.01(t-t_0)^2}$, $t_0=20\,\text{ns}$ is considered. The frequency sweep parameter β=0.01 ns^{-1}. The input pulse is shown in blue. After propagating 110 cm, the pulse gets compressed shown in red. In Fig. 4(b), step shape function is considered. The pulse width is 40 ns and the frequency sweep parameter β=0.02 ns^{-1}. The input pulse and output pulse after propagating 110 cm, are shown in blue and red, respectively. An eleven fold peak power compression is obtained.

4. Conclusion

Helically corrugated waveguide dispersion relation is solved. Pulse compression studied for Gaussian as well as rectangular pulse. Seven fold in former and eleven fold power increase in latter case, is obtained. The numerical results are in agreement with the experimental results of G. Burt et. al.[4].

References

[1] Gaponov-Grekhov A V and Granatstein V L, Eds., *Applications of High-Power Microwaves*. Boston, MA: Artech House, 1994.

[2] Denisov G H, Bratman V L, Cross A W, He W, Phelps A D R, Ronald K, Samsonov S V and Whyte C G 1998 *Phys. Rev. Lett.* **25**, 5680

[3] Burt G, Samsonov S V, Phelps A D R, Bratman V L, Ronald K, Denisov G G, He W, Young A R, Cross A W and Konoplev I V 2005 *IEEE Trans. Plasma Sci.* **33** 661

[4] Burt G, Samsonov S V, Ronald K, Denisov G G, Young A R, Bratman V L, Phelps A D R, Cross A W, Konoplev I V, He W, Thompson J and Whyte C G 2004 *Phys. Rev. E.* **70** 046402

Design of Single disc RF window for High Power Gyrotron

MK Alaria[1], Y Choyal[2] and AK Sinha[1]
[1]Microwave Tubes Area, CSIR-Central Electronics Engineering Research Institute (CEERI), Pilani-333031, Rajasthan, India
[2]Department of physics, Devi Ahilya Vishwavidyalaya, Indore, India

E-mail: mka@ceeri.ernet.in

Abstract. In this paper investigates the design of water edge cooled single-disc CVD-diamond window for 120 GHz, 1MW gyrotron. The design of RF window for 120 GHz, 1MW gyrotron has been carried out using the CST microwave studio. In 120 GHz gyrotron single disc of diameter 90 mm and thickness 2.0 mm CVD diamond window has been used in the simulation. The return loss (S11) and transmission loss (S21) of the 120 GHz gyrotron window have been found - 40.0 dB and -0.02 dB respectively. Thermal analysis of single disc rf window has also been carried out using ANSYS software for high power 120 GHz gyrotron. The temperature range on the disc surface has been found to be 100^0 C -300^0 C.

1. Introduction

The Gyrotron is a high power microwave tube, which emits coherent radiation at approximately the electron cyclotron frequency or its harmonics. Gyrotron is widely used in plasma fusions, ECRH heating, industrial heating and material processing. The need for high power, high frequency rf sources for the magnetic fusion research experiments has provided much of the impetus for the development of present day gyro oscillators. From the last three decades gyrotron oscillators have played a key role in magnetic fusion experiments. Magnetic confined plasma fusion is the most important application of gyrotron oscillator as a high power, high frequency rf source [1, 2]. A 42 GHz, 200kW (CW) gyrotron is being developed for electron cyclotron resonance plasma heating for an Indian TOKAMAK system. A 120 GHz, 1MW Gyrotron is also being developed for International Thermonuclear Experimental Reactor (ITER) for nuclear fusion. The requirement of this special device is increasing with demands for more and more power at high frequency. Therefore, research in this type of fast wave tube, particularly Gyrotron becomes important from both Indian point of view as well as global scenario. RF windows are most critical components of microwave tubes, which are used as a microwave penetrable barrier between the vacuum and a transmission line and also used for fusion tours application. This system separates the ultrahigh vacuum environment inside the tube. The wave-guide type microwave windows are generally preferred for high power microwave tubes due to their higher capacity for handling high peak power. The other functional advantages are broad bandwidth and easy impedance matching with the rest of the transmission line [3]. The desired features of an ideal window are minimum reflection, minimum insertion loss, and high power handling capability, wide bandwidth, excellent mechanical strength and high thermal conductivity. Types of window have been suggested in literature for high power mm wave applications and in gyrotrons,

[1] E-mail: mka@ceeri.ernet.in

these are single disc with edge cooling, single disc with surface cooling by gas, and double disc with surface cooling by liquid. For high power CW gyrotrons at millimeter wavelengths, advanced materials such as sapphire, chemical vapour deposited (CVD) diamond, Silicon nitride composite, BeO, Au-doped silicon etc. have to be used [4]. Design of CVD diamond, sapphire or other dielectric based high-power window operating at more than hundred GHz and capable of handling average power of one up to two megawatts.

2. Electrical design of RF window

The window diameter 90 mm, disc thickness 2.0 mm, length of the circular waveguide is 70 mm have been used for the design of single disc chemical vapour deposition (CVD) window at 120 GHz. The design of single disc CVD window has been carried out using CST Microwave studio [5]. The material of disc is CVD and dielectric constant (ε_r) of the CVD is 4.7, loss tangent is 115×10^{-5}. Figure1 shows the schematic diagram of edge cooled window for high power gyrotron. Figure 2 shows the model of single disc RF window used in this analysis. Table 1 shows the electrical design parameters of single disc window for 120 GHz Gyrotron. The scope of PACVD diamond, sapphire or other dielectric based high-power window operating at more than hundred GHz and capable of handling average power of one mega watt minimum.

Table 1. Electrical design parameters of RF window.

Parameters	Value
Disc Diameter (R)	90 mm
Disc Thickness (d)	2.0 mm
Waveguide length	90 mm

THE GRAPHIC ON THIS PAGE WAS NOT AVAILABLE ON THE E-MEDIA VERSION OF THIS PAPER.

Figure 1. Schematic diagram of single disc edge-cooled window

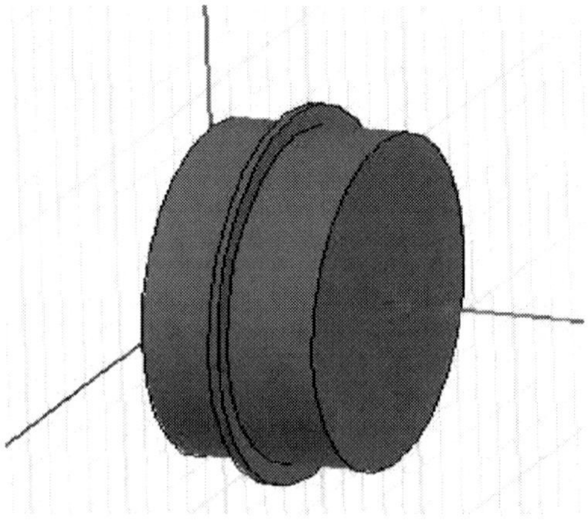

Figure 2. Model of single disc CVD window

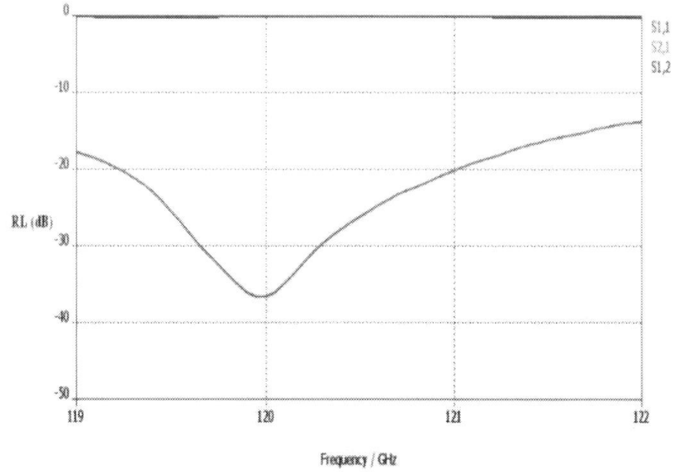

Figure 3. Return loss performance of the single disc window

3. Thermal design of RF window

The aim of the thermal analysis is to assess the temperature distribution, axial stress, radial stress and thermal expansions in the RF window for the 120 GHz Gyrotron during extreme case of operation, i.e. at saturation. ANSYS finite element code v.11.1 has been used for 3-D analysis of single disc RF window [6]. A transient thermal analysis is used to determine the temperature distribution as a function of time. Input parameters used in simulation are material property, coolant property, heat flux, film coefficient and bulk temperature. Output parameters are the temperature profile of RF window. Input parameters of window for thermal analysis shown in Table 2. The RF windows boundaries consist of two parallel of faces are of which is isolated and heat flow across the other face. Cooling channel designed around the window outside surface for proper cooling of the window. The coolant flow rate changed from 5 to 15 L/min to see the effect. The structural analysis has been used to determine the displacements and stresses in window caused by heat loads. The material properties required for structural analysis are Young's modulus, Poisson's ratio and coefficient of thermal expansion. The finite element stress analysis of RF window was performed on the inner disc. This disc has higher pressure loading due to the internal vacuum of the Gyrotron tube. The 3-D stress analysis indicates that the double disc is operating close to its maximum stress level. The largest stress contribution comes from the mechanical loading of the discs by the coolant pressure. The large tensile stresses are compensated to a limited extent by the thermal stresses which arise during Gyrotron operation. The maximum radial tensile stress of 152 MPa occurs near the window. The maximum compressive stress is 250 MPa. The output results of the temperature distributions and different stresses on some typical parts are obtained from the ANSYS v.11.1 software. From these results, it is clear that the deformations in the parts are within the safe limit under the operating condition.

4. Results and discussions

The return loss (S_{11}) and transmission loss (S_{21}) of the 120 GHz single disc window have been obtained - 37.3 dB and -0.04 dB respectively. Figure 3 shows the return loss performance of the single disc 120 GHz window. The thermal and structural analyses have been carried out for the estimation of temperature distributions and resulting stress developed on the various components under the saturation condition. Figure 4 shows the temperature distributions on the CVD disc window for different heat flux. Figure 5 shows the transient response of single disc RF window.

Table 2. Input parameters for thermal analysis

Parameters	Value
Loss tangent	0.00115
Bulk temperature	25^0 C
Water flow rate	5 L/min
Film coefficient	7584(W/m^2 ^{0}C)
Input power	1MW

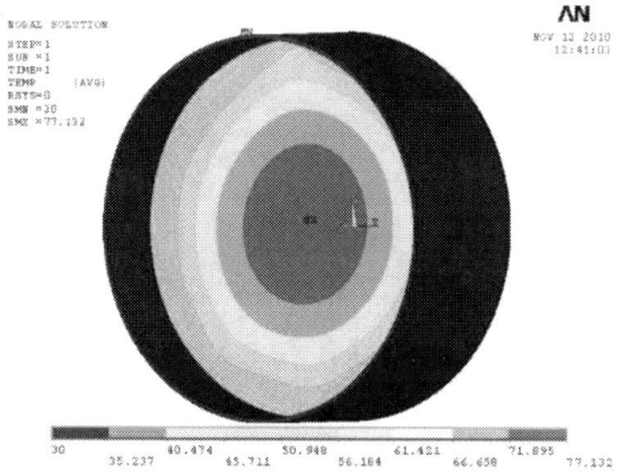

Figure 4. Temperature distribution on disc

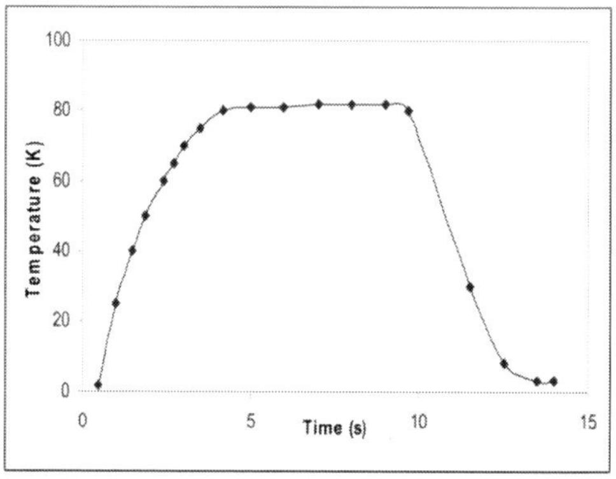

Figure 5. Transient response of single disc RF window

Acknowledgments
This work was carried out under a DST sponsored project. The authors are thankful to the Director, CEERI, Pilani and Dr. SN Joshi, Dr. V Srivastava, Microwave Tubes Area, for all their support and guidance and all other project team members for their support.

References

[1] Gilmour A S 1994 *Microwave Tubes* (Artech House, Boston, London)
[2] Edgecombe C J 1993 Gyroton *Oscillators: Their principles and practice* (Taylor & Francis Ltd London)
[3] Nusinovich G S 2004 *Introduction to the physics of Gyroton* (Maryland, JHU, USA)
[4] Thumm M 2006 *State-of-the-Art of High Power Gyro-Devices and Free Electron Masers Update* (Forschungszentrum Karlsruhe)
[5] CST Microwave Studio Version 9.0, Computer Simulation Technology Darmstadt, Germany
[6] ANSYS Vs 11.1: User manual, (Ansys Inc, USA)

Features of complex networks in a free-software operating system

Rajiv Nair[1], **G Nagarjuna**[2], **Arnab K Ray**[3]

[1] Tata Institute of Social Sciences, V N Purav Marg, Deonar, Mumbai 400088, India
[2] Homi Bhabha Centre for Science Education, Tata Institute of Fundamental Research, V N Purav Marg, Mankhurd, Mumbai 400088, India
[3] Department of Physics, Jaypee University of Engineering and Technology, A B Road, Raghogarh, Guna 473226, Madhya Pradesh, India

E-mail: `rajiv@tiss.edu`, `nagarjun@gnowledge.org`, `arnab.kumar@juet.ac.in`

Abstract. We propose a mathematical model to fit the degree distribution of directed dependency networks in free and open-source software. In this complex system, the intermediate scales of both the in-directed and out-directed dependency networks follow a power-law trend (specifically Zipf's law). Deviations from this feature are found both for the highly linked nodes, and the poorly linked nodes. This is due to finite-size effects in the networks, and the parameters needed to model finite-size behaviour make a quantitative distinction between the in-directed and out-directed networks. We also provide a model to describe the dynamic evolution of the network, and account for its saturation in the long-time limit.

1. Introduction

The last decade has seen the emergence of complex networks [1, 2, 3, 4, 5] across diverse domains like (to name a few) the World Wide Web, the Internet, social, ecological and biological systems, income and wealth distributions, trade and business networks, highway networks, linguistic structures with implications for their syntax and semantics, electronic circuits and the architecture of computer software. A recurring theme in many of these systems is a power-law (scale-free) distribution, a property that is significant for the inherent robustness of the network and its evolutionary aspects. And so it is that power-law features have been found [6] in the dependency networks of free and open-source software (*FOSS*) as well, which have further been shown actually to follow Zipf's law [7]. However, it has also been realised that simple power-law properties do not suffice to provide a complete global model for the networks in a *FOSS*. For any system with a finite size, the power-law trend is not manifested indefinitely, and deviations from this trend appear for both the profusely linked and the sparsely linked nodes in the network. Modelling the finite-size effects (equivalently the saturation properties) in the complex network underlying a *FOSS*, and to study how these effects are related to the dynamic properties of the network are the principal objectives of our work [8].

2. Modelling the degree distributions in a *FOSS* network

When it comes to installing a software package from the *Debian GNU/Linux* repository, many other packages — the "dependencies" — are also called for as prerequisites. This leads to a dependency-based network among all the packages, and each of these packages may be treated as a node in a network of dependency relationships. Each dependency relationship connecting any two packages (nodes) is treated as a link (an edge), and every link establishes a relation between a prior package and a posterior package, whereby the functions defined in the prior package are invoked in the posterior package. As a result, the whole operating system emerges as a functional network, whose fundamental character is determined by the direction of the links.

Networks in a *FOSS* appear in two types: one comprising only the incoming links among the nodes (the in-degree distribution) and the other comprising only the outgoing links among the nodes (the out-degree distribution). To develop any mathematical model that purports to give an accurate global prescription for the degree distribution in both of these networks, it will be necessary first to count the actual number of software packages, ϕ, which are connected by a particular number of links, x, in either kind of network. This gives an unnormalised frequency distribution plot of $\phi \equiv \phi(x)$ versus x. To give continuum description for the power-law feature as well as any finite-size effect in this kind of a frequency distribution, we forward a nonlinear logistic-type equation going as $(x+\lambda)\,\phi'(x) = \alpha\phi\,(1-\eta\phi^{\mu})$, in which $\phi'(x)$ is a first derivative with respect to x, with α being a power-law exponent, μ being a nonlinear saturation exponent, η being a "tuning" parameter for nonlinearity and λ being another parameter that will be instrumental in setting a limiting scale for the poorly connected nodes. The parameter μ crucially controls the advent of scale-free features in the network, while the parameters, η and λ, quantify the finite-size properties in the network on its extremal scales. These two parameters also make a quantitative distinction between the in-directed and out-directed networks.

The integral solution of the foregoing differential equation is obtained as (for $\mu \neq 0$)

$$\phi(x) = \left[\eta + \left(\frac{x+\lambda}{c}\right)^{-\mu\alpha}\right]^{-1/\mu}, \tag{1}$$

in which c is an integration constant (as far as this static model is concerned). It is quite obvious that when $\eta = \lambda = 0$ (with the former condition implying the absence of nonlinearity), there will be a global power-law distribution for the data, going as $\phi(x) = (x/c)^{\alpha}$, regardless of any non-zero value of μ. The situation becomes quite different, however, when both η and λ have non-zero values. In this situation, the network will exhibit a saturation behaviour on extreme scales of x (both low and high). For the high values of x, this can be easily appreciated from the proposed differential equation itself, wherefrom the limiting value of ϕ is obtained as $\phi = \eta^{-1/\mu}$.

The mathematical model implied by Eq. (1) is based on data from two stable *Debian FOSS* releases[1], *Etch* (*Debian GNU/Linux 4.0*) and *Lenny* (*Debian GNU/Linux 5.0*), and it has been tested subsequently on the latest stable release *Squeeze* (*Debian GNU/Linux 6.0*). As mentioned earlier, each release of *Debian FOSS* has two kinds of dependency networks — the in-degree and the out-degree. In fitting the degree distribution of all these networks, the parameter values $\alpha = -2$ and $\mu = -1$, emerge as a universal feature in *all* the cases. While the obvious implication of the former value is that Zipf's law governs the intermediate scales of the dependency networks, the latter value imposes a scale-free character on the network. Carrying out a power series expansion on Eq. (1), it is not difficult to see that a self-contained and natural truncation of the series can only be achieved when $\mu = -1$, i.e. when a single power law operates on all scales of x.

With the values of $\alpha = -2$ and $\mu = -1$, the finite-size properties in the network can be expressed from Eq. (1) as $\phi(x) = \eta + c^2\,(x+\lambda)^{-2}$. The implications of this result are noteworthy. One of these is that the parameter, η, defines a finite lower bound to the discrete count of nodes, regardless of any arbitrarily high value of x, i.e. $\phi \longrightarrow \eta$ as $x \longrightarrow \infty$. The other point worth noting is that when the model is fitted with the data, η makes a quantitative distinction between the in-degree and the out-degree distributions. Going back to the data taken from the *Etch* release, the values of η for the in-degree and the out-degree distributions, respectively, are -8 and 1. For the *Lenny* release, the corresponding values are -15 and 1. The role of η in determining the finite-size properties for high values of x, and in

[1] `http://www.debian.org/releases`

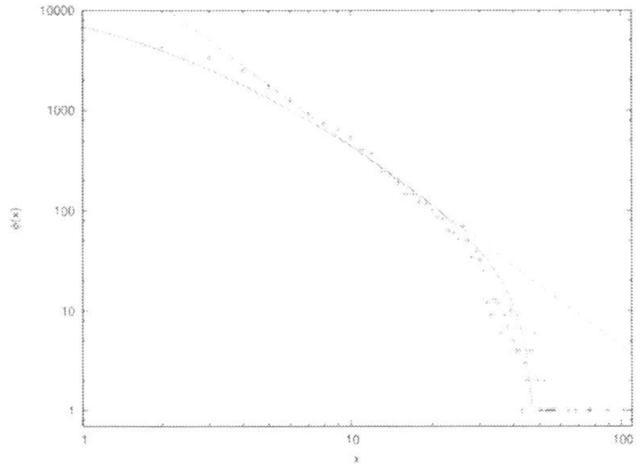

Figure 1. For the network of incoming links in the *Squeeze* release, the degree distribution shows a good fit in the intermediate region with a power-law exponent, $\alpha = -2$ (as indicated by the dotted straight line), which validates Zipf's law. However, for large values of x, there is a saturation behaviour towards a limiting scale that is fitted by the parameter value, $\eta = -28$. On the other hand, when x is small, the fit is good for $\lambda = 2.2$. For this plot, $c \simeq 265$.

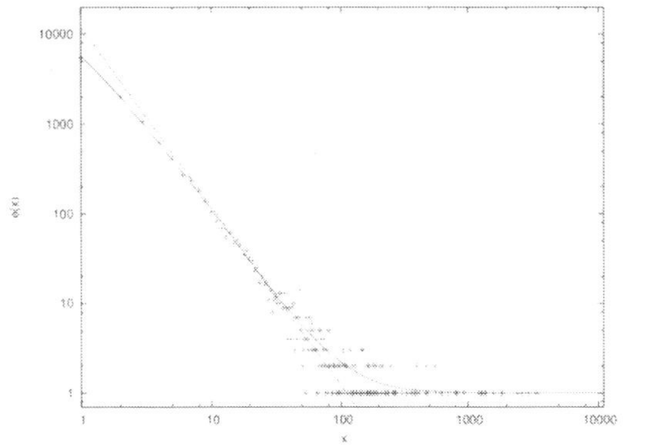

Figure 2. For the network of outgoing links in the *Squeeze* release, the degree distribution of intermediate nodes are again fitted by $\alpha = -2$ (as the dotted straight line shows). However, the saturation behaviour for large values of x is different from that of the network of incoming links. ϕ converges to a limit given by $\eta = 1$. When x is small, the convergence is attained for $\lambda = 0.45$. Thus the sign of η distinguishes the dependency networks. For this plot, $c \simeq 110$.

making a distinction between the two degree distributions is very much in evidence in Figs. 1 & 2, which show the in-degree and the out-degree distributions for *Squeeze*, the latest stable release of *Debian*. It is also interesting to note that for the out-degree distribution, the value of η remains unchanged at 1 with increasing values of x, across all generations of *Debian* releases. Making use of this value, and by requiring the two terms on the right hand side of Eq. (1) to be in rough equipartition with each other, we obtain a scale, $x_{\mathrm{sat}} \sim c$, at which the degree distribution starts saturating to a finite limit.

Considering the other extreme of finite-size effects at small values of x, a noticeable deviation from the power-law solution is also seen. This is especially true for the in-degree distribution in Fig. 1. In the limit of small degree distributions for both the in-directed and out-directed networks, where η ceases to have a quantitative significance, and where $x \sim 1$ (which, in the discrete count of links, is the lowest value that x can assume practically), we can find an upper bound to the number of the very sparsely linked nodes, given as $\phi_{\mathrm{ub}} \simeq c^2 (1+\lambda)^{-2}$, with the full range of ϕ, therefore, going as $\eta \leq \phi \lesssim \phi_{\mathrm{ub}}$.

3. Dynamics and long-time saturation in the network

The *FOSS* network is a dynamically evolving network, undergoing continuous additions (even deletions) and modifications across several generations of *Debian* releases. A realistic model should account for this evolutionary aspect of the network distribution, and at the same time accommodate the finite-size features of the network.

Going by the fact that for the out-degree distributions, the value of η remains fixed at unity, one can reason that the richly connected nodes in this case form the irreducible nucleus of the *FOSS* network. These nodes are by far the most influential in the network and it will be useful to study their dynamic properties. From the perspective of a continuum model, the frequency distribution of the nodes in the network of outgoing links can be seen as a field, $\phi(x,t)$, evolving continuously through time, t, with the saturation in the number of nodes for high values of x, emerging of its own accord from the dynamics. In keeping with this need, an ansatz with a general power-law feature inherent in it, may be framed as $\phi(x,t) = c^{-\alpha}(x+\lambda)^{\alpha} + \varphi(x,t)$ in which $\varphi \longrightarrow \eta$, as $t \longrightarrow \infty$. Under this requirement, we describe

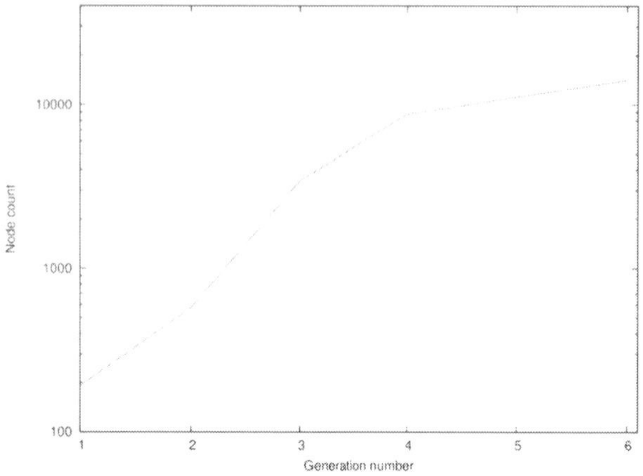

Figure 3. The curve plots the growth of the total number of nodes in the network of outgoing links. Saturation in growth can be seen for the later releases of *Debian*. The generation number plotted along the horizontal axis, effectively marks the time through which the network has evolved. Data for this plot have been taken from all the six releases of *Debian*.

the temporal evolution of the network by a model equation, going as,

$$\tau \frac{\partial \phi}{\partial t} = \frac{\partial \phi}{\partial x} - \frac{\alpha}{c^{\alpha}} (x + \lambda)^{\alpha - 1} , \tag{2}$$

in which τ is a parameter to scale time. The solution of Eq. (2) is obtained by the method of characteristics [9] as $\phi(x,t) = \eta + c^{-2}(x+\lambda)^{\alpha} - c^{-2}[x + \lambda + (t/\tau)]^{\alpha}$, and this, under the condition that $\alpha = -2$, will converge to the expected static distribution for $t \longrightarrow \infty$. The long-time saturation of the total number of nodes in the out-degree distribution can then be evaluated as

$$N_{\rm out}(t) = \int_1^{x_{\rm m}} \phi(x,t)\,{\rm d}x \simeq \eta x_{\rm m} + \frac{c^2}{1+\lambda} - c^2 \left(1 + \lambda + \frac{t}{\tau}\right)^{-1} , \tag{3}$$

in which $x_{\rm m}(\gg 1)$ is the maximum number of links that a node has in the network. Going by the data taken from all the six releases of *Debian*, the first three of which are *Buzz* (*Debian GNU/Linux 1.1*), *Hamm* (*Debian GNU/Linux 2.0*) and *Woody* (*Debian GNU/Linux 3.0*), we see in Fig. 3 that the total number of nodes in the out-degree distribution does indeed saturate to a finite end in accordance with what one can expect from Eq. (3), when $t \longrightarrow \infty$.

Acknowledgments
The authors thank J. K. Bhattacharjee, C. Gershenson, P. Majumdar, S. Sinha, S. Spaeth, R. Stinchcombe, V. M. Yakovenko and S. Zacchiroli for their helpful comments.

References
[1] Strogatz S H 2001 *Nature* **410** 268
[2] Albert R and Barabási A L 2002 *Rev. Mod. Phys.* **74** 47
[3] Newman M, Barabási A L and Watts D J 2006 *The Structure and Dynamics of Networks* (Princeton and Oxford: Princeton University Press)
[4] Barrat A, Barthélemy M and Vespigniani A 2008 *Dynamical Processes on Complex Networks* (Cambridge: Cambridge University Press)
[5] Newman M E J 2011 Complex Systems: A Survey *Preprint* arXiv:1112.1440
[6] LaBelle N and Wallingford E 2004 Inter-Package Dependency Networks in Open-Source Software *Preprint* cs/0411096
[7] Maillart T, Sornette D, Spaeth S and von Krogh G 2008 *Phys. Rev. Lett.* **101** 218701
[8] Nair R, Nagarjuna G and Ray A K 2009 Semantic Structure and Finite-Size Saturation in Scale-Free Dependency Networks of Free Software *Preprint* arXiv:0901.4904
[9] Debnath L 1997 *Nonlinear Partial Differential Equations for Scientists and Engineers* (Boston: Birkhäuser)

Infrared spectra in monitoring biochemical parameters of human blood

S Prabhakar[1], N Jain[2] and R A Singh[3]
[1,3]Department of Physics, Dr. H. S. Gour University, Sagar (M.P.), India, 470 003
[2]Department of Physics, Noble College, Rajakhedi, Sagar (M.P.), India, 470 004

E-mail: [1]s_prabhakar2007@rediffmail.com

Abstract. Infrared spectroscopy is gaining recognition as a promising method. The infrared spectra of selected regions (2000-400cm^{-1}) of blood tissue samples are reported. Present study related to the role of spectral peak fitting in the study of human blood and quantitative interpretations of infrared spectra based on chemometrics. The spectral variations are interpreted in terms of the biochemical and pathological processes involved. The mean RNA/DNA ratio of fitted intensities and analytical area as calculated from the transmittance peaks at 1121cm^{-1}/1020cm^{-1} is found to be 0.911A.U and 2.00A.U. respectively. The ratio of 1659cm^{-1}/1544cm^{-1} (amide-I/amide-II) bands is found to shed light on the change in the DNA content. The ratio of amide-I/amide-II is almost unity (≈1.054) for blood spectra. The deviation from unity is an indication of DNA absorption from the RBC cells. The total phosphate content has found to be 25.09A.U. The level for glycogen/phosphate ratio (areas under peaks 1030cm^{-1}/1082cm^{-1}) is found to be 0.286A.U. The ratio of unsaturated and saturated carbonyl compounds (C=O) in blood samples is in form of esters and the analytical areas under the spectral peaks at 1740cm^{-1} and 1731cm^{-1} for unsaturated esters and saturated esters respectively found to be 0.618A.U.

1. Introduction

The very broad range of mathematical and statistical methods has provided an excellent opportunity for the quantitative analytical descriptions of experimental results and effects in natural sciences. Derived biological markers from IR spectroscopy used in the identification of disease have been extensively investigated [1]. It is strongly believed that in studies related to spectroscopic techniques, both the reliable experimental procedure and detection of spectral peak positions are important. At present, there is no data available in the literature on FT-IR spectra of whole blood, which may be vital in optical study.

In this study, we examined the phosphate, RNA/DNA, carbohydrate and amide-I/amide-II region using a peak fitting program revealing various new spectral peaks in spectra of human blood. The peak fitting procedure consists of modelling an experimental spectrum with a sum of analytical peak functions. The peak fitting is helpful to resolve spectral peaks in spectra which are often overlapped or making direct measurement of their parameters problematic, unreliable or even impossible.

2. Materials and Methods

2.1. Sample preparation for infrared spectroscopy
In the present work, thirty blood samples were collected. The blood was collected on a sterilized glass slide with frosted end. All the blood smears were kept to be dried in clean, dry and airy place at room temperature, so that the moisture content of the smears was removed [2].

2.2. Infrared Spectra
Thick films on the glass slide were removed and smears were grounded with spectroscopic grade KBr in the powder form. FT-IR spectra were recorded using SHIMADZU 8000 series FT-IR spectrophotometer with diffraction technique. 25 scan were performed in the wavenumber region 2000–400cm^{-1} for each sample at the resolution 4 cm^{-1}. Spectral analysis was done using IR solution software. The integrated transmittances were calculated by a band fitting procedure reported earlier [3,4].

3. Observations
The spectra illustrate glycogen bands at 1020cm^{-1} and symmetric phosphate stretching band at 1082cm^{-1}. The region 1300–1000 cm^{-1} belongs to C–O stretching of alcohols, carboxylic acids, ethers and esters but when the sample is biological, the region is dominated by spectral features of phosphodiester [5]. The band at 1369cm^{-1} arise due stretching of C-N of cytosine and guanine [6]. The region 1800 –1600 cm^{-1} belongs to finger print region of C–H phenyl ring substitution overtones [7]. Among several vibrational features, the amide-I band at 1659 cm^{-1} was due primarily to the C = O stretching vibrational bands of the peptide backbone. The amide-II band at 1544 cm^{-1} arises due to coupling of CN stretching and in-plane bending of the N-H group.

4. Curve fitting
The curve or peak fitting [8] is mostly used for resolving the overlapped peaks and finding more precise peak parameters e. g., position, height, width and area.

4.1.1 Initial spectral study

4.1.1 *Baseline Correction.*
Zero baseline correction was used so that the maximum value of transmittance spectrum is 100% without changing the shape of spectrum.

4.1.2 *Smoothening.* Smoothening is used on noisy spectrum to reduce noise level. We used Savitsky-Golay smoothening to make spectrum noise because there is some noise at the ends of spectrum.

4.2 Peak fitting

The spectra generated by infrared spectrophotometer are processed before using to monitor RNA/DNA and Protein/DNA contents e. g., curve fitting. Curve fitting includes baseline correction, curve smoothening, SNR correction, spectral normalization.

The blood samples used in this work are rich in vibrational modes. There is more than one vibration mode in some peaks due to the overlapping of the transmittance peaks. These peaks appeared as doublets or triplets. We used PeakFit Version 4.12 programme to allow band fitting and to detect and locate the hidden peaks. The PeakFit locates the peak centroids up to third derivative of the spectrum described [9]. The weak transmittance bands are resolved by FT-IR spectrometer.

In our study, the spectra were fitted using standard Gaussian peak shape. The Gaussian curve expression with peak amplitude is given by;

$$y = a_0 \exp\left[-\frac{1}{2}\left(\frac{x-a_1}{a_2}\right)^2\right] \tag{1}$$

where; a_0 is the Gaussian amplitude, a_1 is position at the centre and a_2 is the FWHM and the Gaussian curve with peak area is expressed as;

$$y = \frac{A}{\sqrt{2\pi}a_2}\exp\left[-\frac{1}{2}\left(\frac{x-a_1}{a_2}\right)^2\right] \tag{2}$$

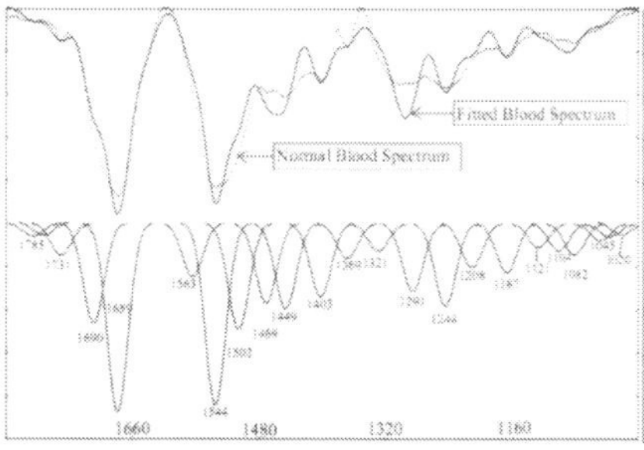

Figure 2a : Statistically processed FT-IR spectrum of blood group A.

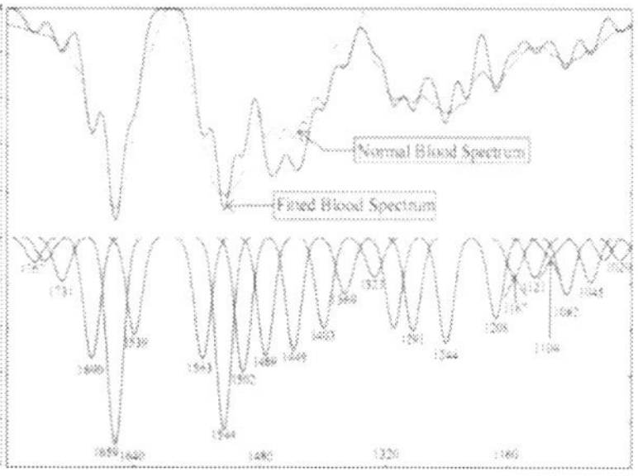

Figure 2b: FT-IR Spectrum of Normal Blood Group B After Peak Fitting.

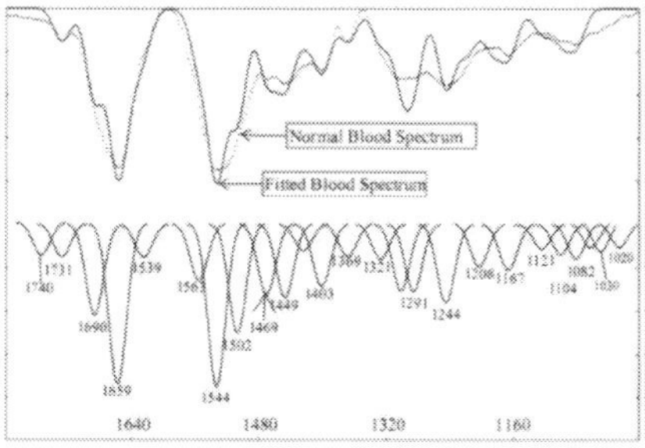

Figure 2c : FT-IR Spectrum of Normal Blood Group AB After Peak Fitting.

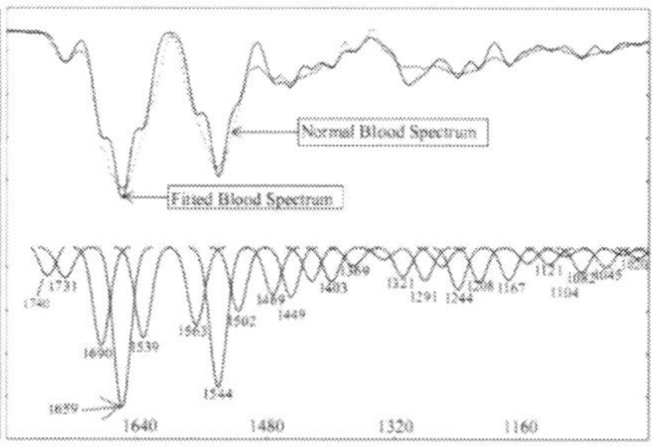

Figure 2d : FT-IR Spectrum of Normal Blood Group O After Peak Fitting.

where; A is the area of Gaussian peak.

The procedure mathematically enhances the resolution of the spectrum, using an iterative statistical curve fitting of symmetric Gaussian peaks to the straight, baseline corrected spectra. The measured IR spectra of thirty blood samples of groups belongs to A, B, AB and O. The statistical spectral peak fitting of blood groups is shown in **Figure 2(a-d).**

5. Results and Discussions

5.1 Hidden Peaks

In the entire blood spectrum, an important peak at 1690 cm^{-1} was hidden due to noise and was not seen, but after processing and peak fitting it clearly observed and appeared as a peak in all the spectra of blood.

The peaks at wavelengths; 1167 cm^{-1}, 1244 cm^{-1}, 1690 cm^{-1} and 1731 cm^{-1} were obtained after peak fitting of blood spectra. The peak at 1167 cm^{-1} arises due to C–O stretching of blood samples [10,11]. Peak at 1244 cm^{-1} arise due to asymmetric PO_2^- stretching [12]. The peaks at 1690cm^{-1} and 1731cm^{-1} arise due to the base carbonyl stretching and ring breathing mode of nucleic acids [13] and fatty acid ester band [14] respectively.

5.2 Biochemical parameters

The analytical area calculated by peak fitting for different peaks have been used to define various biochemical parameters described below;

5.2.1 Total Phosphate Content. The summed phosphate level is obtained under the peaks at 1244 cm^{-1} and 1082 cm^{-1}. The average phosphate content under the peaks 1244cm^{-1} and 1082cm^{-1} was obtained 17.669A.U. and 7.426A.U. respectively. The average summed phosphate level for thirty blood samples was found to be 25.096AU for normal blood.

5.2.2 RNA/DNA Level. The RNA/DNA ratio is one of the biological markers [15]. It is the measured ratio of areas under spectral peaks at 1121cm^{-1} and 1020cm^{-1}. The individual average of 1121cm^{-1} is

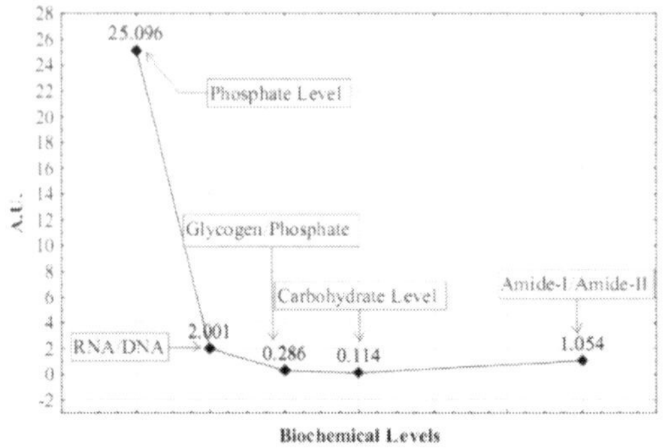

Figure 3: Biochemical levels of different calculated parametrs.

3.939A.U. and is 1.968 for 1020cm^{-1}. In the present study the average of RNA/DNA level of thirty blood spectra was 2.001A.U.

5.2.3 Glycogen/Phosphate Ratio. The spectral assignments show that the region 1350–900cm^{-1} belongs to collagen and glycogen [16]. The average glycogen and phosphate content under the peaks 1030cm^{-1} and 1082cm^{-1} was obtained 2.128A.U. and 7.426A.U. respectively. The average ratio of areas under the peaks 1030cm^{-1}/1082cm^{-1} was found to be 0.286A.U.

5.2.4 Carbohydrates Level. The ratio of 1045cm^{-1}/1544cm^{-1} gives an estimate of carbohydrate concentration and is lower in malignant biological samples. We have also taken this ratio as a variable which could be used to discriminate normal and malignant bloods. The average carbohydrate level was obtained 0.114A.U.

5.2.5 Amide I/II Ratio. The ratio shed light on DNA contents [17]. The contribution of the DNA due to carbonyl group from the bases is quantified by the ratio of integrated transmittance of amide I/II. The contribution of the DNA due to carbonyl group from the bases is quantified by the ratio of amide I/II. It is the measured ratio of areas under spectral peaks at 1659cm^{-1} and 1544cm^{-1}. The individual average of 1659cm^{-1} is 35.524A.U. and is 33.678A.U. for 1544cm^{-1}. The average of amide-I/amide-II was found to be 1.054A.U.

The relative levels of biochemical parameters for normal blood have been shown in **Figure 3**.

6. Conclusion

This technique enables the study of the state of chemical bonds and the relative concentrations of lipids, proteins, carbohydrates, and phosphorylated compounds. Many peaks like; 1167 cm^{-1}, 1244 cm^{-1}, 1690 cm^{-1} and 1731 cm^{-1} which were hidden due to noise are easily seen after spectral processing and peak fitting, it shows the potential role of peak fitting and IR spectral study.

These biochemical parameters derived from IR spectroscopy could be used in identification of different blood related diseases in future.

References

[1] Argov S, Ramesh J, Salman A, Sinelnikov I, Goldstein J, Guterman H, Mordechai S 2002 *Journal of BioMed. Optics* **7(2)** 248

[2] Khanmohammadi M, Ansari M A, Baghaeri Garmarudi A, Hassanzadeh G, Garoosi G 2007 *Cancer Investigation* **25** 397

[3] Salman A, Argov S, Ramesh J, Goldstein J, Igor S, Guterman H, Shaul M 2001 *Cell Mol. Biol.(Paris)* **47(22)** OL159

[4] Lillie R D, Fullmer H M 1976 *Histopathologic Technical and Practical Histochemistry.* (NewYork:McGraw–Hill)

[5] Bruno T J, Svoronos P D N 2006 *CRC Handbook of Fundamental Spectroscopic Correlation Charts* (Boca Rotan:CRC book press)

[6] Dovbeshko G I, Gridina N Y, Kruglova E B, Pashchuk O P 1997 *Talanta* **53** 233

[7] Sahu R K, Zelig U, Huleihel M, Brosh N, Talyshinsky M, Ben-Harosh M, Mordechai S and Kapelushnik 2006 J *Leuk. Res.* **30(6)** 687

[8] Jansson P A 1984 *Deconvulation with Applications in Spectroscopy* (NewYorkY:Academic Press)
[9] Evans M, Hastings N and Peacock B (1993) Statistical Distribution. John Wiley & Sons,
[10] Rigas B, Wong P T T 1992 *Cancer Research* **52** 84
[11] Rigas B, Morgello S, Goldman I S and Wong P T T 1999 Human Colorectal Cancers Display Abnormal Fourier Transform Infrared Spectra: *Proc of the National Academy of Sciences* (USA) **87** 8140
[12] Wood B R, Quinn M A, Burden F R and McNaughton D 1996 *Biospectroscopy* **2** 143
[13] Chiriboga L, Xie P, Yee H, Vigorita V, Zarou D, Zakim D and Diem M 1998 *Biospectroscopy* **4** 47
[14] Yoshida S, Miyazaki M, Sakai K, Takeshita M, Yuasa S, Sato A, Kobayashi T, Watanabe S and Okuyama H 1997 *Biospectroscopy* **3(4)** 281
[15] Andrus P G and Strickland R D 1998 *Biospectroscopy* **4(1)** 37
[16] Bogomolny E, Huleihel M, Suproun Y, Sahu R and Mordechai S 2007*Journal of Biomedical Optics* **12** 2
[17] Benedetti E, Bramanti E, Papineschi F and Rossi I 1997 *Applied Spectroscopy* **51** 792

AUTHOR INDEX

AUTHOR INDEX